Neuroinflammation and Autophagy in Neurodegeneration

Pukar Khanal • B. M. Patil • Rupesh Chikhale
Editors

Neuroinflammation and Autophagy in Neurodegeneration

Mechanistic Insights into Cellular Biology and Advanced Multimodal Therapeutics

Springer

Editors
Pukar Khanal
Pharmacology and Chemical Biology
Emory University School of Medicine
Atlanta, GA, USA

B. M. Patil
Pravara Rural College of Pharmacy
Loni, Maharashtra, India

Rupesh Chikhale
UCL School of Pharmacy
London, UK

ISBN 978-981-95-3944-4 ISBN 978-981-95-3945-1 (eBook)
https://doi.org/10.1007/978-981-95-3945-1

This Springer imprint is published by the registered company Springer Nature Singapore Pte Ltd.
The registered company address is: 152 Beach Road, #21-01/04 Gateway East, Singapore 189721, Singapore

This book is dedicated to all patients and families affected by neurodegenerative disorders, whose resilience inspires the pursuit of deeper understanding and better therapies.

Preface

Neurodegenerative disorders are among the most urgent medical challenges of twenty-first century, imposing a vast burden on patients and healthcare systems worldwide. Despite decades of progress in neuroscience, therapeutic options remain limited, as majority of current interventions offer symptomatic relief, with limited ability to modify the underlying disease trajectory. This gap suggests searching deeper into disease understanding and more innovative treatment strategies. Among multiple neuropathological processes, neuroinflammation and autophagy have gained increasing attention in recent years. Each, in its own right, plays a crucial role in neuronal survival and function, and forms a complex, interdependent axis to shape disease initiation, propagation, and outcomes. This volume was planned to explore that intricate interplay between neuroinflammation and autophagy, not only in terms of basic molecular biology but also in relation to its translational implications.

Organization of this volume is an attempt to guide the reader from foundational principles to disease-specific manifestations and, finally, to innovative therapeutic strategies. The first part provides a mechanistic framework by delineating how neuroinflammation and autophagy function in cellular environment. It highlights the dual nature of these processes, i.e., autophagy, traditionally observed as a protective clearance pathway, can be dysfunctional and contribute to cellular collapse, whereas neuroinflammation under physiological conditions supports repair and defense, and can evolve into chronic and destructive cascade. By examining the altered roles of multiple neurobiological processes, first part builds a comprehensive image of how these pathways converge to shape neuronal vulnerability. The second part applies this mechanistic foundation to individual neurodegenerative diseases. Here, the reader will encounter detailed accounts of how the autophagy–neuroinflammation axis manifests in Alzheimer's disease, Parkinson's disease, amyotrophic lateral sclerosis, Huntington's disease, multiple sclerosis, and frontotemporal dementia. These chapters not only outline the specific molecular disruptions in each condition but also emphasize cross-disease commonalities, underscoring the common pathological nodes that may represent promising therapeutic entry points. At the same time, disease-specific differentiations are described, offering a significant view to balance general principles with unique features of each disorder. The final part turns toward therapy and future directions, with discussions that extend from pharmacological interventions to gene editing, from nanoparticle delivery systems to lifestyle modifications and dietary strategies. In these chapters, the focus shifts from pathogenesis to potential solutions, emphasizing multimodal approaches that combine different strategies to maximize efficacy. The exploration of biomarkers and personalized medicine provides a forward-looking dimension, pointing toward an era in which treatments are tailored to molecular subtypes and individual patient profiles.

The aim to bring this volume was to bridge the gap between mechanistic insight and clinical translation, between cellular biology and therapeutic innovation. Neurodegenerative diseases are complex, multifactorial, and deeply challenging to treat, but by examining the intersections of neuroinflammation and autophagy, we may uncover pathways that lead to more effective interventions. This book does not claim to provide final answers, but it aims to map the current knowledge, highlight promising directions, and encourage continued exploration. If it succeeds in stimulating new research questions, fostering interdisciplinary collaboration, and bringing us closer to therapies that can alter the course of neurodegeneration, then its purpose will have been fulfilled.

Atlanta, GA, USA — Pukar Khanal
Loni, Maharashtra, India — B. M. Patil
London, UK — Rupesh Chikhale

Contents

Editors and Contributors

About the Editors

Pukar Khanal holds a strong academic background and extensive research experience in pharmacological studies. He obtained PhD in Pharmaceutical Sciences from KLE University in India and pursued postdoctoral research at Emory University School of Medicine, USA. His research combines experimental and computational approaches to advance the understanding of complex diseases and to support innovative therapeutic strategies. His research has been published in peer-reviewed journals, including reviews, editorials, and book chapters. He served as an Associate Editor for *Frontiers in Nutrition* and as an Editorial Board Member for multiple journals, including *Pharmacological Research—Reports*, *CNS & Neurological Disorders—Drug Targets*, *Stem Cells International*, and *International Journal of Genomics*. He has received awards, prestigious university gold medals, and research grant. His professional experience encompasses teaching, research supervision, and collaborative projects in various areas of pharmacology. He leverages his interdisciplinary research background to drive innovation, expand scientific understanding, and integrate experimental and computational methods to support translational discoveries.

B. M. Patil is a life member of Indian Pharmaceutical Association, Indian Pharmacological Society, and Association of Pharmacy Teachers of India, served as Senior Professor and Director at Pravara Rural College of Pharmacy, Loni, Maharashtra. With over 35 years of experience in academia and research, he held key positions such as Professor, Principal, and Dean at KLE College of Pharmacy, Belagavi and Hubballi. He made significant contributions to pharmacological research and supervised PhD candidates and MPharm dissertations, while securing research grants from AICTE and industry collaborators. His projects have addressed marker-assisted selection for metabolic and neurodegenerative diseases. His contributions have been recognized with several prestigious honors, including Best

Teacher Award and Prof. M. L. Khorana Memorial Award. In addition to research, he has actively contributed to academic governance, serving on PhD selection and registration committees at Sanjay Ghodawat University and Rajiv Gandhi University of Health Sciences, as well as on boards of pharmacy education at various universities.

Rupesh Chikhale is an affiliated scientist with postdoctoral training from School of Pharmacy, Faculty of Life Sciences, University College London. He specializes in structural biology and drug discovery, with a focus on understanding molecular interactions that drive therapeutic development. His research combines crystallographic and computational approaches to explore complex biological systems, particularly in areas where structural data is scarce or difficult to obtain. He has made significant contributions to the field through high-impact publications in leading international journals and through collaborative drug discovery projects. His expertise in advanced structural biology techniques and data-driven analysis enables him to bridge chemistry and biology in innovative ways, accelerating the identification and design of potential drug candidates.

Contributors

Adil Abalkhail Department of Public Health, College of Applied Medical Sciences, Qassim University, Buraydah, Saudi Arabia

Akil Akhtar Emory Vaccine Center, Department of Microbiology and Immunology, Emory University, Atlanta, GA, USA

Wesam Taher Almagharbeh Medical and Surgical Nursing Department, Faculty of Nursing, University of Tabuk, Tabuk, Saudi Arabia

Abhishek Balmik Department of Pharmacology and Chemical Biology, School of Medicine, Emory University, Atlanta, GA, USA

Abhishek Basu Section on Fibrotic Disorders, National Institute on Alcohol Abuse and Alcoholism, National Institutes of Health, Rockville, MD, USA

Kunal Bhattacharya Pratiksha Institute of Pharmaceutical Sciences, Guwahati, Assam, India

Jashabir Chakraborty Department of Pharmacology, Mata Gujri College of Pharmacy, Mata Gujri University, Kishanganj, Bihar, India

Shuvechha Chakraborty Centre for Integrative Omics Data Science (CIODS), Yenepoya (Deemed to be University), Mangalore, Karnataka, India

Jagdish Chand Department of Pharmaceutical Chemistry, JSS College of Pharmacy, JSS Academy of Higher Education & Research, Ooty, The Nilgiris, Tamil Nadu, India

Silicon Script Sciences Private Limited, Ghorahi, Dang, Nepal

Nongmaithem Randhoni Chanu Pratiksha Institute of Pharmaceutical Sciences, Guwahati, Assam, India

Raushan Kumar Chaudhary Centre for Integrative Omics Data Science (CIODS), Yenepoya (Deemed to be University), Mangalore, Karnataka, India

Dibyajyoti Das Pratiksha Institute of Pharmaceutical Sciences, Guwahati, Assam, India

Anand Desai Department of Pharmacology, Krishna Institute of Pharmacy, Krishna Vishwa Vidyapeeth (Deemed to be University), Karad, Maharashtra, India

Yadu Nandan Dey Pharmacology Division, Dr. B. C. Roy College of Pharmacy & Allied Health Sciences, Durgapur, West Bengal, India

Prarambh S. R. Dwivedi Department of Pharmacology, NGSM Institute of Pharmaceutical Sciences (NGSMIPS), Nitte (Deemed to be University), Derelakatte, Mangalore, Karnataka, India

Hannah Lalengzuali Fanai Department of Pharmaceutical Chemistry, JSS College of Pharmacy, JSS Academy of Higher Education & Research, Ooty, The Nilgiris, Tamil Nadu, India

Atreyee Ganguly Pharmacology Division, Dr. B. C. Roy College of Pharmacy & Allied Health Sciences, Durgapur, West Bengal, India

Lohitha Gujjari Dorothy M. Davis Heart and Lung Research Institute, Division of Pulmonary Critical Care & Sleep Medicine, Department of Internal Medicine, Wexner Medical Center, The Ohio State University, Columbus, OH, USA

Rohan Gupta Department of Biotechnology & Bioengineering, School of Biosciences & Technology, Galgotias University, Greater Noida, Uttar Pradesh, India

Praveen Halagali Department of Pharmaceutics, Manipal College of Pharmaceutical Sciences, Manipal Academy of Higher Education, Manipal, Karnataka, India

Sunil S. Jalalpure Department of Pharmacology, KLE College of Pharmacy, Belagavi, KLE Academy of Higher Education and Research, Deemed-to-be-University, Belagavi, Karnataka, India

Department of Pharmacognosy, KLE College of Pharmacy, Belagavi, KLE Academy of Higher Education and Research, Deemed-to-be-University, Belagavi, Karnataka, India

Sudhir Kumar Department of Microbiology and Immunology, School of Medicine, Emory University, Atlanta, GA, USA

Smita Kumari Division of Medicinal Chemistry and Pharmacognosy, College of Pharmacy, The Ohio State University, Columbus, OH, USA

Rabina Kumpakha Emory University School of Medicine, Atlanta, GA, USA

Shamanand P. Mallapur Department of Pharmacology, KLE College of Pharmacy, Belagavi, KLE Academy of Higher Education and Research, Deemed-to-be-University, Belagavi, Karnataka, India

S. N. Manjula Department of Pharmacology, JSS College of Pharmacy, JSS Academy of Higher Education and Research, Mysuru, Karnataka, India

Kamalika Mazumder Department of Pharmaceutical Chemistry, BCDA College of Pharmacy & Technology, Hridaypur, Barasat, West Bengal, India

Himani Meena Department of Pathology and Genomic Medicine, Houston Methodist Research Institute, Houston, TX, USA

Anuroopa G. Nadh Centre for Integrative Omics Data Science (CIODS), Yenepoya (Deemed to be University), Mangalore, Karnataka, India

Salini P. Nair Department of Pharmacology, JSS College of Pharmacy, JSS Academy of Higher Education and Research, Mysuru, Karnataka, India

Sathgowda Patil Department of Pharmacology, KLE College of Pharmacy, Belagavi, KLE Academy of Higher Education and Research, Deemed-to-be-University, Belagavi, Karnataka, India

Vishal S. Patil Department of Pharmacology, KLE College of Pharmacy, Belagavi, KLE Academy of Higher Education and Research, Deemed-to-be-University, Belagavi, Karnataka, India

Department of Pharmacognosy, KLE College of Pharmacy, Belagavi, KLE Academy of Higher Education and Research, Deemed-to-be-University, Belagavi, Karnataka, India

Mousumi Paul Dr. B.C. Roy College of Pharmacy and Allied Health Sciences, Durgapur, West Bengal, India

Jogendra Singh Pawar The James Cancer Hospital and Solove Research Institute, Comprehensive Cancer Center, The Ohio State University, Columbus, OH, USA

Pavithra Pradeep Prabhu Department of Pharmacognosy, NGSM Institute of Pharmaceutical Sciences (NGSMIPS), Nitte (Deemed to be University), Mangalore, Karnataka, India

Siddhartha Das Pramanik Molecular Endocrinology Laboratory, Department of Biosciences and Bioengineering, Indian Institute of Technology Roorkee, Roorkee, Uttarakhand, India

Rajesh Raju Centre for Integrative Omics Data Science (CIODS), Yenepoya (Deemed to be University), Mangalore, Karnataka, India

Arshad Rizvi Department of Chemistry and Biochemistry, The University of Oklahoma, Norman, OK, USA

Partha Roy Molecular Endocrinology Laboratory, Department of Biosciences and Bioengineering, Indian Institute of Technology Roorkee, Roorkee, Uttarakhand, India

Samir Kumar Samanta Pharmacology Division, Dr. B.C. Roy College of Pharmacy and Allied Health Sciences, Durgapur, West Bengal, India

Asif Shaik Department of Pharmaceutical Chemistry, JSS College of Pharmacy, JSS Academy of Higher Education & Research, Ooty, The Nilgiris, Tamil Nadu, India

Akshay Shamnewadi Department of Pharmaceutics, KLE College of Pharmacy, Hubballi, KLE Academy of Higher Education and Research, Deemed-to-be-University, Belagavi, Karnataka, India

Department of Pharmacology, KLE College of Pharmacy, Belagavi, KLE Academy of Higher Education and Research, Deemed-to-be-University, Belagavi, Karnataka, India

Himanshu Sharma Teerthanker Mahaveer College of Pharmacy, Teerthanker Mahaveer University, Moradabad, Uttar Pradesh, India

Preeti Sharma Pranveer Singh Institute of Technology (Pharmacy), Bhauti, Kanpur, Uttar Pradesh, India

Krishna Shevate Department of Pharmaceutical Chemistry, JSS College of Pharmacy, JSS Academy of Higher Education & Research, Ooty, The Nilgiris, Tamil Nadu, India

Saroj Singhmura Pharmacology Division, Dr. B. C. Roy College of Pharmacy & Allied Health Sciences, Durgapur, West Bengal, India

Aamir Suhail The Gene Lay Institute of Immunology and Inflammation, Brigham and Women's Hospital, Mass General Hospital, and Harvard Medical School, Boston, MA, USA

Ann Romney Center for Neurologic Diseases, Brigham and Women's Hospital, Boston, MA, USA

Broad Institute of MIT and Harvard, Cambridge, MA, USA

Sagarika Deepthy Tallapragada Pharmacology Division, Dr. B. C. Roy College of Pharmacy & Allied Health Sciences, Durgapur, West Bengal, India

Vamshi Krishna Tippavajhala Department of Pharmaceutics, Manipal College of Pharmaceutical Sciences, Manipal Academy of Higher Education, Manipal, Karnataka, India

Prabhat Upadhyay Vincent Center for Reproductive Biology, Department of Obstetrics and Gynaecology, Massachusetts General Hospital, Harvard Medical School, Boston, MA, USA

Anushka Verma Faculty of Pharmaceutical Sciences, Rama University, Madhana, Kanpur, Uttar Pradesh, India

Neha Verma Pranveer Singh Institute of Technology (Pharmacy), Bhauti, Kanpur, Uttar Pradesh, India

Ankita Wal Pranveer Singh Institute of Technology (Pharmacy), Bhauti, Kanpur, Uttar Pradesh, India

Poonam Yadav Department of Pharmacology and Chemical Biology, School of Medicine, Emory University School of Medicine, Emory University, Atlanta, GA, USA

Part I

Neuroinflammation and Autophagy in Neurodegenerative Diseases

Neuroinflammation and Its Impact on Autophagy in Neurodegeneration

1

Kunal Bhattacharya, Nongmaithem Randhoni Chanu, Dibyajyoti Das, and Rabina Kumpakha

Abstract

Neurodegenerative diseases represent a significant and escalating global health challenge, with no curative therapies currently available. Among the converging cellular and molecular mechanisms implicated in neurodegeneration, chronic neuroinflammation and impaired autophagy constitute a pathogenic feedback loop that accelerates proteostatic disruption, mitochondrial dysfunction, and synaptic degeneration. This chapter critically examines current evidence elucidating how microglial and astrocytic inflammatory signaling pathways, predominantly nuclear factor kappa B (NF-κB) and NLRP3 inflammasome activation, impair autophagic flux through key regulatory nodes, including mTORC1 hyperactivation, suppression of Beclin-1, and lysosomal membrane destabilization. In turn, autophagy failure exacerbates neuroinflammatory responses by endorsing the accumulation of dysfunctional organelles and protein aggregates. In addition, disease-specific patterns are explored, with emphasis on amyloid-β-driven pathology in Alzheimer's disease, α-synuclein propagation in Parkinson's disease, and TDP-43 aggregation in amyotrophic lateral sclerosis and frontotemporal dementia. Finally, we discuss emerging therapeutic strategies that combine autophagy-enhancing agents with anti-inflammatory interventions, including advanced drug delivery platforms capable of crossing the blood–brain barrier. By explaining the bidirectional interplay between autophagy and neuroinflammation, this chapter provides a mechanistic framework for the

K. Bhattacharya (✉) · N. R. Chanu · D. Das
Pratiksha Institute of Pharmaceutical Sciences, Guwahati, Assam, India
e-mail: kunal22101994@gmail.com; randhoni27@gmail.com; dibyajyoti184@gmail.com

R. Kumpakha
Emory University School of Medicine, Atlanta, GA, USA
e-mail: rabina.kumpakha@emory.edu

P. Khanal et al. (eds.), *Neuroinflammation and Autophagy in Neurodegeneration*,
https://doi.org/10.1007/978-981-95-3945-1_1

rational design of targeted therapies aimed at interrupting the self-amplifying cycles that drive neurodegeneration.

Keywords

Autophagic flux · Microglia · mTOR · Neurodegeneration · NLRP3 inflammasome · Therapeutic targets

1.1 Introduction

Neurodegenerative disorders affect a substantial portion of the global population, with an estimated 57 million individuals affected by dementia syndromes, of which Alzheimer's disease (AD) accounts for approximately 60–70% of cases [1]. Parkinson's disease (PD) currently affects 8.5 million individuals worldwide and is recognized as the fastest-growing neurological disorder [2]. Although amyotrophic lateral sclerosis (ALS) and frontotemporal dementia (FTD) are less common, yet they contribute disproportionately to the neurodegenerative disease burden. Recent global estimates suggest approximately 330,000 individuals are living with ALS [3], whereas FTD represents 3–26% of all dementia diagnoses [4]. The aggregate socioeconomic impact of these disorders is projected to surpass US$3 trillion annually by 2050, highlighting the critical need for effective disease-modifying interventions. Neuroinflammation represents the brain innate immune response to danger-associated molecular patterns (DAMPs). Age-related changes, proteinopathies, and genetic variants predominantly in genes such as TREM2 can manipulate microglial function, lowering their activation threshold and perpetuating a chronic inflammatory state characterized by sustained tumor necrosis factor-α (TNF-α), interleukin-1β (IL-1β), and reactive oxygen species (ROS) release [5, 6]. Additionally, autophagy plays a critical role in maintaining neuronal proteostasis and bioenergetic homeostasis, whereas the canonical autophagic cascade, involving ULK1-mediated initiation, Beclin-1-dependent nucleation, LC3-II-driven phagophore elongation, and subsequent lysosomal degradation, is highly responsive to nutrient availability and cellular stress. Emerging evidence implicates inflammatory signaling pathways as potent regulators of autophagy and establishes a deleterious feed-forward loop that exacerbates neuronal vulnerability, accelerating neurodegenerative processes (Table 1.1).

1.2 Mechanistic Landscape of Neuroinflammation in Neurodegenerative Diseases

Neuroinflammation is not a monolithic event but a multicellular cascade that involves resident glia, peripheral immune cells, complement factors, and a dense cytokine-chemokine web. These factors increase proteotoxic and metabolic stress, fueling neuronal loss in AD, PD, ALS, and FTD. Under homeostatic conditions,

Table 1.1 Mechanistic intersections between neuroinflammation and autophagy in neurodegeneration

Mechanistic node	Inflammatory effector	Autophagy consequence	Key evidence
mTORC1 hyper-activation	PI3K–Akt signaling triggered by IL-1β and TNF-α	Blocks ULK1 complex, halting autophagosome formation	Maiese (2016) [7]
Transcriptional repression	NF-κB (p65/RelA) suppresses BECN1, ATG5, and MAP1LC3B	Reduces autophagic gene expression	Lei and Klionsky (2023) [8]
Hyperactive secretory autophagy	SKA2 drives pathological cytokine release	Propagates microglial NLRP3 activation	Hartmann et al. (2024) [9]
Inflammasome activation	NLRP3–caspase-1 axis amplifies IL-1β and IL-18	Promotes mitochondrial damage and blocks mitophagy	Xu et al. (2025) [10]
Mitophagy deficit	Accumulated damaged mitochondria and ROS	Sustains DAMP production and microglial activation	Wang et al. (2023) [11]

microglia actively survey the brain parenchyma through highly motile processes and contribute to tissue maintenance by clearing cellular debris through a TREM2-dependent phagocytic mechanism [12]. Importantly, proteomic analyses of cerebrospinal fluid from patients with AD consistently demonstrate a pro-inflammatory cytokine profile [12]. These inflammatory mediators are mTORC1 and concurrently suppress transcription factor EB (TFEB)-dependent lysosomal biogenesis. This results in autophagic clearance impairment, promoting DAMPs accumulation, thereby perpetuating a feed-forward loop of neuroinflammation and autophagic dysfunction. Chronic aggregation of amyloid-β (Aβ), α-synuclein, or hyperphosphorylated tau induces a phenotypic shift in microglia from a homeostatic state toward a pro-inflammatory M1-like profile, which is marked by elevated expression of inducible nitric oxide synthase, NADPH oxidase 2, IL-1β, and TNF-α. Although an anti-inflammatory M2-like phenotype, which is characterized by markers such as arginase-1, insulin-like growth factor-1, and transforming growth factor-β, has been defined, recent studies in AD and ALS reveal a transcriptional continuum of disease-associated microglial states [13]. Sustained disease-associated microglial signaling activates NLRP3 inflammasome to link microglial priming to autophagy failure. Furthermore, astrocytes undergo reactive transformation into at least two principal phenotypic states. First, the neurotoxic A1 phenotype, which is characterized by upregulation of complement component C3 and Serping1, and is induced by a combination of IL-1α, TNF-α, and complement component C1q. Second, the neuroprotective A2 phenotype that is marked by S100A10 and thrombospondins expression, which supports synaptogenesis and neuronal survival [14, 15]. In AD mouse models, plaque-associated astrocytes contribute to complement-mediated synaptic loss under conditions of impaired microglial phagocytic function. The dynamic equilibrium between A1 and A2 astrocytic states thus emerges as a pivotal determinant of neuronal integrity and resilience [16]. Moreover, age-associated disruption of blood–brain barrier (BBB) integrity facilitates peripheral immune cells infiltration, including $CD4^+$ and $CD8^+$ T lymphocytes, B cells, and monocytes, into the central nervous system. In this compartment, these cells enhance α-synuclein-specific Th17 responses, thereby exacerbating nigrostriatal degeneration in PD [17]. Concurrently, reactive astrocytes secrete MMP-9 and VEGF, which further destabilize the BBB and propagate a self-reinforcing cycle of neuroinflammation [18].

In AD, Aβ and hyperphosphorylated tau can elicit lysosomal destabilization in microglia, initiating NLRP3 inflammasome assembly. This process leads to caspase-1 activation and gasdermin-D cleavage, resulting in pyroptotic cell death [10, 19]. Next, the consequent release of mature IL-1β and IL-18 from pyroptotic microglia further exacerbates neuroinflammation and impairs autophagic flux by promoting mTORC1 hyperactivation and nuclear factor kappa B (NF-κB)-dependent transcriptional repression of autophagy-related genes. Furthermore, complement components C1q and C3, which physiologically mediate synaptic pruning during neurodevelopment by tagging weak or redundant synapses for elimination, are abnormally activated in neurodegenerative conditions. Looking into microglial dysfunction, astrocytes are involved in C1q-dependent phagocytic activity as a compensatory mechanism. However, this may result in excessive synaptic loss, disruption of excitatory–inhibitory balance, and progressive cognitive impairment [16, 20]. Therefore, targeting the complement cascade has emerged as a promising disease-modifying strategy, with therapeutic agents such as pegcetacoplan currently undergoing evaluation in clinical trials for indications including geographic atrophy.

1.3 Autophagy in Neurodegeneration

Macroautophagy is the principal degradative pathway for cytoplasmic constituents, which proceeds through a highly conserved, multistep process. Its initiation is regulated by the ULK1 complex, followed by phagophore nucleation mediated by the Beclin-1/VPS34 class III phosphatidylinositol 3-kinase complex. Subsequent phagophore elongation is facilitated by LC3-II conjugation, culminating in the formation of mature, double-membraned autophagosomes. Fusion with lysosomes enables the enzymatic degradation of cargo, and its disruption at any stage results in intracellular accumulation of p62/sequestosome-1-positive protein aggregates, dysfunctional mitochondria, and oxidized lipids, which are the key triggering factors for microglial activation and neuroinflammatory amplification. Autophagic activity is tightly modulated by nutrient and energy status through the reciprocal regulation of mTORC1 and AMPK. Under nutrient-rich conditions, mTORC1 is activated and inhibits ULK1, thereby halting autophagosome biogenesis. Conversely, energy stress characterized by a high AMP/ATP ratio activates AMPK, which phosphorylates Raptor, resulting in mTORC1 inhibition and autophagy activation [21]. Both AMPK and calcineurin also dephosphorylate TFEB, enabling its nuclear translocation and subsequent transcriptional activation of autophagy and lysosomal gene networks. Chronic exposure to inflammatory cytokines skews this regulatory balance toward sustained mTORC1 activation, contributing to the age-related decline in autophagic flux observed in neurodegenerative diseases [22]. Selective autophagy enables the targeted degradation of specific cellular substrates critical for neuronal homeostasis. For example, mitophagy is mediated by the PINK1/Parkin axis and selectively labels depolarized mitochondria for autophagic clearance; pathogenic mutations in PINK1 or PRKN are causally linked to early-onset PD [23]. Aggrephagy facilitates ubiquitylated protein aggregate removal through cargo adaptors such as p62, optineurin, and NBR1. Other forms, including reticulophagy and lysophagy, mitigate endoplasmic reticulum stress and lysosomal damage, respectively. These selective pathways are indispensable in post-mitotic neurons, which lack the proliferative dilution efficacy to eliminate damaged macromolecules.

Moreover, autophagic efficiency declines with age, as evidenced by nearly 40% reduction in autophagosome biogenesis in rodent and human cortex from early adulthood to late life. This decline correlates with increased mTORC1 activity, lysosomal membrane permeabilization, and cathepsin leakage [24]. Therapeutic interventions that enhance autophagy, such as rapamycin, trehalose, or TFEB overexpression, have demonstrated efficacy in multiple AD mouse models, improving cognitive performance and reducing amyloid plaque burden. These findings underscore autophagy as a central therapeutic axis in the management of neurodegenerative disorders.

1.4 Molecular Interplay Bridging Neuroinflammation and Autophagy

Neuroinflammation and autophagy are not parallel phenomena but are functionally intertwined through a tightly regulated, bidirectional feedback loop. Canonical inflammatory signaling pathways, including NF-κB, mTORC1, and the NLRP3 inflammasome, act as potent inhibitors of autophagic flux. Conversely, impaired autophagy promotes the accumulation of DAMPs, dysfunctional organelles, and protein aggregates, which further activate innate immune pathways. This mutual relationship forms a pathological magnification circuit that contributes to neurodegenerative diseases progression (Fig. 1.1).

Involvement of pattern recognition receptors such as TLR4 or the receptor for advanced glycation end-products by Aβ oligomers or high-mobility group box 1 activates the IκB kinase complex, leading to IκBα phosphorylation and degradation. This event liberates NF-κB subunits (p65/p50), which translocate to the nucleus to induce transcription of pro-inflammatory genes including IL-1β, TNF-α, PTGS2, and CYBB [25]. Beyond its pro-inflammatory role, NF-κB also functions as a transcriptional repressor of autophagy by binding κB response elements in the promoters of core autophagy genes such as BECN1, ATG5, and MAP1LC3B, thereby suppressing their expression and autophagosome formation [26]. Chromatin

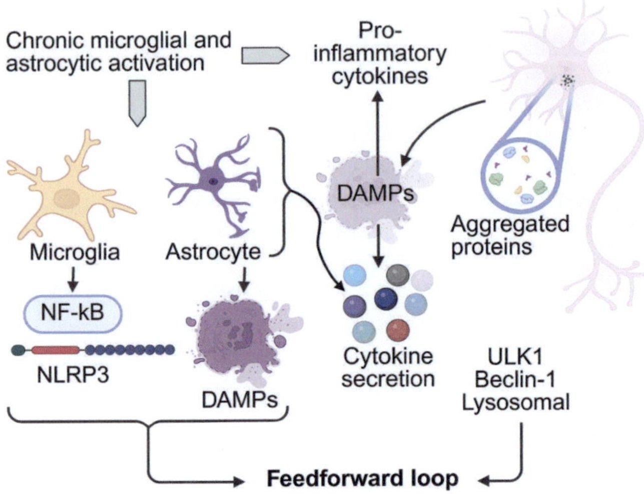

Fig. 1.1 Bidirectional feed-forward loop linking neuroinflammation and autophagy dysfunction in neurodegeneration. Inflammatory signaling through NF-κB and NLRP3 impairs autophagic flux. In turn, autophagy blockade promotes DAMP accumulation, which reactivates microglia and astrocytes, amplifying the cycle. (Created using BioRender. https://www.biorender.com/)

immunoprecipitation studies in AD mouse models confirmed reduced Beclin-1 promoter occupancy concurrent with increased RelA (p65) binding, which is a molecular signature of an autophagy-deficient, inflammation-dominant state.

ULK1 is the upstream initiator of autophagy and is subjected to reciprocal regulation by mTORC1 and AMPK. AMPK is activated by elevated AMP/ATP ratios and phosphorylates ULK1 at Ser^{555} to promote autophagy initiation, whereas mTORC1 phosphorylates ULK1 at Ser^{757} to inhibit its activity [27]. Pro-inflammatory cytokines such as TNF-α and IL-1β promote activation of the PI3K–Akt–Rheb axis, leading to sustained mTORC1 signaling and suppression of autophagy. Notably, mTORC1 and AMPK localize to the lysosomal membrane, forming a molecular "toggle switch" that integrates nutrient and inflammatory cues. Chronic inflammation skews this switch toward mTORC1 dominance, contributing to autophagic insufficiency in neurodegenerative settings.

ROS and reactive nitrogen species, generated by NADPH oxidase 2 and inducible nitric oxide synthase, oxidize lysosomal membrane proteins such as LAMP-1 and LAMP-2, leading to lysosomal membrane permeabilization. The resulting leakage of cathepsins into the cytosol activates caspase-3 and reinforces NF-κB signaling [28]. In parallel, oxidative carbonylation impairs cathepsin B and D activity, stalling cargo degradation and augmenting proteotoxic stress [29]. Moreover, it has been demonstrated that elevated cathepsin B activity accelerates Aβ-mediated neurotoxicity and cognitive decline in 5×FAD mice [30]. Furthermore, NLRP3 inflammasome activation facilitates adaptor protein ASC and caspase-1 assembly, leading to cleavage of gasdermin-D and membrane pore formation. This pyroptotic response not only results in IL-1β release but also induces lysosomal disruption and cytosolic leakage of galectin-3 [31]. This can result in a sustained autophagy suppression through transcriptional repression of lysosomal biogenesis, linking inflammasome activation directly to autophagic arrest. Interestingly, extracellular α-synuclein fibrils also activate microglial TLR2, initiating MyD88-dependent NF-κB signaling and subsequent IL-1β secretion, which suppresses neuronal autophagy. Intracellularly, α-synuclein interacts with cargo adaptor p62 but fails to progress through the autophagy–lysosome pathway, leading to clogging of the autophagic system. This defective clearance further propagates proteotoxic stress and inflammatory activation [32, 33].

1.5 Pathophysiological Consequences of Neuroinflammation and Autophagy Crosstalk

Autophagic dysfunction leads to the accumulation of p62-positive protein aggregates, including hyperphosphorylated tau, TDP-43, and α-synuclein. These aggregates act as nucleation sites for further misfolding and physically impair synaptic architecture. Concurrently, the aberrant reactivation of developmental complement pathways, specifically C1q and C3, results in the mis-targeting of excitatory synapses. Moreover, inflammation promotes A1 astrocyte activation, leading to dysregulated synapse elimination and impaired cognition [34–36]. Further, ineffective clearance of dysfunctional mitochondria due to impaired PINK1/Parkin-mediated mitophagy leads to depolarized mitochondria. This not only escalates intracellular ROS levels but also promotes mitochondrial DNA release, which is a potent DAMP and direct activator of the NLRP3 inflammasome, propagating a vicious cycle of inflammation and cellular stress [37–39]. Likewise, neuroinflammatory signaling upregulates hepcidin expression, downregulating iron efflux transporter ferroportin. The resulting intracellular accumulation of ferrous iron fuels Fenton reactions that catalyze lipid peroxidation, and neurons with high metabolic demand become susceptible to ferroptosis, which is regulated necrosis characterized by iron-dependent oxidative damage to cellular membranes [40–44]. These interconnected pathological processes converge to establish a self-amplifying circuit. Pro-inflammatory cytokines activate the mTORC1 and NF-κB signaling pathways, which in turn suppress autophagic flux. Impaired autophagy promotes the accumulation of DAMPs, further activating microglia and sustaining a pro-inflammatory, neurotoxic milieu. Disrupting this reciprocal interaction between inflammation and autophagy is, therefore, a critical therapeutic objective in neurodegenerative disease intervention.

1.6 Snapshots Across Major Neurodegenerative Diseases

In AD, Aβ plaques chronically activate TREM2 on plaque-associated microglia, and loss-of-function variants or age-related downregulation of TREM2 shift these cells from phagocytic "barrier" phenotypes to dystrophic states, which are incapable of clearing Aβ or LC3-associated tau [45–47]. Tau-seeding models demonstrate that TREM2 knockout increases phosphorylated tau dissemination, underscoring its role in constraining tau spread through autophagy–endolysosomal mechanisms [48]. Concurrently, Aβ-induced NLRP3 activation drives caspase-1-mediated gasdermin-D cleavage and lysosomal membrane permeabilization. The resulting cathepsin leakage and luminal acidification defects suppress tau degradation and exacerbate cytokine release [49]. Similarly, in PD, α-synuclein fibrils robustly stimulate NLRP3 inflammasome activity in microglia, promoting IL-1β maturation and dopaminergic neuron loss [50]. Genetic or pharmacological blockade of NLRP3 attenuates fibril pathology and motor deficits in pre-formed fibril-seeded mice [51]. Furthermore, downstream p38 kinase impairs TFEB function by phosphorylation and nuclear exclusion, reducing lysosomal gene transcription. Additionally, either p38 or NLRP3 inhibition restores TFEB nuclear localization, revives autophagic flux, and preserves substantia nigra integrity [52]. PD risk alleles in GBA1 lower glucocerebrosidase activity, causing lipid-laden, alkalized lysosomes that synergistically block α-synuclein clearance and autophagosome–lysosome fusion, explaining the disproportionately severe synucleinopathy in glucocerebrosidase-PD cohorts [53, 54]. Furthermore, in both ALS and FTD, hexanucleotide repeat expansions in C9orf72 translate into arginine-rich dipeptide-repeat proteins that bind coatomer protein complex I and stall autophagosome-to-lysosome trafficking, leading to swollen LC3-positive vacuoles and cytotoxic stress granules in motor neurons [55]. Simultaneously, insoluble TDP-43 aggregates sequester small heat-shock protein HSPB8, which is a key chaperone that, together with BAG3, drives p62-cargo delivery to nascent autophagosomes; HSPB8 depletion exacerbates truncated TDP-43 accumulation and accelerates neurodegeneration in zebrafish and mouse models [56]. Recent human-iPSC microglia harboring progranulin deficiency further reveal lysosomal alkalinization, cytoplasmic TDP-43 mislocalization, and runaway NF-κB activation, which solidify the convergence of autophagic failure and neuroinflammation in FTD pathobiology [57].

1.7 Therapeutic Implications in Targeting the Neuroinflammation–Autophagy Axis

Emerging pharmacological strategies suggest that attenuating maladaptive inflammatory signaling or restoring impaired autophagy can mitigate neurodegenerative progression in preclinical models. However, a translationally promising approach can integrate both modalities with advanced delivery platforms designed to effectively traverse the BBB. Here, we look into some lead compounds and mechanistic strategies that demonstrate potential for clinical development (Fig. 1.2).

Anti-inflammatory strategies that secondarily restore autophagy with NLRP3 blockade using small molecule MCC950 (also called CP-456,773) reverse cognitive decline and normalize LC3-II/Beclin-1 levels in APP/PS1 or Aβ-injected rodents by shutting down NLRP3–caspase-1 signaling, which otherwise cleaves lysosomal gasdermin-D and stalls autophagic flux [58]. Likewise, another molecule,

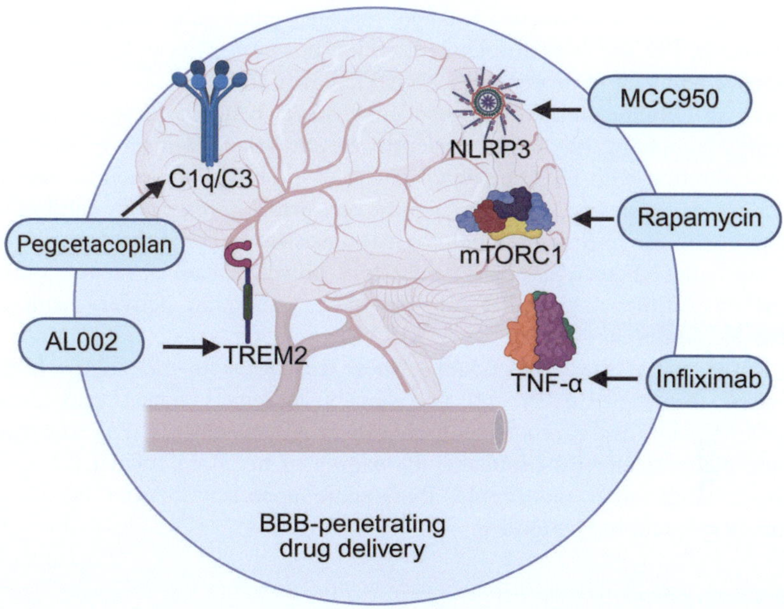

Fig. 1.2 Therapeutic targets across the neuroinflammation–autophagy axis. Key targets involved in neuroinflammation and autophagy are shown with corresponding therapeutic agents. (Created using BioRender. https://www.biorender.com/)

Dapansutrile (OLT1177), also lowers α-synuclein-triggered IL-1β and reduces nigral loss in MPTP-lesioned mice, supporting its entry into a first-in-human PD program [59]. Furthermore, microglial modulating two therapeutic antibodies targeting TREM2 activation are currently under investigation. First, Ab2 TVD-Ig/αTfR is a tetravalent agonist engineered with a transferrin receptor-binding single-chain domain to facilitate BBB transcytosis. In 5×FAD mice models, this construct enhanced SYK phosphorylation, induced disease-associated microglia-like transcriptional signature, and significantly reduced amyloid plaque burden [60]. Second, AL002 is a humanized IgG1 agonist with an Fc-silenced backbone, demonstrating central nervous system penetration and a dose-dependent decrease in cerebrospinal fluid soluble TREM2 with an acceptable safety profile in clinical evaluation. Additionally, its murine analog, AL002c, has shown comparable efficacy in reducing fibrillar amyloid deposition and neuritic dystrophy in 5×FAD mice expressing human TREM2 [61]. Furthermore, direct autophagy enhancement with mTOR-independent boosters such as disaccharide trehalose activates TFEB through Akt inhibition, thereby upregulating lysosomal and autophagy genes without suppressing mTORC1; chronic oral dosing ameliorates proteinopathy in multiple AD and PD disease models [62–65]. Additionally, small-molecule enhancer of rapamycin (SMER)-28 allosterically activates ULK1/Vps34, promotes neurite outgrowth, and protects primary neurons from excitotoxic death [66]. Next, lysosomal stabilizers such as N-acetyl-l-cysteine augment glutathione, intercept ROS signaling, and prevent cathepsin leakage; repeated intravenous dosing raises brain glutathione in PD and is well-tolerated, which positions N-acetyl-l-cysteine as an inexpensive adjuvant to preserve lysosomal integrity [67].

Moreover, combinatorial regimens may have better effects as rapamycin reactivates autophagy but may transiently elevate pro-inflammatory IL-6 levels. Co-administration with FDA-approved anti-TNF-α antibodies, such as infliximab, may mitigate systemic cytokine spill-over while preserving inhibition of phosphorylated S6 (p-S6). Epidemiological data also indicate that individuals on long-term anti-TNF therapy for inflammatory bowel disease have a significantly reduced risk of developing PD [68–70]. Next-generation delivery with brain-targeted exosomes engineered with RVG-Lamp2b or transferrin peptides ferry small RNAs across the BBB. Loading these vesicles with miR-124, which is a microRNA that represses NF-κB and boosts autophagy genes, may attenuate neuroinflammation and promote neurogenesis in AD models [71–73]. The modular exosome platform enables simultaneous loading of trehalose and ULK1 agonists, offering a single-shot, multimodal therapeutic approach tailored to individual inflammatory–autophagy profiles.

1.8 Conclusion and Perspectives

Neuroinflammation and autophagy are interconnected in a bidirectional, self-reinforcing cycle that exacerbates proteotoxic burden and neuronal degeneration across neurodegenerative diseases. Thus, therapeutic efficacy depends on recognizing microglia, astrocytes, and neurons that activate distinct autophagy–inflammatory pathways, necessitating cell-type-specific intervention strategies. Current multi-omics analyses indicate that targeting a single component of this loop, such as inflammasome activation, often produces only temporary benefits, highlighting the importance of combinatorial therapies that concurrently suppress inflammation and restore autophagic function. However, multiple key questions remain unresolved. One major challenge is defining the temporal inflection point at which initially protective neuroinflammatory responses become chronically detrimental. Another is the development of reliable biomarkers, such as cerebrospinal fluid LC3-II or galectin-3, to monitor autophagic efficiency and inflammatory load in real time. Finally, the potential of artificial intelligence to identify small molecules that co-regulate NF-κB and autophagy pathways is being explored, with early candidates emerging for targets once considered pharmacologically intractable. As longitudinal studies increasingly integrate fluid biomarkers, single-cell epigenetic profiling, and computational drug discovery, the field advances toward precision-based, combinatorial therapies that recalibrate neuroimmune dynamics by maintaining neuronal proteostasis throughout aging.

References

1. World Health Organization (WHO). Dementia – key facts. Fact-sheet, updated 31 Mar 2025 [Internet]. Geneva: WHO; 2025 [cited 2025 Jul 09]. Available from: https://www.who.int/news-room/fact-sheets/detail/dementia.
2. World Health Organization (WHO). Parkinson disease – key facts. Fact-sheet, 9 Aug 2023 [Internet]. Geneva: WHO; 2023 [cited 2025 Jul 09]. Available from: https://www.who.int/news-room/fact-sheets/detail/parkinson-disease.
3. Wolfson C, Gauvin DE, Ishola F, Oskoui M. Global prevalence and incidence of amyotrophic lateral sclerosis: a systematic review. Neurology. 2023;101(6):e613. https://doi.org/10.1212/WNL.0000000000207474.
4. Henríquez F, Cabello V, Baez S, Souza LC, Lillo P, Olavarría L, et al. Multidimensional clinical assessment in frontotemporal dementia and its spectrum in Latin America and the Caribbean: a narrative review and a glance at future challenges. Front Neurol. 2022;12:768591. https://doi.org/10.3389/fneur.2021.768591.
5. Shi Q, Gutierrez RA, Bhat MA. Microglia, Trem2, and neurodegeneration. Neuroscientist. 2025;31(2):159–76. https://doi.org/10.1177/10738584241254118.
6. Gao C, Jiang J, Tan Y, Chen S. Microglia in neurodegenerative diseases: mechanism and potential therapeutic targets. Signal Transduct Target Ther. 2023;8(1):1–37. https://doi.org/10.1038/s41392-023-01588-0.
7. Maiese K. Targeting molecules to medicine with mTOR, autophagy and neurodegenerative disorders. Br J Clin Pharmacol. 2016;82(5):1245–66. https://doi.org/10.1111/bcp.12804.
8. Lei Y, Klionsky DJ. Transcriptional regulation of autophagy and its implications in human disease. Cell Death Differ. 2023;30(6):1416. https://doi.org/10.1038/s41418-023-01162-9.
9. Hartmann J, Bajaj T, Otten J, Klengel C, Ebert T, Gellner A, et al. SKA2 regulated hyperactive secretory autophagy drives neuroinflammation-induced neurodegeneration. Nat Commun. 2024;15(1):1–20. https://doi.org/10.1038/s41467-024-46953-x.
10. Xu W, Huang Y, Zhou R. NLRP3 inflammasome in neuroinflammation and central nervous system diseases. Cell Mol Immunol. 2025;22(4):341–55. https://doi.org/10.1038/s41423-025-01275-w.
11. Wang S, Long H, Hou L, Feng B, Ma Z, Wu Y, et al. The mitophagy pathway and its implications in human diseases. Signal Transduct Target Ther. 2023;8(1):1–28. https://doi.org/10.1038/s41392-023-01503-7.
12. AmeliMojarad M, AmeliMojarad M. The neuroinflammatory role of microglia in Alzheimer's disease and their associated therapeutic targets. CNS Neurosci Ther. 2024;30(7):e14856. https://doi.org/10.1111/cns.14856.
13. Adamu A, Li S, Gao F, Xue G. The role of neuroinflammation in neurodegenerative diseases: current understanding and future therapeutic targets. Front Aging Neurosci. 2024;16:1347987. https://doi.org/10.3389/fnagi.2024.1347987.
14. Santiago-Balmaseda A, Aguirre-Orozco A, Valenzuela-Arzeta IE, Villegas-Rojas MM, Pérez-Segura I, Jiménez-Barrios N, et al. Neurodegenerative diseases: unraveling the heterogeneity of astrocytes. Cells. 2024;13(11):921. https://doi.org/10.3390/cells13110921.
15. Santos DE, Silva Lima SA, Moreira LS, Costa SL, Schitine S. New perspectives on heterogeneity in astrocyte reactivity in neuroinflammation. Brain Behav Immun Health. 2025;44:100948. https://doi.org/10.1016/j.bbih.2025.100948.
16. Dejanovic B, Wu T, Tsai M, Graykowski D, Gandham VD, Rose CM, et al. Complement C1q-dependent excitatory and inhibitory synapse elimination by astrocytes and microglia in Alzheimer's disease mouse models. Nat Aging. 2022;2(9):837–50. https://doi.org/10.1038/s43587-022-00281-1.
17. Lau K, Kotzur R, Richter F. Blood–brain barrier alterations and their impact on Parkinson's disease pathogenesis and therapy. Transl Neurodegener. 2024;13:37. https://doi.org/10.1186/s40035-024-00430-z.

18. Roodveldt C, Bernardino L, Oztop-Cakmak O, Dragic M, Fladmark KE, Ertan S, et al. The immune system in Parkinson's disease: what we know so far. Brain. 2024;147(10):3306. https://doi.org/10.1093/brain/awae177.
19. Jha D, Bakker ENTP, Kumar R. Mechanistic and therapeutic role of NLRP3 inflammasome in the pathogenesis of Alzheimer's disease. J Neurochem. 2024;168(10):3574–98. https://doi.org/10.1111/jnc.15788.
20. Zelek WM, Tenner AJ. Complement therapeutics in neurodegenerative diseases. Immunobiology. 2025;230(4):153089. https://doi.org/10.1016/j.imbio.2025.153089.
21. Kim H. Contrasting views on the role of AMPK in autophagy. Bioessays. 2024;46(3): e2300211. https://doi.org/10.1002/bies.202300211.
22. Chen H, Gong S, Zhang H, Chen Y, Liu Y, Hao J, et al. From the regulatory mechanism of TFEB to its therapeutic implications. Cell Death Discov. 2024;10(1):1–11. https://doi.org/10.1038/s41420-024-01850-6.
23. Sever T, Erbaş O. The role of the PINK1-Parkin pathway in mitophagy and its implications in neurodegenerative disorders. J Exp Basic Med Sci. 2024;5(4):264–70.
24. Palmer JE, Wilson N, Son SM, Obrocki P, Wrobel L, Rob M, et al. Autophagy, aging, and age-related neurodegeneration. Neuron. 2025;113(1):29–48. https://doi.org/10.1016/j.neuron.2024.09.015.
25. Trocoli A, Djavaheri-Mergny M. The complex interplay between autophagy and NF-κB signaling pathways in cancer cells. Am J Cancer Res. 2011;1(5):629. https://pmc.ncbi.nlm.nih.gov/articles/PMC3189824/.
26. Cao Z, Tian K, Ran Y, Zhou H, Zhou L, Ding Y, et al. Beclin-1: a therapeutic target at the intersection of autophagy, immunotherapy, and cancer treatment. Front Immunol. 2024;15: 1506426. https://doi.org/10.3389/fimmu.2024.1506426.
27. Smiles WJ, Ovens AJ, Kemp BE, Galic S, Petersen J, Oakhill JS. New developments in AMPK and mTORC1 cross-talk. Essays Biochem. 2024;68(3):321–36. https://doi.org/10.1042/EBC20240007.
28. Scott O, Saran E, Freeman SA. The spectrum of lysosomal stress and damage responses: from mechanosensing to inflammation. EMBO Rep. 2025;26(6):1425–39. https://doi.org/10.1038/s44319-025-00405-9.
29. Liu H, Liu T, Wu Y, Li A, Ren D, Cheng M, et al. Broadening horizons: exploring the cathepsin family as therapeutic targets for Alzheimer's disease. Aging Dis. 2024;16(4):1987. https://doi.org/10.14336/AD.2024.0456.
30. Siddiqui AA, Merquiol E, Hirbawi J, Boocholez H, Cohen I, Yan Y, et al. Cathepsin B promotes Aβ proteotoxicity by modulating aging regulating mechanisms. Nat Commun. 2024;15(1):1–14. https://doi.org/10.1038/s41467-024-52540-x.
31. McManus RM, Latz E. NLRP3 inflammasome signalling in Alzheimer's disease. Neuropharmacology. 2024;252:109941. https://doi.org/10.1016/j.neuropharm.2024.109941.
32. Dutta D, Jana M, Majumder M, Mondal S, Roy A, Pahan K. Selective targeting of the TLR2/MyD88/NF-κB pathway reduces α-synuclein spreading in vitro and in vivo. Nat Commun. 2021;12(1):1–19. https://doi.org/10.1038/s41467-021-25767-1.
33. Zhang X, Yu H, Feng J. Emerging role of microglia in inter-cellular transmission of α-synuclein in Parkinson's disease. Front Aging Neurosci. 2024;16:1411104. https://doi.org/10.3389/fnagi.2024.1411104.
34. Foster AD, Flynn LL, Cluning C, Cheng F, Davidson JM, Lee A, et al. P62 overexpression induces TDP-43 cytoplasmic mislocalisation, aggregation and cleavage and neuronal death. Sci Rep. 2021;11:11474. https://doi.org/10.1038/s41598-021-90822-2.
35. Brady OA, Meng P, Zheng Y, Mao Y, Hu F. Regulation of TDP-43 aggregation by phosphorylation and p62/SQSTM1. J Neurochem. 2011;116(2):248–59. https://doi.org/10.1111/j.1471-4159.2010.07098.x.
36. Zhang H, Jin Q, Li J, Wang J, Li M, Yin Q, et al. Astrocyte-derived complement C3 facilitated microglial phagocytosis of synapses in Staphylococcus aureus-associated neurocognitive deficits. PLoS Pathog. 2025;21(4):e1013126. https://doi.org/10.1371/journal.ppat.1013126.

37. Wu W, Xu H, Wang Z, Mao Y, Yuan L, Luo W, et al. PINK1-Parkin-mediated mitophagy protects mitochondrial integrity and prevents metabolic stress-induced endothelial injury. PLoS One. 2015;10(7):e0132499. https://doi.org/10.1371/journal.pone.0132499.
38. Barodia SK, Creed RB, Goldberg MS. Parkin and PINK1 functions in oxidative stress and neurodegeneration. Brain Res Bull. 2016;133:51. https://doi.org/10.1016/j.brainresbull.2016.12.004.
39. Zhao H, Zhang Y, Ren Y, Wang W. PINK1/Parkin-mediated mitophagy ameliorates mitochondrial dysfunction in lacrimal gland acinar cells during aging. Invest Ophthalmol Vis Sci. 2024;65(13):12. https://doi.org/10.1167/iovs.65.13.12.
40. Vela D. The dual role of hepcidin in brain iron load and inflammation. Front Neurosci. 2018;12:740. https://doi.org/10.3389/fnins.2018.00740.
41. Raha AA, Ghaffari SD, Henderson J, Chakraborty S, Allinson K, Friedland RP, et al. Hepcidin increases cytokines in Alzheimer's disease and down's syndrome dementia: implication of impaired iron homeostasis in neuroinflammation. Front Aging Neurosci. 2021;13:653591. https://doi.org/10.3389/fnagi.2021.653591.
42. Zhang DL, Rouault TA. How does hepcidin hinder ferroportin activity? Blood. 2018;131(8):840–2. https://doi.org/10.1182/blood-2018-01-824151.
43. Braughler JM, Duncan LA, Chase RL. The involvement of iron in lipid peroxidation. Importance of ferric to ferrous ratios in initiation. J Biol Chem. 1986;261(22):10282–9.
44. Li S, Huang Y. Ferroptosis: an iron-dependent cell death form linking metabolism, diseases, immune cell and targeted therapy. Clin Transl Oncol. 2021;24(1):1. https://doi.org/10.1007/s12094-021-02669-8.
45. Fracassi A, Marcatti M, Tumurbaatar B, Woltjer R, Moreno S, Taglialatela G. TREM2-induced activation of microglia contributes to synaptic integrity in cognitively intact aged individuals with Alzheimer's neuropathology. Brain Pathol. 2022;33(1):e13108. https://doi.org/10.1111/bpa.13108.
46. Lin M, Yu JX, Zhang WX, et al. Roles of TREM2 in the pathological mechanism and the therapeutic strategies of Alzheimer's disease. J Prev Alzheimers Dis. 2024;11:1682–95. https://doi.org/10.14283/jpad.2024.164.
47. Thomas AL, Lehn MA, Janssen EM, Hildeman DA, Chougnet CA. Naturally-aged microglia exhibit phagocytic dysfunction accompanied by gene expression changes reflective of underlying neurologic disease. Sci Rep. 2022;12(1):1–10. https://doi.org/10.1038/s41598-022-21920-y.
48. Li Y, Xu H, Wang H, Yang K, Luan J, Wang S. TREM2: potential therapeutic targeting of microglia for Alzheimer's disease. Biomed Pharmacother. 2023;165:115218. https://doi.org/10.1016/j.biopha.2023.115218.
49. Zhu C, Xu S, Jiang R, Yu Y, Bian J, Zou Z. The gasdermin family: emerging therapeutic targets in diseases. Signal Transduct Target Ther. 2024;9(1):1–33. https://doi.org/10.1038/s41392-024-01801-8.
50. Pike AF, Varanita T, Herrebout MAC, Plug BC, Kole J, Musters RJP, et al. α-Synuclein evokes NLRP3 inflammasome-mediated IL-1β secretion from primary human microglia. Glia. 2021;69(6):1413–28. https://doi.org/10.1002/glia.23970.
51. Yu J, Zhao Z, Li Y, Chen J, Huang N, Luo Y. Role of NLRP3 in Parkinson's disease: specific activation especially in dopaminergic neurons. Heliyon. 2024;10(7):e28838. https://doi.org/10.1016/j.heliyon.2024.e28838.
52. Chen J, Mao K, Yu H, et al. p38-TFEB pathways promote microglia activation through inhibiting CMA-mediated NLRP3 degradation in Parkinson's disease. J Neuroinflammation. 2021;18:295. https://doi.org/10.1186/s12974-021-02349-y.
53. Zhang X, Wu H, Tang B, Guo J. Clinical, mechanistic, biomarker, and therapeutic advances in GBA1-associated Parkinson's disease. Transl Neurodegener. 2024;13(1):48. https://doi.org/10.1186/s40035-024-00437-6.

54. Koros C, Bougea A, Alefanti I, Simitsi AM, Papagiannakis N, Pachi I, et al. A global perspective of GBA1-related Parkinson's disease: a narrative review. Genes. 2024;15(12): 1605. https://doi.org/10.3390/genes15121605.
55. Diab R, Pilotto F, Saxena S. Autophagy and neurodegeneration: unraveling the role of C9ORF72 in the regulation of autophagy and its relationship to ALS-FTD pathology. Front Cell Neurosci. 2023;17:1086895. https://doi.org/10.3389/fncel.2023.1086895.
56. Crippa V, Cicardi ME, Ramesh N, Seguin SJ, Ganassi M, Bigi I, et al. The chaperone HSPB8 reduces the accumulation of truncated TDP-43 species in cells and protects against TDP-43-mediated toxicity. Hum Mol Genet. 2016;25(18):3908–24. https://doi.org/10.1093/hmg/ddw232.
57. Sung W, Noh MY, Nahm M, Kim YS, Ki CS, Kim YE, et al. Progranulin haploinsufficiency mediates cytoplasmic TDP-43 aggregation with lysosomal abnormalities in human microglia. J Neuroinflammation. 2024;21(1):47. https://doi.org/10.1186/s12974-024-03039-1.
58. Naeem A, Prakash R, Kumari N, Khan MA, Khan AQ, Uddin S, et al. MCC950 reduces autophagy and improves cognitive function by inhibiting NLRP3-dependent neuroinflammation in a rat model of Alzheimer's disease. Brain Behav Immun. 2024;116:70–84. https://doi.org/10.1016/j.bbi.2023.11.031.
59. Amo-Aparicio J, Daly J, Højen JF, Dinarello CA. Pharmacologic inhibition of NLRP3 reduces the levels of α-synuclein and protects dopaminergic neurons in a model of Parkinson's disease. J Neuroinflammation. 2023;20(1):147. https://doi.org/10.1186/s12974-023-02830-w.
60. Zhao P, Xu Y, Fan X, Li L, Li X, Arase H, et al. Discovery and engineering of an anti-TREM2 antibody to promote amyloid plaque clearance by microglia in 5XFAD mice. MAbs. 2022;14 (1):2107971. https://doi.org/10.1080/19420862.2022.2107971.
61. Long H, Simmons A, Mayorga A, Burgess B, Nguyen T, Budda B, et al. Preclinical and first-in-human evaluation of AL002, a novel TREM2 agonistic antibody for Alzheimer's disease. Alzheimers Res Ther. 2024;16(1):235. https://doi.org/10.1186/s13195-024-01599-1.
62. Benito-Cuesta I, Ordoñez-Gutierrez L, Wandosell F. Trehalose reduces the secreted beta-amyloid levels in primary neurons independently of autophagy induction. Metabolites. 2021;11(7):421. https://doi.org/10.3390/metabo11070421.
63. Rusmini P, Cortese K, Crippa V, Cristofani R, Cicardi ME, Ferrari V, et al. Trehalose induces autophagy via lysosomal-mediated TFEB activation in models of motoneuron degeneration. Autophagy. 2019;15(4):631–51. https://doi.org/10.1080/15548627.2018.1535292.
64. Palmieri M, Pal R, Nelvagal HR, Lotfi P, Stinnett GR, Seymour ML, et al. MTORC1-independent TFEB activation via Akt inhibition promotes cellular clearance in neurodegenerative storage diseases. Nat Commun. 2017;8(1):1–19. https://doi.org/10.1038/ncomms14338.
65. Sarkar S, Davies JE, Huang Z, Tunnacliffe A, Rubinsztein DC. Trehalose, a novel mTOR-independent autophagy enhancer, accelerates the clearance of mutant huntingtin and alpha-synuclein. J Biol Chem. 2007;282(8):5641–52. https://doi.org/10.1074/jbc.M609532200.
66. Kirchenwitz M, Stahnke S, Grunau K, Melcher L, van Ham M, Rottner K, et al. The autophagy inducer SMER28 attenuates microtubule dynamics mediating neuroprotection. Sci Rep. 2022;12(1):17805. https://doi.org/10.1038/s41598-022-20563-3.
67. Coles LD, Tuite PJ, Öz G, Mishra UR, Kartha RV, Sullivan KM, et al. Repeated-dose oral N-acetylcysteine in Parkinson's disease: pharmacokinetics and effect on brain glutathione and oxidative stress. J Clin Pharmacol. 2018;58(2):158–67. https://doi.org/10.1002/jcph.1008.
68. Zhang Y, He X, Wu X, Lei M, Wei Z, Zhang X, et al. Rapamycin upregulates glutamate transporter and IL-6 expression in astrocytes in a mouse model of Parkinson's disease. Cell Death Dis. 2017;8(2):e2611. https://doi.org/10.1038/cddis.2016.491.
69. Cappoli N, Mezzogori D, Tabolacci E, Coletta I, Navarra P, Pani G, et al. The mTOR kinase inhibitor rapamycin enhances the expression and release of pro-inflammatory cytokine interleukin 6 modulating the activation of human microglial cells. EXCLI J. 2019;18:779. https://doi.org/10.17179/excli2019-1715.

70. Peter I, Dubinsky M, Bressman S, Park A, Lu C, Chen N, et al. Anti-tumor necrosis factor therapy and incidence of Parkinson disease among patients with inflammatory bowel disease. JAMA Neurol. 2018;75(8):939–46. https://doi.org/10.1001/jamaneurol.2018.0605.
71. Williams A, Branscome H, Kashanchi F, Batrakova EV. Targeting of extracellular vesicle-based therapeutics to the brain. Cells. 2025;14(7):548. https://doi.org/10.3390/cells14070548.
72. Mehdizadeh S, Mamaghani M, Hassanikia S, Pilehvar Y, Ertas YN. Exosome-powered neuropharmaceutics: unlocking the blood-brain barrier for next-gen therapies. J Nanobiotechnol. 2025;23:329. https://doi.org/10.1186/s12951-025-03352-8.
73. Évora A, Garcia G, Rubi A, Vitis ED, Matos AT, Vaz AR, et al. Exosomes enriched with miR-124-3p show therapeutic potential in a new microfluidic triculture model that recapitulates neuron–glia crosstalk in Alzheimer's disease. Front Pharmacol. 2025;16:1474012. https://doi.org/10.3389/fphar.2025.1474012.

Mechanisms of Autophagy and Its Dysregulation in Neurodegenerative Diseases

2

Saroj Singhmura, Sagarika Deepthy Tallapragada, Atreyee Ganguly, Jashabir Chakraborty, Abhishek Basu, and Yadu Nandan Dey

Abstract

Clinical evidence suggests that most of the neurodegenerative diseases are closely related to autophagy failure. This chapter decodes the evolving molecular interplay of autophagy, highlighting newly uncovered mechanisms and their disruption in neurodegeneration. It includes understanding canonical and non-canonical autophagy pathways, including roles of ATG9 vesicles and the regulatory effect of liquid–liquid phase separation on autophagosome assembly. This chapter also explores autophagy dysregulation in neurodegenerative disorders, from failures at endoplasmic reticulum–mitochondria contact to lysosomal system collapse and unexpected outcomes like disrupted lipid homeostasis and retrotransposon activation. With a rising provocative question, can targeted, organelle-specific autophagy enhancement restore function in the face of global pathway failure?

Keywords

Autophagy · Neurodegeneration · LINE-1 activation · Neuroinflammation

S. Singhmura · S. D. Tallapragada · A. Ganguly · Y. N. Dey (✉)
Pharmacology Division, Dr. B. C. Roy College of Pharmacy & Allied Health Sciences, Durgapur, West Bengal, India
e-mail: sarojsinghmura@bcrcp.org; sagarika.d@bcrcp.org; atreyee.ganguly@bcrcp.org; yadunandan.dey@bcrcp.org

J. Chakraborty
Department of Pharmacology, Mata Gujri College of Pharmacy, Mata Gujri University, Kishanganj, Bihar, India
e-mail: jashabir@mgcopharmacy.com

A. Basu
Section on Fibrotic Disorders, National Institute on Alcohol Abuse and Alcoholism, National Institutes of Health, Rockville, MD, USA
e-mail: abhishek.basu@nih.gov

P. Khanal et al. (eds.), *Neuroinflammation and Autophagy in Neurodegeneration*,
https://doi.org/10.1007/978-981-95-3945-1_2

2.1 Introduction

In neurodegenerative diseases like Alzheimer's disease (AD), amyotrophic lateral sclerosis (ALS), Parkinson's disease (PD), or Huntington's disease (HD), autophagy failure is considered one of the causes for disease progression [1–3]. The postmortem of AD brain has revealed autophagosome marker LC3 accumulation, confirming its colocalization with hyperphosphorylated tau [4]. Additionally, elevated tetraspanin levels, such as CD63 and CD81, are the outcome of autophagic flux failure and compromised lysosomal clearance [5]. It was also observed that FKBP52, which is a co-chaperone involved in protein folding, was reduced in PD [6]. Further, in postmortem PD brain, it was shown that elevated toll-like receptor-2 expression in an area of α-synuclein aggregation also increased the expression of autophagy receptor SQSTM1/p62 [7], followed by altered LAMP-1 expression in postmortem substantia nigra neurons [8]. These clinical observations illustrate a clear and consistent pattern that autophagic failure and lysosomal processes are not merely a consequence but a contributing factor in the progression of neurodegenerative diseases [2].

Autophagy was first described by Christian De Duve in 1963 [9], and this cellular recycling process is involved in degrading and removing damaged cytoplasmic components [10]. Macroautophagy, microautophagy, and chaperone-mediated autophagy (CMA) are mechanistically different, but their endpoint is delivering intracellular cargo to the lysosome. Macroautophagy is characterized in mammalian cells [11] and is involved in degrading larger protein aggregates and damaged organelles. In response to stress, a phagophore is formed from the lipid bilayer of plasma membrane, endoplasmic reticulum, and/or trans-Golgi, endosomes, or mitochondria [12–15]. Availability of cellular energy and AMPK activation also regulate autophagy switching. Intracellular protein aggregates are engulfed by an expanding phagophore, forming a double-membraned autophagosome that sequesters cargo. The autophagosome is then transported along microtubules to the microtubule-organizing center, where it fuses with endosomes to form amphisomes. These mature and fuse with lysosomes, where the cargo is degraded by acidic proteases. Finally, lysosomal membrane proteins export the resulting breakdown products back into the cytoplasm for reuse [16]. Next, microautophagy is similar to macroautophagy but is less understood in mammals and involves direct lysosomal engulfment via pinocytosis, remaining active even at rest [17]. CMA is distinct from other two autophagy pathways and mediates selective degradation of cytosolic proteins containing a KFERQ-like motif, through recognition by the chaperone hsc70 and its co-chaperones. These substrates are then delivered to lysosomes, where the receptor LAMP-2A binds them for translocation and degradation [18].

Autophagy regulators act through various pathways, including AMPK, mTOR, and PI3K/Akt, and some are independent of mTOR. AMPK activation promotes autophagy by inhibiting mTORC1 and activating ULK1. Compounds such as urolithin A, artemisinin, justicidin A, berberine, stigmasterol, arctigenin, and rutaecarpine are promising AMPK agonists with anti-AD effects [19]. Inhibition of mTORC1/mTORC2 activates autophagy and promotes neurotoxic proteins such as Aβ, tau, and α-synuclein clearance. Rapamycin and its analogs (everolimus, temsirolimus), and selective compounds like NR1 and DL001, effectively inhibit mTORC1. Moreover, curcumin and cubeben promote autophagy by suppressing PI3K/Akt, leading to mTORC1 inhibition [20]. mTOR-independent autophagy is regulated by pathways such as cAMP/Epac/PLCε, Ca^{2+}/calpain, and inositol/IP_3. Lithium, L-690,330, carbamazepine, and valproic acid enhance autophagy via the inositol/IP_3 pathway, whereas verapamil, loperamide, and related compounds inhibit calpain by reducing Ca^{2+} influx. Calpastatin, calpeptin, and NF449 block autophagy-suppressing feedback through cAMP signaling. Next, Trehalose, SMERs, and Tat-Beclin-1 directly induce autophagy independent of mTOR [21].

2.2 Molecular Mechanisms of Autophagy

2.2.1 Canonical and Non-canonical Autophagy Pathways

LC3 lipidation on the autophagosome membrane is a key step in cargo recruitment and fusion with lysosomes for degradation and recycling. Autophagosome formation involves ~18 essential ATG proteins. Beyond canonical autophagy, ATG proteins also function in membrane trafficking, signaling, organelle communication, and non-canonical pathways (Fig. 2.1). Canonical autophagy is Atg5/Atg7 dependent, which originates from the endoplasmic reticulum or mitochondria-associated membrane and primarily targets ubiquitinated proteins and organelles via p62/SQSTM1. In contrast, alternative autophagy is Atg5/Atg7-independent, derives membranes from trans-Golgi and late endosomes, and is crucial for mitophagy under stress [22–24]. Autophagosome assembly relies on liquid–liquid phase separation to concentrate autophagy proteins at membranes. Under stress, prion-like proteins such as TDP-43 and FUS localize to stress granules. In ALS, mutations promote aggregate formation, disrupting mTORC1 signaling and impairing ATG13 function in autophagy initiation [25].

2.2.2 Organelle-Specific Autophagy in Neurons

Autophagy is a catabolic process that clears protein aggregates and damaged organelles to maintain cellular homeostasis. Under stress conditions such as energy deprivation, oxidative stress, or hypoxia, organelle-specific autophagy, such as mitophagy, pexophagy, reticulophagy, ribophagy, lysophagy, and nucleophagy, is activated to remove dysfunctional organelles [11]. Axonal and dendritic regional degenerations are most abundantly identifiable early pathogenesis of many neurodegenerative disorders. The unique structural features of axons and dendrites may be vulnerable due to protein aggregates and damaged organelles [26]. Moreover, dynein motors mediate retrograde transport of autophagic vacuoles from axon terminals to soma. Mutations or inhibition of dynein impair this transport, disrupting

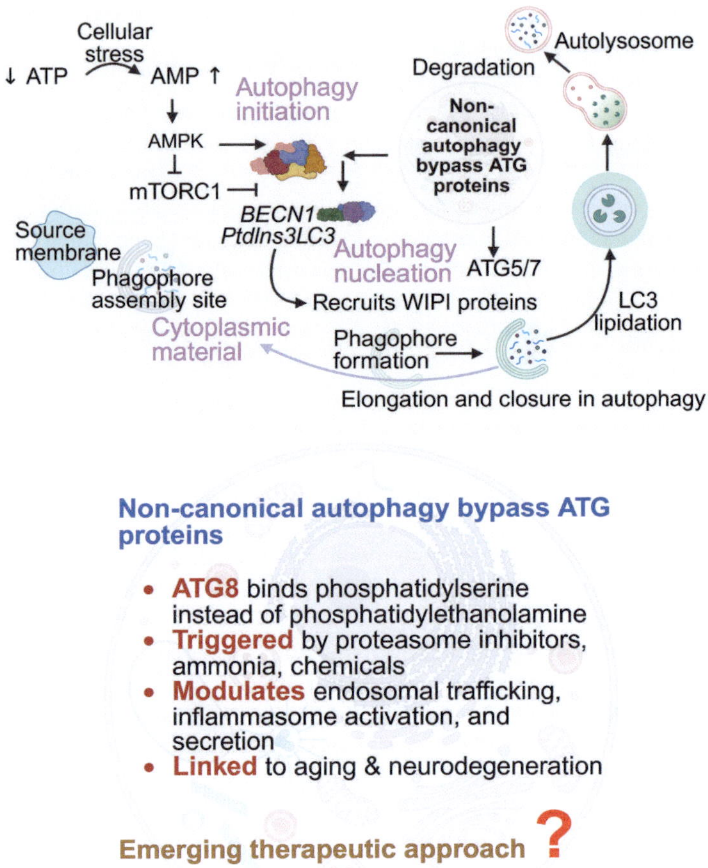

Fig. 2.1 Non-canonical autophagy pathways and therapeutic potential. Non-canonical autophagy bypasses classical ATG proteins, involves ATG8-phosphatidylserine conjugation, and is linked to cellular stress responses, aging, and neurodegeneration, highlighting its promise as an emerging therapeutic target. (Created using BioRender. https://www.biorender.com/)

autophagic clearance. This defect is a key pathogenic mechanism in Charcot-Marie-Tooth disease, which is an inherited neuropathy affecting peripheral nerve signal transmission [27]. Dynein is the primary minus-end motor responsible for transporting cargoes along microtubules. In GFP-LC3-labeled neurons, dynein co-migrates with LC3-positive puncta in axons. This retrograde movement is significantly reduced by dynein inhibition using EHNA or by expressing the dominant-negative inhibitor CC1, which disrupts dynein-dynactin interaction [28]. Next, lysophagy is the selective autophagic removal of acidic organelles like lysosomes, which are sequestered when their membranes are damaged. Lysosomal damage can result from factors such as amyloid-β (Aβ) deposition or crystal (e.g., silica, monosodium urate) accumulation [29]. Moreover, mitochondria are essential for cellular function and survival and rely on quality control mechanisms to maintain their integrity and interactions with other organelles, and their dysfunction has been recognized as a hallmark of aging and neurodegenerative diseases [30]. In PD, dopaminergic neuron loss is associated with Lewy bodies and α-synuclein aggregates, which impair autophagy and lysosomal function. Tau pathology and Aβ deposition further disrupt cellular clearance mechanisms. These changes compromise mitochondrial quality control, increase reactive oxygen species (ROS), and promote a cycle of neurodegeneration [31].

2.2.3 Epigenetic Regulation of Autophagy

Epigenetic change was first introduced by Waddington in 1939, and has now been expanded for autophagy regulation, involving more than 30 associated proteins and signaling molecules. Although autophagy was once considered purely cytoplasmic, recent evidence reveals significant nuclear involvement, including roles for transcription factors, histone modifications, and microRNAs [32]. Next, microRNAs are short, single-stranded non-coding RNAs involved in various physiological and pathological processes. Approximately 2000 microRNAs, each ~21 nt in length, act as key regulators of post-transcriptional gene expression and are evolutionarily conserved. Among them, miR-124 is the most abundant in brain and neurons and plays critical roles in differentiation, cell proliferation, fate determination, neuron migration, apoptosis, synaptic plasticity, and cognition [32]. Previously, it has been shown that the primary function of miR-124 is to restrain neural stem/precursor cell proliferation and promote neuronal differentiation [33, 34]. Overexpression of miR-124 may stimulate neural stem/precursor cell premature differentiation by repressing STAT3 [35]. Moreover, miR-124 promotes neurite sprouting, elongation, morphological adaptation, and changes in dendritic spine density [36]. Moreover, low miR-124 expression in brain led to impaired cognition. Interestingly, inhibiting miR-124 has been reported to improve spatial memory more than working memory. Further, it is suggested that miR-124 is a potential biomarker and therapeutic target for neurodegenerative diseases [37–42]. In AD, miR-124 expression is typically suppressed, and its restoration or exogenous supplementation has shown neuroprotective effects [43, 44]. Cerebral hypoperfusion promotes Aβ-peptide accumulation, which inhibits miR-124 and upregulates BACE1 via the EPAC-rap1 pathway [45]. Microvascular dysfunction further impairs Aβ clearance, disrupts miR-124 regulation, increases C1pl3 expression, and reduces microvessel density, accelerating AD progression [46]. Although exogenous miR-124 can be beneficial, its overexpression may impair learning and memory by downregulating PTPN1, which can be reversed using miR-124 inhibitors or PTPN1 blockade [47]. In PD, miR-124-loaded nanoparticles enhanced neural stem cell differentiation and neuronal migration [48]. Similarly, miR-124 overexpression promotes neuroprotection by downregulating p62/p-p38, enhancing autophagy, and preventing dopaminergic neuron loss [49, 50]. DNA methylation is closely linked to aging and longevity across species [51]. Transcription factor EB (TFEB) is a key regulator of autophagy and lysosomal biogenesis, which contributes to aging by promoting proteostasis, mitochondrial function, energy metabolism, and cell regeneration; modulating inflammation; and inhibiting senescence. Its expression is epigenetically regulated through DNA methylation. DNA methylation represses gene expression by blocking transcription factor binding or recruiting repressor proteins [52, 53]. Aberrant methylation is common in tumors involving tumor suppressor silencing or proto-oncogene activation through various epigenetic mechanisms, including histone modifications and non-coding RNAs [54]. DNA methyltransferases catalyze methyl group transfer using S-adenosylmethionine, which is facilitated by covalent intermediate formation at cytosine's C6 and activation of C5 nucleophilicity [55]. Studies confirm that DNA methylation influences learning and memory in neurodegenerative diseases like AD [56, 57]. Age-related changes in CpG methylation affect both adults and children, with hypermethylation frequently targeting developmentally relevant gene promoters, some of which are also methylated in cancer, suggesting shared epigenetic mechanisms [58]. Autophagy induction is associated with chromatin remodeling, including DNA methylation and histone modifications, which may be heritable and contribute to transgenerational effects of environmental factors like famine [59].

2.3 Dysregulation in Neurodegenerative Conditions

Neurons rely on efficient protein turnover and organelle recycling to maintain homeostasis. The endo-lysosomal and autophagy pathways are essential to this process, mediating misfolded proteins and damaged organelles. Dysfunction in these systems is a hallmark of several neurodegenerative diseases and lysosomal storage disorders [1, 60, 61]. Autophagy is vital for neuronal function and organelle turnover, and its initiation is critical for autophagosome biogenesis. Emerging evidence shows that disruptions at this early stage via kinase inhibition, receptor mislocalization, ubiquitin-mediated degradation, or epigenetic alterations can activate neurodegenerative pathways [62, 63]. Some key mechanisms and their pathological implications are detailed below.

The ULK1 kinase complex (ULK1, ATG13, FIP200) integrates nutrient and energy signals to initiate autophagy. In neurons, ULK1 is inhibited by mTOR and E3 ubiquitin ligases such as RPM1, which target ULK1 for degradation, reducing axonal autophagy and affecting synaptic maintenance [64]. The human homolog PAM/MYCBP2 shows similar effects, indicating a conserved mechanism limiting protective autophagy [65]. Autophagy initiation also depends on omegasome formation, requiring proper ATG9 localization. In PD, α-synuclein overexpression mislocalizes ATG9, while VPS35 mutations disrupt WASH complex-mediated ATG9A trafficking, both impairing autophagy and promoting α-synuclein accumulation [66, 67]. Chronic mTOR hyperactivation in AD, HD, and tauopathies inhibits ULK1/ATG13, blocking autophagy and leading to protein aggregate buildup

[63, 68]. Concurrently, AMPK activity declines, further reducing ULK1 activation [69]. Deficiencies in BECN1 and ATG16L1 impair clearance of Aβ and tau and affect non-canonical autophagy [70], whereas disrupted FUNDC1-mediated mitophagy increases mitochondrial stress in AD [69]. Reticulophagy dysfunction contributes to endoplasmic reticulum stress, misfolded protein accumulation, and apoptosis in neurodegeneration [71, 72]. In AD, reduced BECN1 leads to Aβ aggregation, NLRP3 inflammasome activation, and microglial dysfunction [72]. Similar mechanisms in PD also contribute to dopaminergic neuronal loss [64]. Moreover, dysregulated lncRNAs and miRNAs alter autophagy gene expression, affecting autophagy initiation in both AD and PD [67, 68].

Lysosomes are vital for cellular function, functioning as signaling hubs and autophagy regulators, and dysregulated lysosomal pathways are commonly linked to neurodegenerative diseases [73]. This is especially evident in microglia, promoting neuroinflammation and reducing degradation efficacy [1]. Proper lysosomal biogenesis depends on trafficking through the Golgi-M6P and endosomal pathways. In addition, mutations affecting GNPTAB or M6P tagging result in enzyme missorting [18]. Retromer mutations (e.g., VPS35-D620N) can disrupt WASH complex function and lysosomal enzyme sorting, exacerbating proteins, for example, α-synuclein accumulation [1]. Though autophagosome formation may remain intact or elevated in early AD, fusion with lysosomes often fails, causing vesicle accumulation [17]. Aggregation-prone proteins like α-synuclein and mutant Tau can block CMA by interfering with LAMP-2A [61, 74]. Genetic risk factors intersect with lysosomal enzyme activity pathways such as GBA mutations impairing glucocerebrosidase in PD, PGRN haploinsufficiency affecting acidification and cathepsin maturation in ALS and FTD, and AD-linked genes (APOE, SORL1, PLD3) disrupt enzyme activity [60]. In lysosomal storage disorders (e.g., Niemann-Pick), lipid accumulation impairs calcium handling and vesicle trafficking, reflecting age-related neurodegeneration [1, 75]. Emerging tools may track lysosomal dysfunction, such as pH sensors to reveal defects in neurons and microglia [76], whereas CRISPRi screens link glycosphingolipid accumulation to lysosomal stress and oxidative damage [73].

2.4 Novel Consequences Beyond Protein Aggregation

Lipophagy regulates intracellular lipid droplets and is activated during starvation through lipolysis, microlipophagy, and macrolipophagy. Perilipins (PLIN2, PLIN3) on lipid droplets are degraded by HSC70, permitting lipolysis or autophagic degradation [77]. In HD, hypoxia and neuronal stress disrupt lipid metabolism and mitochondrial function, elevating triglycerides and oxidative stress. Lipid peroxidations with 12/15-lipoxygenase and dysregulated sphingolipid metabolism contribute to mitochondrial damage. Sphingolipids are tightly regulated through synthesis, recycling, and autophagy, but are imbalanced with age, impairing membrane function and promoting neurodegenerative diseases. Under stress (e.g., oxidative stress, TNF-α, DNA damage), neutral sphingomyelinase activates the

ERK-MAPK pathway, increasing ceramide, which is a key mediator of apoptosis and inflammation. Additionally, acid sphingomyelinase also contributes to ceramide accumulation. Salvage pathways recycle complex sphingolipids to ceramide, and ceramide synthase converts sphingosine back to ceramide. Although moderate ceramide levels support autophagy, prolonged elevation triggers apoptosis and neuronal loss. In HD, striatal ceramide accumulation under hypoxia and metabolic stress exacerbates neuroinflammation and disease progression. Sphingosine-1-phosphate counterbalances ceramide—high favors survival, while low promotes neuronal death [78]. In AD, deficient autophagy in microglia leads to inflammasome activation, cytokine release, and tau mislocalization [79, 80]. Impaired Aβ clearance further reflects autophagy deficits [81]. Additionally, LINE-1 retrotransposons are activated in AD microglia. Increased LINE-1 activity alters microglial function, leading to abnormal cytokine production, disrupted morphology, impaired phagocytosis, and excessive IL-10 expression. LINE-1 insertions may drive somatic mutations in AD, suggesting that epigenetic regulation of LINE-1 could offer therapeutic potential [82].

2.5 Deep Dive into Emerging Cross-Disease Mechanisms

PD is a neurodegenerative disorder marked by dopaminergic neuron loss, α-synuclein aggregation, and impaired protein clearance. Mitochondrial dysfunction, mostly defective mitophagy, and neuroinflammation are key contributors. Mitophagy involves PINK1 accumulation on depolarized mitochondrial membranes, which activates the E3 ubiquitin ligase PARKIN to tag mitochondria for autophagic degradation [83–85]. Mutations in PINK1 and PARKIN impair this process, promoting Lewy body formation and early-onset PD [86]. Dysfunctional mitophagy leads to mitochondrial DNA leakage, triggering NLRP3 inflammasome activation. This induces caspase-1-mediated maturation of IL-1β and IL-18, driving neuroinflammation and dopaminergic neurodegeneration. In neuronal demyelination, CMA is crucial for proteostasis, which relies on HSPA8 to unfold proteins with a consensus motif, directing them to lysosomal receptor LAMP-2A for degradation. Additionally, IFN-γ is a key factor in impairing oligodendrocyte function, which reduces myelination and induces endoplasmic reticulum stress, triggering unfolded protein response and compensatory autophagy activation [87]. The gut–brain axis is a bidirectional neuroimmune-endocrine system linking gut microbiota with central nervous system (CNS), influencing brain function and behavior through neural (e.g., vagus nerve), immune (e.g., cytokines), hormonal, and microbial signaling. Microbial metabolites like short-chain fatty acids, specifically butyrate, act as key messengers modulating inflammation, gene expression, and neural activity [88]. Butyrate is primarily absorbed in the colon but exerts systemic effects, including CNS [89], and acts as a potent epigenetic modulator, promoting gene expression by inducing histone hyperacetylation, and may activate nuclear translocation and transcriptional activity.

2.6 Case Study: Novel Genetic Drivers

TBK1 is one of the key genetic factors in ALS and FTD. Haploinsufficiency typically occurs due to loss-of-function mutations, which reduce TBK1 levels and disrupt autophagy, immune signaling, and neuroinflammation [90]. TBK1 phosphorylates autophagy receptors like OPTN, enhancing their efficacy to recognize and clear damaged organelles and protein aggregates. In TBK1 deficiency, OPTN is hypophosphorylated, impairing cargo binding and LC3 recruitment, leading to defective mitophagy and autophagic flux. This results in mitochondrial dysfunction, oxidative stress, and neurodegeneration, highlighting TBK1 as a critical therapeutic target [91, 92]. Next, VPS13C also regulates lipid transfer and lysosomal homeostasis, and its mutations impair cholesterol transport, causing its accumulation in lysosomes, disrupting pH balance, enzyme activity, and autophagosome–lysosome fusion. This "upstream arrest" in autophagy leads to the accumulation of damaged mitochondria and proteins, contributing to oxidative stress and neurodegeneration in PD [93]. Most VPS13C mutations are autosomal recessive and may be exacerbated by environmental or epigenetic factors.

2.7 Conclusion

To sum up, this chapter offers a comprehensive overview of autophagy mechanisms and their dysregulation, highlighting their critical role in neurodegenerative diseases. It explores both canonical and non-canonical autophagy pathways, with precise attention to recent insights into ATG9 function, and organelle-specific autophagy in neurons. Additionally, it explored the effect of epigenetic regulators, including microRNAs and DNA methylation, on autophagy. It concludes that impairments in autophagy initiation and lysosomal degradation are central to proteostasis disruption and neuroinflammation, underscoring the therapeutic potential of targeting these processes in neurodegeneration.

References

1. Malik BR, Maddison DC, Smith GA, Peters OM. Autophagic and endo-lysosomal dysfunction in neurodegenerative disease. Mol Brain. 2019;12(1):100. https://doi.org/10.1186/s13041-019-0504-x.
2. Guo F, Liu X, Cai H, Le W. Autophagy in neurodegenerative diseases: pathogenesis and therapy. Brain Pathol. 2018;28(1):3–13. https://doi.org/10.1111/bpa.12545.
3. Giovedì S, Ravanelli MM, Parisi B, Bettegazzi B, Guarnieri FC. Dysfunctional autophagy and endolysosomal system in neurodegenerative diseases: relevance and therapeutic options. Front Cell Neurosci. 2020;14:602116. https://doi.org/10.3389/fncel.2020.602116.
4. Piras A, Collin L, Grüninger F, Graff C, Rönnbäck A. Autophagic and lysosomal defects in human tauopathies: analysis of post-mortem brain from patients with familial Alzheimer disease, corticobasal degeneration and progressive supranuclear palsy. Acta Neuropathol Commun. 2016;4(1):22. https://doi.org/10.1186/s40478-016-0292-9.

5. Guix FX, Sannerud R, Berditchevski F, et al. Tetraspanin 6: a pivotal protein of the multiple vesicular body determining exosome release and lysosomal degradation of amyloid precursor protein fragments. Mol Neurodegener. 2017;12(1):25. https://doi.org/10.1186/s13024-017-0165-0.
6. Meduri G, Guillemeau K, Dounane O, et al. Caspase-cleaved Tau-D421 is colocalized with the immunophilin FKBP52 in the autophagy-endolysosomal system of Alzheimer's disease neurons. Neurobiol Aging. 2016;46:124–37. https://doi.org/10.1016/j.neurobiolaging.2016.06.017.
7. Dzamko N, Gysbers A, Perera G, et al. Toll-like receptor 2 is increased in neurons in Parkinson's disease brain and may contribute to alpha-synuclein pathology. Acta Neuropathol. 2017;133(2):303–19. https://doi.org/10.1007/s00401-016-1648-8.
8. Murphy KE, Gysbers AM, Abbott SK, et al. Lysosomal-associated membrane protein 2 isoforms are differentially affected in early Parkinson's disease. Mov Disord. 2015;30(12):1639–47. https://doi.org/10.1002/mds.26141.
9. Ktistakis NT. In praise of M. Anselmier who first used the term "autophagie" in 1859. Autophagy. 2017;13(12):2015–7. https://doi.org/10.1080/15548627.2017.1367473.
10. Parzych KR, Klionsky DJ. An overview of autophagy: morphology, mechanism, and regulation. Antioxid Redox Signal. 2014;20(3):460–73. https://doi.org/10.1089/ars.2013.5371.
11. Anding AL, Baehrecke EH. Cleaning house: selective autophagy of organelles. Dev Cell. 2017;41(1):10–22. https://doi.org/10.1016/j.devcel.2017.02.016.
12. Axe EL, Walker SA, Manifava M, et al. Autophagosome formation from membrane compartments enriched in phosphatidylinositol 3-phosphate and dynamically connected to the endoplasmic reticulum. J Cell Biol. 2008;182(4):685–701. https://doi.org/10.1083/jcb.200803137.
13. Simonsen A, Tooze SA. Coordination of membrane events during autophagy by multiple class III PI3-kinase complexes. J Cell Biol. 2009;186(6):773–82. https://doi.org/10.1083/jcb.200907014.
14. Hailey DW, Rambold AS, Satpute-Krishnan P, et al. Mitochondria supply membranes for autophagosome biogenesis during starvation. Cell. 2010;141(4):656–67. https://doi.org/10.1016/j.cell.2010.04.009.
15. Ravikumar B, Moreau K, Jahreiss L, Puri C, Rubinsztein DC. Plasma membrane contributes to the formation of pre-autophagosomal structures. Nat Cell Biol. 2010;12(8):747–57. https://doi.org/10.1038/ncb2078.
16. Mizushima N. Autophagy: process and function. Genes Dev. 2007;21(22):2861–73. https://doi.org/10.1101/gad.1599207.
17. Koh JY, Kim HN, Hwang JJ, Kim YH, Park SE. Lysosomal dysfunction in proteinopathic neurodegenerative disorders: possible therapeutic roles of cAMP and zinc. Mol Brain. 2019;12(1):18. https://doi.org/10.1186/s13041-019-0439-2.
18. Akaaboune SR, Wang Y. Golgi defect as a major contributor to lysosomal dysfunction. Front Cell Dev Biol. 2024;12:1386149. https://doi.org/10.3389/fcell.2024.1386149.
19. Zhao Z, Yan J, Huang L, Yang X. Phytochemicals targeting Alzheimer's disease via the AMP-activated protein kinase pathway, effects, and mechanisms of action. Biomed Pharmacother. 2024;173:116373. https://doi.org/10.1016/j.biopha.2024.116373.
20. Heras-Sandoval D, Pérez-Rojas JM, Hernández-Damián J, Pedraza-Chaverri J. The role of PI3K/AKT/mTOR pathway in the modulation of autophagy and the clearance of protein aggregates in neurodegeneration. Cell Signal. 2014;26(12):2694–701. https://doi.org/10.1016/j.cellsig.2014.08.019.
21. Sarkar S. Regulation of autophagy by mTOR-dependent and mTOR-independent pathways: autophagy dysfunction in neurodegenerative diseases and therapeutic application of autophagy enhancers. Biochem Soc Trans. 2013;41(5):1103–30. https://doi.org/10.1042/BST20130134.
22. Feng H, Wang N, Zhang N, Liao H. Alternative autophagy: mechanisms and roles in different diseases. Cell Commun Signal. 2022;20(1):43. https://doi.org/10.1186/s12964-022-00851-1.

23. Millard E, Tooze SA. ATG9 not just an autophagy related protein. J Mol Biol. 2025;437(18): 169288. https://doi.org/10.1016/j.jmb.2025.169288.
24. Holzer E, Martens S, Tulli S. The role of ATG9 vesicles in autophagosome biogenesis. J Mol Biol. 2024;436(15):168489. https://doi.org/10.1016/j.jmb.2024.168489.
25. Noda NN, Wang Z, Zhang H. Liquid–liquid phase separation in autophagy. J Cell Biol. 2020;219(8):e202004062. https://doi.org/10.1083/jcb.202004062.
26. Yang Y, Coleman M, Zhang L, Zheng X, Yue Z. Autophagy in axonal and dendritic degeneration. Trends Neurosci. 2013;36(7):418–28. https://doi.org/10.1016/j.tins.2013.04.001.
27. Murphy SM, Laura M, Fawcett K, et al. Charcot–Marie–Tooth disease: frequency of genetic subtypes and guidelines for genetic testing. J Neurol Neurosurg Psychiatry. 2012;83(7):706–10. https://doi.org/10.1136/jnnp-2012-302451.
28. Maday S, Wallace KE, Holzbaur ELF. Autophagosomes initiate distally and mature during transport toward the cell soma in primary neurons. J Cell Biol. 2012;196(4):407–17. https://doi.org/10.1083/jcb.201106120.
29. Hasegawa J, Maejima I, Iwamoto R, Yoshimori T. Selective autophagy: lysophagy. Methods. 2015;75:128–32. https://doi.org/10.1016/j.ymeth.2014.12.014.
30. Picca A, Guerra F, Calvani R, et al. Generation and release of mitochondrial-derived vesicles in health, aging and disease. J Clin Med. 2020;9(5):1440. https://doi.org/10.3390/jcm9051440.
31. Hou X, Watzlawik JO, Fiesel FC, Springer W. Autophagy in Parkinson's disease. J Mol Biol. 2020;432(8):2651–72. https://doi.org/10.1016/j.jmb.2020.01.037.
32. Wang Y, Zhang H. Regulation of autophagy by mTOR signaling pathway. In: Qin ZH, editor. Autophagy: biology and diseases: basic science. Springer Singapore; 2019. p. 67–83. https://doi.org/10.1007/978-981-15-0602-4_3.
33. Mokabber H, Najafzadeh N, Mohammadzadeh Vardin M. *miR-124* promotes neural differentiation in mouse bulge stem cells by repressing *Ptbp1* and *Sox9*. J Cell Physiol. 2019;234(6): 8941–50. https://doi.org/10.1002/jcp.27563.
34. Visvanathan J, Lee S, Lee B, Lee JW, Lee SK. The microRNA miR-124 antagonizes the anti-neural REST/SCP1 pathway during embryonic CNS development. Genes Dev. 2007;21(7): 744–9. https://doi.org/10.1101/gad.1519107.
35. Yang Y, Ye Y, Kong C, et al. MiR-124 enriched exosomes promoted the M2 polarization of microglia and enhanced hippocampus neurogenesis after traumatic brain injury by inhibiting TLR4 pathway. Neurochem Res. 2019;44(4):811–28. https://doi.org/10.1007/s11064-018-02714-z.
36. Schouten M, Aschrafi A, Bielefeld P, Doxakis E, Fitzsimons CP. microRNAs and the regulation of neuronal plasticity under stress conditions. Neuroscience. 2013;241:188–205. https://doi.org/10.1016/j.neuroscience.2013.02.065.
37. Tuoc TC, Pavlakis E, Tylkowski MA, Stoykova A. Control of cerebral size and thickness. Cell Mol Life Sci. 2014;71(17):3199–218. https://doi.org/10.1007/s00018-014-1590-7.
38. Zhang WH, Jiang L, Li M, Liu J. MicroRNA-124: an emerging therapeutic target in central nervous system disorders. Exp Brain Res. 2023;241(5):1215–26. https://doi.org/10.1007/s00221-022-06524-2.
39. Malmevik J, Petri R, Klussendorf T, et al. Identification of the miRNA targetome in hippocampal neurons using RIP-seq. Sci Rep. 2015;5(1):12609. https://doi.org/10.1038/srep12609.
40. Dutta R, Chomyk AM, Chang A, et al. Hippocampal demyelination and memory dysfunction are associated with increased levels of the neuronal microRNA miR-124 and reduced AMPA receptors. Ann Neurol. 2013;73(5):637–45. https://doi.org/10.1002/ana.23860.
41. Nguyen LH, Diao HJ, Chew SY. MicroRNAs and their potential therapeutic applications in neural tissue engineering. Adv Drug Deliv Rev. 2015;88:53–66. https://doi.org/10.1016/j.addr.2015.05.007.
42. Ambrogini P, Albertini MC, Betti M, et al. Neurobiological correlates of alpha-tocopherol antiepileptogenic effects and microRNA expression modulation in a rat model of kainate-induced seizures. Mol Neurobiol. 2018;55(10):7822–38. https://doi.org/10.1007/s12035-018-0946-7.

43. Jin XF, Wu N, Wang L, Li J. Circulating microRNAs: a novel class of potential biomarkers for diagnosing and prognosing central nervous system diseases. Cell Mol Neurobiol. 2013;33(5): 601–13. https://doi.org/10.1007/s10571-013-9940-9.
44. Woldemichael BT, Mansuy IM. Micro-RNAs in cognition and cognitive disorders: potential for novel biomarkers and therapeutics. Biochem Pharmacol. 2016;104:1–7. https://doi.org/10.1016/j.bcp.2015.11.021.
45. Zhang X, Huang X, Fang C, et al. miR-124 regulates the expression of BACE1 in the hippocampus under chronic cerebral hypoperfusion. Mol Neurobiol. 2017;54(4):2498–506. https://doi.org/10.1007/s12035-016-9845-y.
46. Li AD, Tong L, Xu N, et al. miR-124 regulates cerebromicrovascular function in APP/PS1 transgenic mice via C1ql3. Brain Res Bull. 2019;153:214–22. https://doi.org/10.1016/j.brainresbull.2019.09.002.
47. Wang X, Liu D, Huang HZ, et al. A novel microRNA-124/PTPN1 signal pathway mediates synaptic and memory deficits in Alzheimer's disease. Biol Psychiatry. 2018;83(5):395–405. https://doi.org/10.1016/j.biopsych.2017.07.023.
48. Saraiva C, Paiva J, Santos T, Ferreira L, Bernardino L. MicroRNA-124 loaded nanoparticles enhance brain repair in Parkinson's disease. J Control Release. 2016;235:291–305. https://doi.org/10.1016/j.jconrel.2016.06.005.
49. Wang H, Ye Y, Zhu Z, Mo L, Lin C, Wang Q, Wang H, Gong X, He X, Lu G, Lu F, Zhang S. MiR-124 regulates apoptosis and autophagy process in MPTP model of Parkinson's disease by targeting to Bim. Brain Pathol. 2016;26(2):167–76.
50. Yao L, Zhu Z, Wu J, et al. MicroRNA-124 regulates the expression of p62/p38 and promotes autophagy in the inflammatory pathogenesis of Parkinson's disease. FASEB J. 2019;33(7): 8648–65. https://doi.org/10.1096/fj.201900363R.
51. Sedivy JM, Banumathy G, Adams PD. Aging by epigenetics—a consequence of chromatin damage? Exp Cell Res. 2008;314(9):1909–17. https://doi.org/10.1016/j.yexcr.2008.02.023.
52. Todde V, Veenhuis M, van der Klei IJ. Autophagy: principles and significance in health and disease. Biochim Biophys Acta (BBA) Mol Basis Dis. 2009;1792(1):3–13. https://doi.org/10.1016/j.bbadis.2008.10.016.
53. Moore LD, Le T, Fan G. DNA methylation and its basic function. Neuropsychopharmacology. 2013;38(1):23–38. https://doi.org/10.1038/npp.2012.112.
54. Qiao Q, Wang Y, Zhang R, Pang Q. Autophagy related DNA methylation signature predict clinical prognosis and immune microenvironment in low-grade glioma. Transl Cancer Res. 2022;11(7):2157–74. https://doi.org/10.21037/tcr-22-310.
55. Johnson AA, Akman K, Calimport SRG, Wuttke D, Stolzing A, de Magalhães JP. The role of DNA methylation in aging, rejuvenation, and age-related disease. Rejuvenation Res. 2012;15(5):483–94. https://doi.org/10.1089/rej.2012.1324.
56. Xu X. DNA methylation and cognitive aging. Oncotarget. 2015;6(16):13922–32. https://doi.org/10.18632/oncotarget.4215.
57. Pellegrini C, Pirazzini C, Sala C, et al. A meta-analysis of brain DNA methylation across sex, age, and Alzheimer's disease points for accelerated epigenetic aging in neurodegeneration. Front Aging Neurosci. 2021;13:639428. https://doi.org/10.3389/fnagi.2021.639428.
58. Horvath S, Zhang Y, Langfelder P, et al. Aging effects on DNA methylation modules in human brain and blood tissue. Genome Biol. 2012;13(10):R97. https://doi.org/10.1186/gb-2012-13-10-r97.
59. González-Rodríguez P, Füllgrabe J, Joseph B. The hunger strikes back: an epigenetic memory for autophagy. Cell Death Differ. 2023;30(6):1404–15. https://doi.org/10.1038/s41418-023-01159-4.
60. Udayar V, Chen Y, Sidransky E, Jagasia R. Lysosomal dysfunction in neurodegeneration: emerging concepts and methods. Trends Neurosci. 2022;45(3):184–99. https://doi.org/10.1016/j.tins.2021.12.004.

61. Mançano ASF, Pina JG, Froes BR, Sciani JM. Autophagy-lysosomal pathway impairment and cathepsin dysregulation in Alzheimer's disease. Front Mol Biosci. 2024;11:1490275. https://doi.org/10.3389/fmolb.2024.1490275.
62. Oettinger D, Yamamoto A. Autophagy dysfunction and neurodegeneration: where does it go wrong? J Mol Biol. 2025;437(18):169219. https://doi.org/10.1016/j.jmb.2025.169219.
63. Barmaki H, Nourazarian A, Shademan B, khaki-khatibi F. The autophagy paradox: a new hypothesis in neurodegenerative disorders. Neurochem Int. 2024;179:105827. https://doi.org/10.1016/j.neuint.2024.105827.
64. Nixon RA, Rubinsztein DC. Mechanisms of autophagy–lysosome dysfunction in neurodegenerative diseases. Nat Rev Mol Cell Biol. 2024;25(11):926–46. https://doi.org/10.1038/s41580-024-00757-5.
65. Crawley O, Grill B. Autophagy in axonal and presynaptic development. Curr Opin Neurobiol. 2021;69:139–48. https://doi.org/10.1016/j.conb.2021.03.011.
66. Crawley O, Opperman KJ, Desbois M, et al. Autophagy is inhibited by ubiquitin ligase activity in the nervous system. Nat Commun. 2019;10(1):5017. https://doi.org/10.1038/s41467-019-12804-3.
67. Mohseni M, Behzad G, Farhadi A, Behroozi J, Mohseni H, Valipour B. MicroRNAs regulating autophagy: opportunities in treating neurodegenerative diseases. Front Neurosci. 2024;18:1397106. https://doi.org/10.3389/fnins.2024.1397106.
68. Meng T, Lin S, Zhuang H, et al. Recent progress in the role of autophagy in neurological diseases. Cell Stress. 2019;3(5):141–61. https://doi.org/10.15698/cst2019.05.186.
69. Shi D, Guo X, Ning Z, et al. Review of FUNDC1-mediated mitochondrial autophagy in Alzheimer's disease. Front Aging Neurosci. 2025;17:1544241. https://doi.org/10.3389/fnagi.2025.1544241.
70. Ajoolabady A, Aslkhodapasandhokmabad H, Henninger N, et al. Targeting autophagy in neurodegenerative diseases: from molecular mechanisms to clinical therapeutics. Clin Exp Pharmacol Physiol. 2021;48(7):943–53. https://doi.org/10.1111/1440-1681.13500.
71. Zhao XY, Xu DE, Wu ML, Liu JC, Shi ZL, Ma QH. Regulation and function of endoplasmic reticulum autophagy in neurodegenerative diseases. Neural Regen Res. 2025;20(1):6–20. https://doi.org/10.4103/NRR.NRR-D-23-00995.
72. Liu X, Ye M, Ma L. The emerging role of autophagy and mitophagy in tauopathies: from pathogenesis to translational implications in Alzheimer's disease. Front Aging Neurosci. 2022;14:1022821. https://doi.org/10.3389/fnagi.2022.1022821.
73. Chen C, Chen Y, Sidransky E. Technological advances expand our knowledge of lysosomal dysfunction in neurodegeneration. Neural Regen Res. 2023;18(3):539. https://doi.org/10.4103/1673-5374.346490.
74. Root J, Merino P, Nuckols A, Johnson M, Kukar T. Lysosome dysfunction as a cause of neurodegenerative diseases: lessons from frontotemporal dementia and amyotrophic lateral sclerosis. Neurobiol Dis. 2021;154:105360. https://doi.org/10.1016/j.nbd.2021.105360.
75. Peng W, Minakaki G, Nguyen M, Krainc D. Preserving lysosomal function in the aging brain: insights from neurodegeneration. Neurotherapeutics. 2019;16(3):611–34. https://doi.org/10.1007/s13311-019-00742-3.
76. Quick JD, Silva C, Wong JH, et al. Lysosomal acidification dysfunction in microglia: an emerging pathogenic mechanism of neuroinflammation and neurodegeneration. J Neuroinflammation. 2023;20(1):185. https://doi.org/10.1186/s12974-023-02866-y.
77. Khawar MB, Abbasi MH, Rafiq M, Naz N, Mehmood R, Sheikh N. A decade of mighty lipophagy: what we know and what facts we need to know? Oxidative Med Cell Longev. 2021;2021(1):5539161. https://doi.org/10.1155/2021/5539161.
78. Burtscher J, Pepe G, Maharjan N, et al. Sphingolipids and impaired hypoxic stress responses in Huntington disease. Prog Lipid Res. 2023;90:101224. https://doi.org/10.1016/j.plipres.2023.101224.
79. 2021 Alzheimer's disease facts and figures. Alzheimers Dement. 2021;17(3):327–406. https://doi.org/10.1002/alz.12328.

80. Yang X, Wen J, Yang H, et al. Functional characterization of Alzheimer's disease genetic variants in microglia. Nat Genet. 2023;55(10):1735–44. https://doi.org/10.1038/s41588-023-01506-8.
81. Boche D, Nicoll JAR. The role of the immune system in clearance of Abeta from the brain. Brain Pathol. 2008;18(2):267–78. https://doi.org/10.1111/j.1750-3639.2008.00134.x.
82. Roy N, Haq I, Ngo JC, et al. Elevated expression of the retrotransposon LINE-1 drives Alzheimer's disease-associated microglial dysfunction. Acta Neuropathol. 2024;148(1):75. https://doi.org/10.1007/s00401-024-02835-6.
83. Jin SM, Youle RJ. PINK1- and Parkin-mediated mitophagy at a glance. J Cell Sci. 2012;125(4): 795–9. https://doi.org/10.1242/jcs.093849.
84. Narendra D, Walker JE, Youle R. Mitochondrial quality control mediated by PINK1 and Parkin: links to Parkinsonism. Cold Spring Harb Perspect Biol. 2012;4(11):a011338. https://doi.org/10.1101/cshperspect.a011338.
85. Pickrell AM, Youle RJ. The roles of PINK1, Parkin, and mitochondrial fidelity in Parkinson's disease. Neuron. 2015;85(2):257–73. https://doi.org/10.1016/j.neuron.2014.12.007.
86. Kitada T, Asakawa S, Hattori N, et al. Mutations in the parkin gene cause autosomal recessive juvenile parkinsonism. Nature. 1998;392(6676):605–8. https://doi.org/10.1038/33416.
87. Shen D, Liu K, Wang H, Wang H. Autophagy modulation in multiple sclerosis and experimental autoimmune encephalomyelitis. Clin Exp Immunol. 2022;209(2):140–50. https://doi.org/10.1093/cei/uxac017.
88. Mazzoli R, Pessione E. The neuro-endocrinological role of microbial glutamate and GABA signaling. Front Microbiol. 2016;7:1934. https://doi.org/10.3389/fmicb.2016.01934.
89. Kalkan AE, BinMowyna MN, Raposo A, et al. Beyond the gut: unveiling butyrate's global health impact through gut health and dysbiosis-related conditions: a narrative review. Nutrients. 2025;17(8):1305. https://doi.org/10.3390/nu17081305.
90. Oakes JA, Davies MC, Collins MO. TBK1: a new player in ALS linking autophagy and neuroinflammation. Mol Brain. 2017;10(1):5. https://doi.org/10.1186/s13041-017-0287-x.
91. Richter B, Sliter DA, Herhaus L, et al. Phosphorylation of OPTN by TBK1 enhances its binding to Ub chains and promotes selective autophagy of damaged mitochondria. Proc Natl Acad Sci. 2016;113(15):4039–44. https://doi.org/10.1073/pnas.1523926113.
92. Stamatakou E, Wróbel L, Hill SM, et al. Mendelian neurodegenerative disease genes involved in autophagy. Cell Discov. 2020;6(1):24. https://doi.org/10.1038/s41421-020-0158-y.
93. Lesage S, Drouet V, Majounie E, et al. Loss of VPS13C function in autosomal-recessive parkinsonism causes mitochondrial dysfunction and increases PINK1/Parkin-dependent mitophagy. Am J Hum Genet. 2016;98(3):500–13. https://doi.org/10.1016/j.ajhg.2016.01.014.

Glial Cells, Cytokines, and Inflammasomes in Neurodegeneration

3

Vishal S. Patil, Anand Desai, and Himani Meena

Abstract

Neurodegenerative disorders are driven by complex interactions among glial cells, inflammatory cytokines, and inflammasomes, resulting in progressive neuronal degeneration. Microglia and astrocytes, traditionally considered passive supportive elements, are now recognized as dynamic regulators capable of exerting either neuroprotective or neurotoxic effects depending on the pathological context. Pro-inflammatory cytokines, including interleukin-1β and tumor necrosis factor-α, amplify neuroinflammatory cascades, compromise synaptic integrity, and promote neuronal hyperexcitability. Concurrently, chemokines facilitate the infiltration of peripheral immune cells into the central nervous system, thereby exacerbating inflammatory responses. The NOD-like receptor family pyrin domain-containing 3 (NLRP3) inflammasome represents a central molecular hub, sensing pathological stimuli such as misfolded protein aggregates and mitochondrial dysfunction, and initiating inflammatory signaling that accelerates disease progression. Therapeutic approaches under investigation

V. S. Patil (✉)
Department of Pharmacology, KLE College of Pharmacy, Belagavi, KLE Academy of Higher Education and Research, Deemed-to-be-University, Belagavi, Karnataka, India

Department of Pharmacognosy, KLE College of Pharmacy, Belagavi, KLE Academy of Higher Education and Research, Deemed-to-be-University, Belagavi, Karnataka, India
e-mail: vishalpatil@klepharm.edu

A. Desai
Department of Pharmacology, Krishna Institute of Pharmacy, Krishna Vishwa Vidyapeeth (Deemed to be University), Karad, Maharashtra, India

H. Meena
Department of Pathology and Genomic Medicine, Houston Methodist Research Institute, Houston, TX, USA

P. Khanal et al. (eds.), *Neuroinflammation and Autophagy in Neurodegeneration*,
https://doi.org/10.1007/978-981-95-3945-1_3

include modulation of glial reactivity, selective inhibition of pro-inflammatory cytokine signaling, and pharmacological blockade of inflammasome activation. Advances in combinatorial pharmacotherapy, genome-editing technologies, and nanoparticle-based drug delivery platforms offer the potential for targeted and more effective interventions. Furthermore, high-resolution molecular profiling of glial cell subpopulations is anticipated to elucidate their functional heterogeneity in disease, thereby informing the design of precision therapies that preserve neuronal integrity while mitigating chronic neuroinflammation.

Keywords

Cytokine · Glial cells · Inflammasomes · Neurodegeneration · Microglia · Neuroinflammation

3.1 Introduction

The progressive loss of neurons in the brain and spinal cord is a hallmark of neurodegenerative diseases, which include Huntington's disease (HD), Parkinson's disease (PD), Alzheimer's disease (AD), frontotemporal dementia (FTD), amyotrophic lateral sclerosis (ALS), and multiple sclerosis (MS) [1]. Since neuronal injury is thought to be the key driver for disease progression, research has traditionally concentrated mostly on neurons, the cells in the nervous system that send signals [2]. Recent research has changed this perspective, emphasizing the critical role that glial cells, particularly microglia and astrocytes, play in the onset and progression of various illnesses [3, 4]. Glial cells are non-neuronal components of the central nervous system (CNS) that provide structural, metabolic, and immunological support to neurons. Microglia act as the resident immune cells of the CNS, constantly surveying the microenvironment for signs of injury, infection, or homeostatic disturbance. Astrocytes contribute to neuronal function by regulating cerebral blood flow, maintaining the integrity of the blood–brain barrier (BBB), recycling neurotransmitters, and sustaining metabolic balance [5]. Under physiological conditions, both microglia and astrocytes preserve neural homeostasis through protective and regulatory functions. In contrast, during aging, toxic exposure, or pathological insults, these cells undergo reactive transformation and release pro-inflammatory mediators along with neurotoxic factors. This state of gliosis is a major driver of sustained neuroinflammation and contributes to the progression of neurodegenerative pathology [6, 7].

A key mechanism driving chronic neuroinflammation is inflammasome activation. These cytosolic protein complexes act as sensors of cellular stress and damage [8]. Among them, the NOD-like receptor family pyrin domain-containing 3 (NLRP3) inflammasome is the most extensively studied in neurodegenerative disorders. Pathological stimuli such as misfolded proteins (e.g., amyloid-β in Alzheimer's disease), oxidative stress, and cellular debris trigger its assembly. Activated NLRP3 then mediates the cleavage of pro-caspase-1 into the active enzyme caspase-1, initiating downstream inflammatory signaling. Caspase-1 processes the inactive precursors pro-interleukin-1β (pro-IL-1β) and pro-interleukin-18 (pro-IL-18) into their biologically active forms, IL-1β and IL-18, which are subsequently secreted into the extracellular space [9, 10]. These cytokines act as potent mediators of inflammation, amplifying immune signaling and influencing neighboring cells. Once released, IL-1β and IL-18 activate surrounding microglia and astrocytes, enhance oxidative stress, and disrupt neuronal homeostasis. This establishes a self-perpetuating inflammatory cycle that exacerbates neuronal injury and impairs the brain's capacity for repair [11].

Microglia respond to cytokines and damage-associated signals, and their initial activation is protective, facilitating debris clearance and tissue repair [12]. In chronic conditions, however, sustained activation leads to excessive production of pro-inflammatory mediators such as tumor necrosis factor-alpha (TNF-α), IL-6, and IL-1β, along with reactive oxygen species (ROS) and proteolytic enzymes. Prolonged microglial activity also promotes aberrant synaptic pruning and contributes to bystander tissue injury. These pathological changes underlie the cognitive and motor deficits characteristic of many neurodegenerative diseases [13]. Astrocytes likewise play a central role in sustaining this inflammatory milieu. Under the influence of cytokines released by activated microglia or inflammasome signaling, astrocytes can transform into a neurotoxic phenotype termed A1 astrocytes. In this state, they lose their homeostatic and trophic functions while releasing cytotoxic molecules that accelerate neuronal injury and cell death [12]. Under normal conditions, astrocytes contribute to neuronal homeostasis by detoxifying the brain and maintaining a stable microenvironment [14]. In their reactive state, however, astrocytes exacerbate neuroinflammation and promote disease progression. Inflammasomes function as central regulators of this response, sensing danger signals and initiating cytokine release that engages both microglia and astrocytes [15]. This interaction establishes a self-perpetuating cycle of glial activation, inflammation, and cellular injury, ultimately accelerating neuronal loss. Additional signaling pathways, including NF-κB, MAPK, and JAK/STAT, further amplify these neuroinflammatory processes [16].

Insights into the glial–inflammasome–cytokine cycle have enabled the development of novel therapeutic strategies for neurodegenerative disorders, targeting specific components of the inflammasome pathway. Pharmacological interventions such as NLRP3 inhibitors, caspase-1 antagonists, and IL-1β neutralizing antibodies are under evaluation in preclinical models and early-phase clinical trials [17]. Additional approaches aim to modulate microglial and astrocytic activity to restore their homeostatic and neuroprotective functions while limiting neurotoxic responses. These strategies mark a paradigm shift from neuron-centric therapies toward interventions that also regulate neuroimmune signaling and disrupt the self-perpetuating cycle of chronic neuroinflammation [18]. This chapter explores in detail how glial cells, inflammasomes, and cytokines interact in neurodegenerative diseases (Fig. 3.1) and also explains how these components form a network that sustains inflammation and leads to neuronal loss.

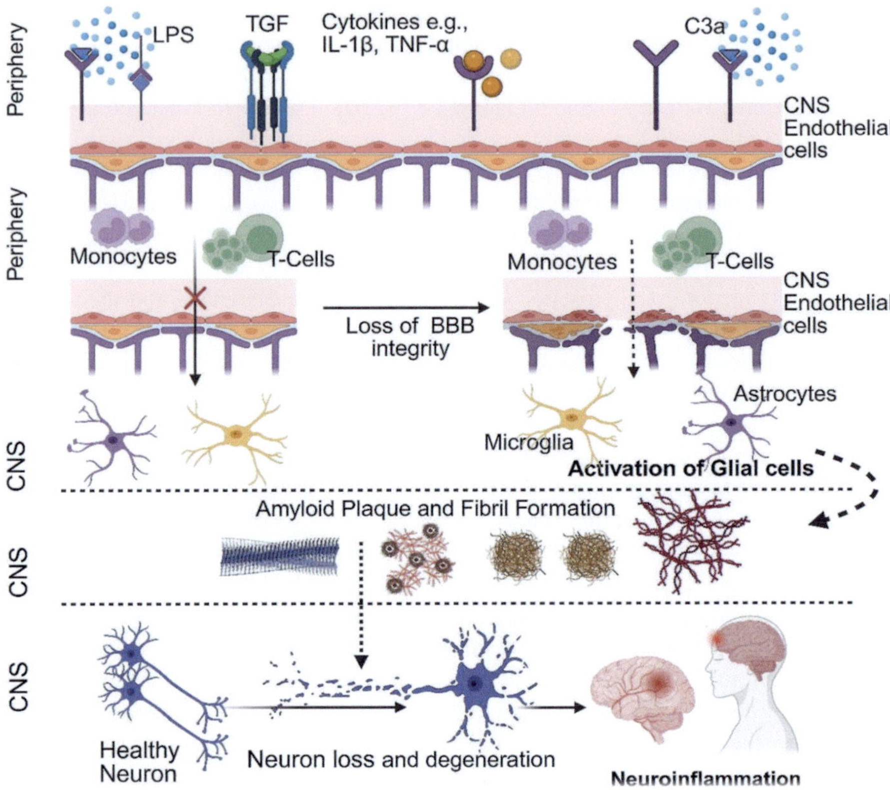

Fig. 3.1 Pathological BBB disruption and glial activation drive neuroinflammation and neurodegeneration. Peripheral inflammatory signals (e.g., LPS, TGF-β, IL-1β, TNF-α, C3a) compromise CNS endothelial cells, leading to BBB breakdown. This permits monocyte and T-cell infiltration into the CNS, which, along with local cues, activates microglia and astrocytes. Activated glia contribute to amyloid plaques and fibrils accumulation. These pathological changes drive progressive neuronal damage, synaptic loss, and degeneration. (Created using BioRender. https://www.biorender.com/)

3.2 Glial Cells: Guardians Turned Adversaries

Glial cells, which were once thought to play only a supporting role in the CNS, are now understood to be essential for both normal brain function and neurological diseases. Although neurons have traditionally received most attention in neuroscience, research has shown that glial cells actively contribute to maintaining the brain's internal environment [19]. They regulate synaptic communication, provide metabolic support, maintain BBB integrity, and coordinate immune responses in the CNS [20]. The main types of glial cells are microglia, astrocytes, oligodendrocytes, and ependymal cells, and each carries out specific functions that are crucial for homeostatic brain function. Importantly, glial cells can shift from a protective to a harmful state under stress or disease conditions [21]. This shift contributes to prolonged inflammation, neuronal injury, and worsening neurodegenerative processes. Understanding this transformation is key to identifying how glial cells influence the course of diseases such as AD, PD, and others [22].

3.2.1 Microglia: From Homeostatic Surveillance to Pathogenic Activation

In healthy CNS, microglia maintain a ramified (i.e., highly branched with multiple primary and secondary processes) morphology and perform continuous surveillance. They modulate synaptic architecture through complement-dependent pruning (via C1q and C3), contributing to neural circuit refinement and plasticity [23]. However, in pathological conditions, microglia transition to an activated phenotype in response to pattern recognition receptor (PRR) signaling. Key receptors such as toll-like receptor 4 (TLR4), triggering receptor expressed on myeloid cells 2 (TREM2), and CD33 detect aggregated proteins, including Aβ, hyperphosphorylated tau, and α-synuclein, triggering downstream inflammatory cascades [24]. Persistent stimulation leads to a dysregulated inflammatory state characterized by pro-inflammatory cytokines (e.g., IL-1β, TNF-α) release, reactive oxygen species (ROS), and impaired phagocytosis. This phagocytic dysfunction contributes to toxic protein aggregates and facilitates disease progression [25]. In AD, microglial PRR activation by Aβ and tau not only triggers inflammation but also compromises clearance mechanisms, promoting plaque persistence [26].

Transcriptional profiling has identified distinct disease-associated phenotypes such as disease-associated microglia (DAM) and senescence-associated microglia (SAM). DAM is marked by homeostatic gene downregulation, followed by upregulation of lipid metabolism and phagocytosis-related pathways, reflecting a shift toward maladaptive immune responses. SAM phenotypes, often seen in aging, exhibit diminished surveillance capacity and enhanced pro-inflammatory signatures [27]. Another pathogenic mechanism involves synaptic stripping, whereas activated microglia erroneously tag and eliminate functional synapses through complement activation, an early event in AD and other neurodegenerative disorders [28]. Furthermore, in conditions such as MS, microglia contribute to demyelination and inhibit oligodendrocyte maturation, although a reparative role may also emerge following phenotypic reprogramming [29]. Thus, microglia exhibit a dual character as critical for CNS defense and maintenance, yet capable of driving neuroinflammatory and neurodegenerative cascades when dysregulated.

3.2.2 Astrocytes: Metabolic Support Versus Neurotoxic Reactivity

Astrocytes are key regulators of the CNS microenvironment, supporting neuronal metabolism, neurotransmitter homeostasis, and BBB integrity [30]. Under physiological conditions, they facilitate glutamate clearance through EAAT2, participate in the astrocyte–neuron lactate shuttle, and modulate extracellular ion balance. However, in response to injury or chronic neurodegeneration, astrocytes undergo reactive astrogliosis and are marked by GFAP upregulation, morphological hypertrophy, and functional reprogramming [31]. This reactive state may diverge into distinct phenotypes. A1 astrocytes, induced by pro-inflammatory cues, acquire a neurotoxic profile, releasing factors that promote neuronal and oligodendrocyte apoptosis. In contrast, A2 astrocytes are often associated with ischemic conditions, expressing neuroprotective and regenerative molecules [32].

In AD, astrocytes exhibit impaired Aβ clearance and reduced EAAT2 expression, contributing to glutamate excitotoxicity [33]. Elevated intracellular Ca^{2+} levels promote tonic GABA release, disrupting synaptic plasticity and memory consolidation [34]. These dysfunctions are exacerbated by chronic inflammatory signaling, particularly IL-1β, TNF-α, and mitochondrial ROS. Astrocyte-mediated BBB disruption is another hallmark of neurodegenerative pathology [35]. MMP-9 secretion and mislocalization of AQP4 from perivascular endfeet compromise barrier integrity, facilitating peripheral immune cell infiltration [36]. In PD, astrocytic uptake of α-synuclein leads to dysfunction and secondary microglial activation, amplifying dopaminergic neuron loss [37]. In diseases like ALS and HD, astrocytes display reduced metabolic support due to mitochondrial impairment and glutamate transport deficits. In ALS, loss of EAAT2 function and lactate production exacerbate motor neuron death, whereas in HD, mutant huntingtin accumulation disrupts K^+ buffering and Ca^{2+} signaling [38]. Furthermore, astrocytes also participate in innate immune responses through TLR2/3/4 activation, initiating cytokine release and contributing to sustained neuroinflammation. Although glial cells contribute to BBB repair and tissue remodeling, their sustained activation is frequently associated with deleterious effects on neuronal integrity and function [39].

3.3 Cytokine Networks: Amplifiers of Neuroinflammation

Neuroinflammation is a hallmark of various neurodegenerative disorders and is orchestrated by a complex network of cytokines and chemokines. Although acute inflammation is part of the CNS defense mechanism, chronic activation of cytokine signaling disrupts neuronal homeostasis, exacerbates tissue damage, and accelerates disease progression. Glial cells, particularly microglia and astrocytes, are the primary sources and targets of these mediators in the brain [40, 41].

3.3.1 Pro-Inflammatory Cytokine Cascades

The progression of neurodegeneration is closely linked to elevated pro-inflammatory cytokine levels such as IL-1β and TNF-α. These cytokines act through conserved signaling pathways. To detail, IL-1β binds to IL-1 receptor, activating MyD88/NF-κ B axis, whereas TNF-α signals through TNFR1, initiating TRADD/Caspase-8- cascade, leading to inflammation, apoptosis, or necroptosis depending on cell type, duration of stimulation, and presence of specific regulatory molecules [40, 42]. In AD, both cytokines are elevated in response to Aβ plaque deposition, contributing to microglial activation and a sustained inflammatory microenvironment. Furthermore, IL-1β promotes microglial neurotoxic mediators' production and facilitates peripheral immune cells recruitment into CNS [43]. Similarly, TNF-α not only induces inflammation but also enhances AMPA receptor trafficking, disrupts synaptic plasticity, and impairs long-term depression, hindering cognitive processes such as learning and memory. Another pleiotropic cytokine, IL-6, is consistently elevated in the cerebrospinal fluid and serum of AD patients, which is associated with impaired cognition and disease severity [44]. Furthermore, IL-17 is produced by Th17 cells and innate lymphoid populations, disrupting BBB and enhancing neuroinflammation by activating glial cells and promoting leukocyte infiltration. Although IL-17 is essential for host defense, its chronic expression contributes to CNS pathology in disorders such as MS and PD [45].

3.3.2 Anti-Inflammatory Failure and Resolution Pathways

The resolution of neuroinflammation relies on anti-inflammatory cytokines such as IL-10 and TGF-β. These molecules normally suppress glial activation and restore tissue homeostasis [43]. However, during chronic neurodegeneration, their efficacy is diminished due to receptor desensitization and intracellular negative regulators, such as SMAD7 in the TGF-β pathway, leading to a failure in resolving inflammation [46]. Recent interest has also emerged around specialized pro-resolving mediators such as lipoxins, resolvins, and maresins, derived from omega-3 polyunsaturated fatty acids. These bioactive lipids counteract pro-inflammatory signaling, promote phagocytosis of cellular debris, and aid in tissue repair [47]. Although specialized pro-resolving mediators have shown promise in peripheral inflammatory disorders, their therapeutic utility in neuroinflammation remains underexplored and represents a promising area for future investigation [48].

3.3.3 Chemokines: Orchestrators of Immune Cell Trafficking

Chemokines, also known as chemotactic cytokines, regulate leukocyte migration and CNS immune surveillance. Their dysregulation facilitates the recruitment of peripheral immune cells (e.g., monocytes, T cells) into inflamed CNS regions [49]. Chemokines are categorized based on conserved cysteine motifs into CXC, CC, CX3C, and XC subfamilies. Among them, CX3CL1 (fractalkine) and CCL2 (MCP-1) are known to trigger neuroinflammation. CX3CL1 is constitutively expressed by neurons and signals through CX3CR1, predominantly on microglia, maintaining neuroimmune homeostasis [12, 49]. However, in neurodegenerative diseases, altered CX3CL1–CX3CR1 signaling impairs microglial function, exacerbating neurotoxicity. Furthermore, CCL2 binds to CCR2 and facilitates monocyte infiltration into the CNS. Elevated CCL2 levels are also implicated in

MS, PD, and AD, where they contribute to BBB breakdown, microglial activation, and neuroinflammation [12, 50]. In PD, astrocyte-derived CCL2 promotes dopaminergic neuron vulnerability [51]. In MS, its overexpression correlates with disease activity and lesion burden, whereas TNF-α, VEGF, and LPS further enhance its production [52]. Although many chemokines exacerbate inflammation, some exhibit neuroprotective functions based on the timing of their release, receptor specificity, and surrounding cellular environment. For example, precisely regulated chemokine signaling can promote neural stem cell proliferation and modulate synaptic activity. Thus, chemokines and their receptors represent potential diagnostic biomarkers and therapeutic targets in neuroinflammatory and neurodegenerative disorders [40, 53].

3.4 Inflammasomes as Molecular Hubs of Neurodegeneration

Inflammasomes are cytosolic multiprotein complexes that are critical to innate immune responses and serve as molecular platforms for the activation of caspase-1, which catalyzes the proteolytic maturation of pro-inflammatory cytokines IL-1β and IL-18, as well as the induction of pyroptosis [54]. Among the various inflammasome complexes identified (e.g., NLRP1, NLRP2, NLRC4, AIM2), the NLRP3 inflammasome is most extensively studied due to its wide range of stimuli and central role in neuroinflammation [55]. Activation of NLRP3 follows a two-step mechanism. First, priming via NF-κB-mediated upregulation of NLRP3 and pro-IL-1β in response to pattern recognition receptors (e.g., TLRs), and, second, activation triggered by intracellular events such as potassium efflux, mitochondrial ROS, lysosomal rupture, and the presence of aggregated proteins like Aβ and α-synuclein [56]. This activation leads to the recruitment of the adaptor protein Apoptosis-associated Speck-like protein containing a CARD (ASC) and caspase-1, culminating in mature cytokines and pyroptotic cell death initiation [57].

3.4.1 NLRP3 Inflammasome Activation and Regulation

In AD, both Aβ oligomers and hyperphosphorylated tau proteins function as danger-associated molecular patterns (DAMPs), activating TLR4 and TREM2 signaling pathways [40]. These events facilitate the assembly of the NLRP3 inflammasome within microglia. A key component of this complex is the adaptor protein ASC, which mediates the recruitment and activation of pro-caspase-1 [58]. Upon activation, ASC aggregates into oligomeric structures known as ASC specks, which are released extracellularly and taken up by neighboring cells, thereby perpetuating a self-amplifying, prion-like inflammatory cascade. This sustained neuroinflammation not only disrupts neuronal homeostasis but also impairs Aβ clearance mechanisms [9, 59]. Moreover, activation of the NLRP3 inflammasome drives a phenotypic shift in microglia from a reparative, M2-like state to a pro-inflammatory M1-like profile, further promoting amyloid accumulation and disease progression (Fig. 3.2) [60]. Mouse models deficient in NLRP3 or caspase-1 exhibit improved cognitive

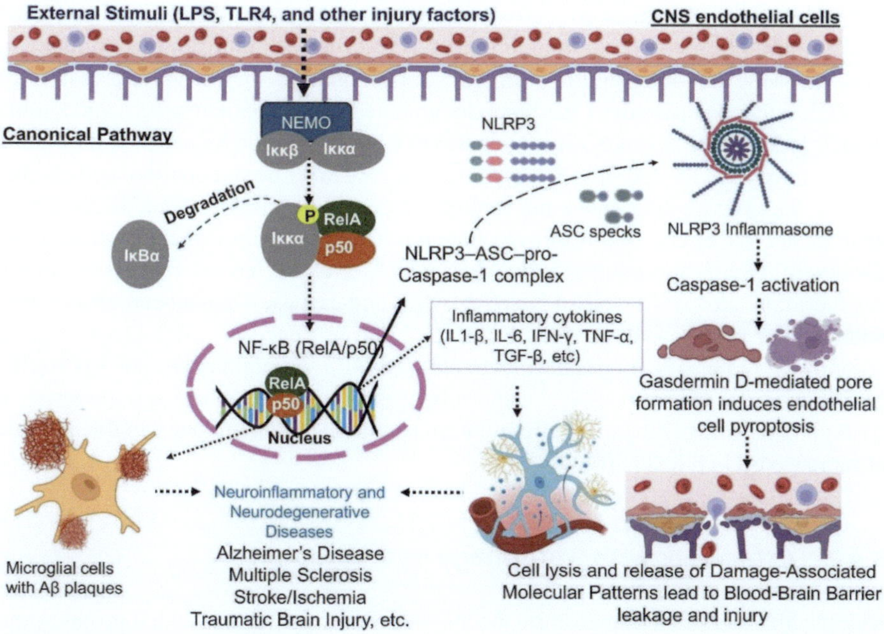

Fig. 3.2 NF-κB and NLRP3 inflammasome-mediated neuroinflammation and BBB disruption. External stimuli (e.g., LPS, TLR4 activation) trigger canonical NF-κB signaling via IKK complex activation, leading to RelA/p50 nuclear translocation and proinflammatory gene expression. NF-κB also primes the NLRP3 inflammasome, which activates caspase-1, resulting in cytokine release and gasdermin-D-mediated pyroptosis of CNS endothelial cells. This process disrupts the BBB, promotes DAMP release, and contributes to neuroinflammatory and neurodegenerative diseases such as AD, MS, stroke, and traumatic brain injury. (Created using BioRender. https://www.biorender.com/)

outcomes and reduced Aβ burden [61]. Additionally, INPP5D (SHIP1) and CALHM2, genes implicated in AD, modulate microglial inflammatory signaling through NLRP3 regulation. A decline in SHIP1 activity also exacerbates inflammasome activation, as evidenced by elevated caspase-1 cleavage and IL-1β secretion [62].

In PD, misfolded α-synuclein aggregates stimulate NLRP3 activation through microglial PRRs, primarily TLR2 and TLR4 [40]. Mitochondrial dysfunction and ROS production caused by α-syn accumulation trigger NLRP3 inflammasome assembly, promoting neuroinflammation and dopaminergic neuronal loss in the substantia nigra [63]. The exosome-mediated spread of α-syn and ASC specks contributes to disease propagation. The MPTP model of PD also shows upregulation of other inflammasomes, including AIM2 and NLRP1, though their roles remain less defined. Similarly, HD caused by polyglutamine expansions in mutant huntingtin is marked by chronic neuroinflammation, with elevated levels of IL-1β, IL-6, and TNF-α [64]. Mutant huntingtin disrupts mitochondrial integrity and impairs autophagic flux, which can potentiate NLRP3 activation. Microglial activation driven by these cytokines exacerbates neurotoxicity, and disruption of the BBB permits infiltration of peripheral immune cells, further amplifying inflammation [65]. In MS, an autoimmune demyelinating disease, increased expression of NLRP3, ASC, and caspase-1 has been documented in both patient tissues and animal models [66]. NLRP3 activation contributes to CNS infiltration by peripheral immune cells through IL-1β and IL-18 signaling, exacerbating demyelination and axonal injury. Targeting inflammasome using caspase-1 inhibitors can demonstrate neuroprotective effects and improve clinical outcomes in experimental autoimmune encephalomyelitis models [67]. In ALS, which is characterized by motor neuron degeneration, elevated levels of NLRP3, ASC, and caspase-1 have been observed in both human tissues and transgenic models. The presence of TAR DNA-binding protein 43 (TDP-43) aggregates activates NLRP3 inflammasome in microglia, contributing to neurotoxicity. Furthermore, serum IL-18 levels are elevated in ALS patients, suggesting systemic inflammasome involvement, even in the absence of increased IL-1β levels [68, 69].

3.4.2 Cross-Talk with Autophagy and Mitophagy

Autophagy serves as a homeostatic mechanism that degrades misfolded proteins and damaged organelles, thus functioning as a negative regulator of inflammasome activation [70]. Under physiological conditions, autophagic flux limits NLRP3 activation by targeting inflammasome components such as NLRP3 and ASC for lysosomal degradation. However, in neurodegeneration, autophagy is frequently impaired, leading to inflammasome persistence and chronic inflammation [71]. IL-1β, in turn, can suppress autophagy, establishing a feed-forward loop that amplifies neuroinflammation [72]. Mitochondrial dysfunction represents a key link between inflammasome activation and neurodegeneration. The release of mitochondrial DNA through voltage-dependent anion channels into the cytosol serves as a potent trigger for NLRP3 [73]. Additionally, mutations in mitophagy-regulating genes such as PINK1 and Parkin impair the clearance of dysfunctional mitochondria, leading to ROS accumulation and subsequent activation of the inflammasome. These mechanisms are predominantly relevant in PD pathogenesis, where defective mitophagy due to PINK1/Parkin loss contributes to dopaminergic neurodegeneration through sustained NLRP3 activity. Together, the inflammasome–autophagy–mitophagy axis forms a critical interface in neurodegenerative disorders pathogenesis, underscoring its potential as a therapeutic target to halt or reverse disease progression [40, 73, 74].

3.5 Neurodegenerative Consequences of Glial-Driven Inflammation

Persistent glial activation in the CNS is a key driver of progressive neurodegeneration. Although microglia and astrocytes initially serve protective roles, their chronic inflammatory responses can disrupt synaptic communication, impair network function, and promote neuronal death [12, 40]. These glia-mediated alterations not only compromise brain function but also contribute to the self-perpetuating cycles that worsen disease pathology [75].

3.5.1 Synaptic and Network Dysfunction

Neuroinflammatory cytokines released by glial cells have a direct effect on synaptic physiology [76]. In detail, microglia-derived TNF-α enhances the insertion of calcium-permeable AMPA receptors at neuronal synapses. This process increases calcium influx, placing neurons under excitotoxic stress, which is a known cause to trigger neuronal damage in disorders like AD and ALS [77]. Furthermore, elevated IL-1β levels disrupt excitatory and inhibitory signaling balance in neuronal networks, impairing synaptic plasticity, contributing to cognitive deficits observed in AD and related disorders [78]. These alterations in network activity often precede overt neuronal loss, highlighting their importance as early indicators of neurodegenerative disease.

3.5.2 Feed-Forward Loops in Disease Progression

A hallmark of neurodegeneration is feed-forward inflammatory loops formation, where pathological protein aggregates and immune activation mutually increase disease progression [79]. In AD, Aβ and hyperphosphorylated tau trigger risk signals to activate NLRP3 inflammasome in microglia. This leads to IL-1β release, which in turn upregulates β-secretase 1, which is responsible for Aβ production, accelerating plaque formation and neuroinflammation, driving disease progression [9, 80]. In neuronal death, different inflammatory pathways trigger distinct forms of degeneration. For example, apoptosis is mediated by caspase-3, leading to a controlled and typically non-inflammatory cell death [4]. However, pyroptosis, which is a highly inflammatory form of cell death, occurs when gasdermin-D forms pores in the cell membrane, leading to cell rupture and pro-inflammatory contents release [73]. Interestingly, pyroptosis has been linked to neurodegenerative diseases such as ALS and FTD, especially due to protein aggregates like TDP-43 that activate inflammasomes in microglia and maintain inflammatory responses [40]. Together, these events show how inflammation driven by glial cells disrupts neural networks, interferes with brain activity, and speeds up neurodegeneration, shifting the CNS from a state of repair to one of ongoing damage and decline.

3.6 Therapeutic Strategies

3.6.1 Targeting Glial Cell Polarization

Glial cells, including microglia and astrocytes, also play key roles in regulating neuroinflammation and neurodegeneration [40]. Modulating their activation states, commonly referred to as polarization, offers a promising approach to influence the progression of neurodegenerative diseases [73]. In detail, the TREM2 receptor is a critical modulator of microglial function, regulating processes such as phagocytosis, survival, proliferation, and inflammatory responses. Loss-of-function mutations in TREM2 are strongly linked to increased AD risk and other neurodegenerative conditions [81]. Agonistic antibodies targeting TREM2 have been developed to potentiate its neuroprotective effects. For example, mouse-specific AL002a and human-specific AL002c monoclonal antibodies enhance microglial phagocytic activity and anti-inflammatory signaling [82]. These antibodies stimulate downstream pathways, including SYK and AKT, to improve pathogenic protein aggregate clearance, such as Aβ and apoptotic neuronal debris [40]. Another promising candidate, CGX101, binds to the TREM2 extracellular domain, promoting microglial proliferation and enhancing phagocytosis, with preclinical evidence indicating cognitive improvements in AD models [83]. Such strategies aim to shift microglia from pro-inflammatory to a reparative phenotype to restore homeostasis and mitigate chronic neuroinflammation [84]. This approach exhibits potential applications across multiple neurodegenerative diseases characterized by protein aggregation and chronic immune activation, including PD and FTD.

Astrocytes provide metabolic and trophic support to neurons and undergo reactive transformations during neurodegeneration that exacerbate pathology. Reactive astrocytes can contribute to excitotoxicity, BBB disruption, and synaptic dysfunction [35]. Pharmacological enhancement of astrocytic glutamate uptake is a key target to counter excitotoxic damage. A β-lactam antibiotic, ceftriaxone, has been shown to upregulate the astrocytic glutamate transporter EAAT2, promoting excess extracellular glutamate clearance and preventing excitotoxic neuronal injury [85]. Additionally, connexin-43 hemichannel blockers can limit harmful intercellular signaling and neuroinflammation through astrocytic networks [86]. By modulating astrocyte function, these interventions aim to reduce neurotoxic inflammation and preserve synaptic integrity, offering complementary benefits to microglial-targeted therapies.

3.6.2 Cytokine and Inflammasome Inhibition

Neuroinflammatory cascades driven by cytokines and inflammasomes constitute critical pathological mechanisms in neurodegenerative diseases [12]. Targeting these signaling axes can attenuate inflammation-induced neuronal damage. For example, IL-1β is a pro-inflammatory cytokine essential to neuroinflammation augmentation. Agents such as anakinra, a recombinant IL-1 receptor antagonist, and canakinumab, a monoclonal antibody against IL-1β, can inhibit IL-1β signaling. These agents are currently undergoing clinical evaluation in AD and PD to disrupt detrimental IL-1β-mediated pathways that exacerbate neurodegeneration [87]. Next, the NLRP3 inflammasome is a multiprotein complex that regulates caspase-1 activation and subsequent IL-1β and IL-18 maturation, playing a pivotal role in neuroinflammation [88]. Its dysregulation is implicated in multiple neurodegenerative disorders, making it a prime pharmacological target. MCC950 is the most extensively studied specific NLRP3 inhibitor, and it effectively inhibits both canonical and non-canonical activation pathways of NLRP3, without affecting other inflammasomes like AIM2 or NLRC4. Importantly, MCC950 crosses the BBB and demonstrates neuroprotective effects in AD and PD preclinical models by reducing neuroinflammation and promoting clearance of protein aggregates [89, 90]. Other inhibitors, such as CY-09, directly bind to the ATPase domain of NLRP3, preventing its oligomerization and assembly. CY-09 and related analogues have shown efficacy in in vitro and animal models, though their application across neurodegenerative diseases requires further validation [91]. Additional compounds, including OLT1177, tranilast, oridonin, JC171, and MNS, target NLRP3 through distinct molecular mechanisms [92]. Some broad-spectrum inflammasome inhibitors like BAY 11-7082 and parthenolide modulate ATPase activity but lack specificity, potentially affecting other inflammasomes and increasing off-target effects [93].

3.6.3 Multimodal Approaches and Nanotechnology-Based Delivery

Due to the complex, multifactorial nature of neurodegenerative diseases, monotherapies often fall short of comprehensive disease modification. Multimodal strategies integrating diverse mechanisms of action and personalized medicine approaches offer promising avenues for effective intervention [94]. Synergistic targeting of multiple inflammatory and degenerative pathways may enhance therapeutic outcomes. For example, combining NLRP3 inhibitors with anti-TNF-α biologics and autophagy enhancers [95] aims to simultaneously suppress inflammasome-driven inflammation, neutralize pro-inflammatory cytokines, and promote cellular clearance of toxic proteins. Advances in gene editing technologies provide novel opportunities for precise modulation of neuroinflammatory pathways. Next, CRISPR-Cas9-based approaches have been developed to silence key inflammasome components such as ASC or IL-1β in glial cells [96]. Moreover, overcoming the restrictive BBB remains a major hurdle in CNS drug delivery. However, nanoparticles engineered for BBB penetration can deliver therapeutic agents such as small interfering RNAs against pro-inflammatory cytokines with high specificity [97] to minimize off-target toxicity and enhance bioavailability, facilitating efficient modulation of neuroinflammation.

3.7 Conclusion

The complex interactions among glial cells, cytokine signaling, and inflammasome activation form a central mechanism underlying neuroinflammation and neurodegeneration in various disorders. Although glial cells play essential roles in maintaining CNS homeostasis, their dysregulated activation, driven by pro-inflammatory cytokines and inflammasome pathways, can worsen neuronal dysfunction and loss. Therapeutic approaches that selectively target this glial–cytokine–inflammasome axis offer considerable potential but must balance critical neuroprotective immune functions, reducing chronic inflammatory damage. Advances in single-cell omics facilitate detailed characterization of glial diversity and dynamic states in disease environments, enabling more precise interventions. Additionally, developing therapies that simultaneously inhibit inflammasome activation and enhance autophagy may disrupt harmful feedback mechanisms and support neuroprotection to advance precision medicine strategies for neurodegenerative diseases.

References

1. De Marchi F, Munitic I, Vidatic L, Papić E, Rački V, Nimac J, Jurak I, Novotni G, Rogelj B, Vuletic V, Liscic RM, Cannon JR, Buratti E, Mazzini L, Hecimovic S. Overlapping neuroimmune mechanisms and therapeutic targets in neurodegenerative disorders. Biomedicines. 2023;11(10):2793. https://doi.org/10.3390/biomedicines11102793.
2. Li S, Lei Z, Sun T. The role of microRNAs in neurodegenerative diseases: a review. Cell Biol Toxicol. 2023;39:53–83. https://doi.org/10.1007/s10565-022-09761-x.
3. Rim C, You MJ, Nahm M, et al. Emerging role of senescent microglia in brain aging-related neurodegenerative diseases. Transl Neurodegener. 2024;13:10. https://doi.org/10.1186/s40035-024-00402-3.
4. Cai L, Fan Q, Pang R, et al. Microglia programmed cell death in neurodegenerative diseases and CNS injury. Apoptosis. 2025;30:446–65. https://doi.org/10.1007/s10495-024-02041-5.
5. Gonçalves de Andrade E, Šimončičová E, Carrier M, Vecchiarelli HA, Robert MÈ. Microglia fighting for neurological and mental health: on the central nervous system frontline of COVID-19 pandemic. Front Cell Neurosci. 2021;15:647378. https://doi.org/10.3389/fncel.2021.647378.
6. Zhao R. Exercise mimetics: a novel strategy to combat neuroinflammation and Alzheimer's disease. J Neuroinflammation. 2024;21:40. https://doi.org/10.1186/s12974-024-03031-9.
7. Dias-Carvalho A, Sá SI, Carvalho F, et al. Inflammation as common link to progressive neurological diseases. Arch Toxicol. 2024;98:95–119. https://doi.org/10.1007/s00204-023-03628-8.
8. Chen Y, Ye X, Escames G, et al. The NLRP3 inflammasome: contributions to inflammation-related diseases. Cell Mol Biol Lett. 2023;28:51. https://doi.org/10.1186/s11658-023-00462-9.
9. Van Zeller M, Dias D, Sebastião AM, Valente CA, Wu Z. NLRP3 inflammasome: a starring role in amyloid-β- and tau-driven pathological events in Alzheimer's disease. J Alzheimers Dis. 2021;83(3):939–61. https://doi.org/10.3233/JAD-210268.
10. Sharma B, Satija G, Madan A, et al. Role of NLRP3 inflammasome and its inhibitors as emerging therapeutic drug candidate for Alzheimer's disease: a review of mechanism of activation, regulation, and inhibition. Inflammation. 2023;46:56–87. https://doi.org/10.1007/s10753-022-01730-0.

11. Singh A, Tiwari V, Roy S. Multifaceted role of oxidative stress in neurological disorders. Mol Biol Rep. 2025;52:640. https://doi.org/10.1007/s11033-025-10760-4.
12. Gao C, Jiang J, Tan Y, et al. Microglia in neurodegenerative diseases: mechanism and potential therapeutic targets. Signal Transduct Target Ther. 2023;8:359. https://doi.org/10.1038/s41392-023-01588-0.
13. Biswas K. Microglia mediated neuroinflammation in neurodegenerative diseases: a review on the cell signaling pathways involved in microglial activation. J Neuroimmunol. 2023;383: 578180. https://doi.org/10.1016/j.jneuroim.2023.578180.
14. Ahadiat SA, Hosseinian Z. Astrocytes' innate role in neurodegenerative disorders. Bull Natl Res Cent. 2023;47:105. https://doi.org/10.1186/s42269-023-01083-0.
15. Kumari S, et al. The impact of cytokines in neuroinflammation-mediated stroke. Cytokine Growth Factor Rev. 2024;78:105–19. https://doi.org/10.1016/j.cytogfr.2024.06.002.
16. Stoian A, Muntean C, Babă D-F, Manea A, Dénes L, Simon-Szabó Z, et al. NF-κB and JAK/STAT signaling pathways as crucial regulators of neuroinflammation and astrocyte modulation in spinal cord injury. Cells. 2024;13(7):581. https://doi.org/10.3390/cells13070581.
17. Schwaid AG, Spencer KB. Strategies for targeting the NLRP3 inflammasome in the clinical and preclinical space. J Med Chem. 2020;64(1):101–22. https://doi.org/10.1021/acs.jmedchem.0c01307.
18. Yong HYF, Rawji KS, Ghorbani S, et al. The benefits of neuroinflammation for the repair of the injured central nervous system. Cell Mol Immunol. 2019;16:540–6. https://doi.org/10.1038/s41423-019-0223-3.
19. Rahman MM, et al. Emerging role of neuron-glia in neurological disorders: at a glance. Oxidative Med Cell Longev. 2022;2022(1):3201644. https://doi.org/10.1155/2022/3201644.
20. Schiller M, Ben-Shaanan TL, Rolls A. Neuronal regulation of immunity: why, how and where? Nat Rev Immunol. 2021;21:20–36. https://doi.org/10.1038/s41577-020-0387-1.
21. Bitar L, Puig B, Oertner TG, et al. Changes in neuroimmunological synapses during cerebral ischemia. Transl Stroke Res. 2025;16:1369–82. https://doi.org/10.1007/s12975-024-01286-1.
22. Kwon HS, Koh SH. Neuroinflammation in neurodegenerative disorders: the roles of microglia and astrocytes. Transl Neurodegener. 2020;9:42. https://doi.org/10.1186/s40035-020-00221-2.
23. Gomez-Arboledas A, Acharya MM, Tenner AJ. The role of complement in synaptic pruning and neurodegeneration. Immunotargets Ther. 2021;10:373–86. https://doi.org/10.2147/itt.s305420.
24. Rosciszewski G, Cadena V, Murta V, et al. Toll-like receptor 4 (TLR4) and triggering receptor expressed on myeloid cells-2 (TREM-2) activation balance astrocyte polarization into a proinflammatory phenotype. Mol Neurobiol. 2018;55:3875–88. https://doi.org/10.1007/s12035-017-0618-z.
25. Bennett JM, et al. Inflammation–nature's way to efficiently respond to all types of challenges: implications for understanding and managing "the epidemic" of chronic diseases. Front Med. 2018;5:316. https://doi.org/10.3389/fmed.2018.00316.
26. Leng F, Edison P. Neuroinflammation and microglial activation in Alzheimer disease: where do we go from here? Nat Rev Neurol. 2021;17:157–72. https://doi.org/10.1038/s41582-020-00435-y.
27. Malvaso A, Gatti A, Negro G, Calatozzolo C, Medici V, Poloni TE. Microglial senescence and activation in healthy aging and Alzheimer's disease: systematic review and neuropathological scoring. Cells. 2023;12(24):2824. Published 2023 Dec 12. https://doi.org/10.3390/cells12242824.
28. Carpanini SM, Torvell M, Bevan RJ, et al. Terminal complement pathway activation drives synaptic loss in Alzheimer's disease models. Acta Neuropathol Commun. 2022;10:99. https://doi.org/10.1186/s40478-022-01404-w.
29. Zveik O, Rechtman A, Ganz T, et al. The interplay of inflammation and remyelination: rethinking MS treatment with a focus on oligodendrocyte progenitor cells. Mol Neurodegener. 2024;19:53. https://doi.org/10.1186/s13024-024-00742-8.

30. Che J, Sun Y, Deng Y, et al. Blood-brain barrier disruption: a culprit of cognitive decline? Fluids Barriers CNS. 2024;21:63. https://doi.org/10.1186/s12987-024-00563-3.
31. Todd AC, Hardingham GE. The regulation of astrocytic glutamate transporters in health and neurodegenerative diseases. Int J Mol Sci. 2020;21(24):9607. https://doi.org/10.3390/ijms21249607.
32. Beard E, Lengacher S, Dias S, Magistretti PJ, Finsterwald C. Astrocytes as key regulators of brain energy metabolism: new therapeutic perspectives [published correction appears in Front Physiol. 2022;13:867827. https://doi.org/10.3389/fphys.2022.867827]. Front Physiol. 2022;12: 825816. Published 2022. https://doi.org/10.3389/fphys.2021.825816.
33. Takahashi K, Foster JB, Lin CLG. Glutamate transporter EAAT2: regulation, function, and potential as a therapeutic target for neurological and psychiatric disease. Cell Mol Life Sci. 2015;72:3489–506. https://doi.org/10.1007/s00018-015-1937-8.
34. Padamsey Z, Foster WJ, Emptage NJ. Intracellular CA2+ release and synaptic plasticity: a tale of many stores. Neuroscientist. 2018;25(3):208–26. https://doi.org/10.1177/1073858418785334.
35. Lawrence JM, Schardien K, Wigdahl B, et al. Roles of neuropathology-associated reactive astrocytes: a systematic review. Acta Neuropathol Commun. 2023;11:42. https://doi.org/10.1186/s40478-023-01526-9.
36. Zhu B, Cao A, Chen C, Zhou W, Luo W, Gui Y, Wang Q, Xu Z, Wang J. MMP-9 inhibition alleviates postoperative cognitive dysfunction by improving glymphatic function via regulating AQP4 polarity. Int Immunopharmacol. 2023;126:111215. https://doi.org/10.1016/j.intimp.2023.111215.
37. Saramowicz K, Siwecka N, Galita G, Kucharska-Lusina A, Rozpędek-Kamińska W, Majsterek I. Alpha-synuclein contribution to neuronal and glial damage in Parkinson's disease. Int J Mol Sci. 2023;25(1):360. https://doi.org/10.3390/ijms25010360.
38. Dittlau KS, Freude K. Astrocytes: the stars in neurodegeneration? Biomolecules. 2024;14(3): 289. https://doi.org/10.3390/biom14030289.
39. Abarca-Merlin DM, Martínez-Durán JA, Medina-Pérez JD, Rodríguez-Santos G, Alvarez-Arellano L. From immunity to neurogenesis: toll-like receptors as versatile regulators in the nervous system. Int J Mol Sci. 2024;25(11):5711. https://doi.org/10.3390/ijms25115711.
40. Zhang W, Xiao D, Mao Q, et al. Role of neuroinflammation in neurodegeneration development. Signal Transduct Target Ther. 2023;8:267. https://doi.org/10.1038/s41392-023-01486-5.
41. Jung YJ, Tweedie D, Scerba MT, Greig NH. Neuroinflammation as a factor of neurodegenerative disease: thalidomide analogs as treatments. Front Cell Dev Biol. 2019;7:313. https://doi.org/10.3389/fcell.2019.00313.
42. Druszczyńska M, Godkowicz M, Kulesza J, Wawrocki S, Fol M. Cytokine receptors—regulators of antimycobacterial immune response. Int J Mol Sci. 2022;23(3):1112. https://doi.org/10.3390/ijms23031112.
43. Thakur S, Dhapola R, Sarma P, et al. Neuroinflammation in Alzheimer's disease: current progress in molecular signaling and therapeutics. Inflammation. 2023;46:1–17. https://doi.org/10.1007/s10753-022-01721-1.
44. Bourgognon J, Cavanagh J. The role of cytokines in modulating learning and memory and brain plasticity. Brain Neurosci Adv. 2020;4:239821282097980. https://doi.org/10.1177/2398212820979802.
45. Milovanovic J, Arsenijevic A, Stojanovic B, Kanjevac T, Arsenijevic D, Radosavljevic G, Milovanovic M, Arsenijevic N. Interleukin-17 in chronic inflammatory neurological diseases. Front Immunol. 2020;11:947. https://doi.org/10.3389/fimmu.2020.00947.
46. Stolfi C, Troncone E, Marafini I, Monteleone G. Role of TGF-Beta and SMAD7 in gut inflammation, fibrosis and cancer. Biomolecules. 2020;11(1):17. https://doi.org/10.3390/biom11010017.
47. Kwon Y. Immuno-resolving ability of resolvins, protectins, and maresins derived from Omega-3 fatty acids in metabolic syndrome. Mol Nutr Food Res. 2019;64(4):1900824. https://doi.org/10.1002/mnfr.201900824.

48. Shang P, Zhang Y, Ma D, Hao Y, Wang X, Xin M, Zhang Y, Zhu M, Feng J. Inflammation resolution and specialized pro-resolving lipid mediators in CNS diseases. Expert Opin Ther Targets. 2019;23(11):967–86. https://doi.org/10.1080/14728222.2019.1691525.
49. Hughes CE, Nibbs RJB. A guide to chemokines and their receptors. FEBS J. 2018;285(16): 2944–71. https://doi.org/10.1111/febs.14466.
50. Subbarayan MS, Joly-Amado A, Bickford PC, Nash KR. CX3CL1/CX3CR1 signaling targets for the treatment of neurodegenerative diseases. Pharmacol Ther. 2021;231:107989. https://doi.org/10.1016/j.pharmthera.2021.107989.
51. Wang C, Yang T, Liang M, et al. Astrocyte dysfunction in Parkinson's disease: from the perspectives of transmitted α-synuclein and genetic modulation. Transl Neurodegener. 2021;10: 39. https://doi.org/10.1186/s40035-021-00265-y.
52. Khan Z, Mehan S, Gupta GD, Narula AS. Immune system dysregulation in the progression of multiple sclerosis: molecular insights and therapeutic implications. Neuroscience. 2024;548:9–26. https://doi.org/10.1016/j.neuroscience.2024.04.004.
53. Wojcieszak J, Kuczyńska K, Zawilska JB. Role of chemokines in the development and progression of Alzheimer's disease. J Mol Neurosci. 2022;72:1929–51. https://doi.org/10.1007/s12031-022-02047-1.
54. Huang Y, Xu W, Zhou R. NLRP3 inflammasome activation and cell death. Cell Mol Immunol. 2021;18:2114–27. https://doi.org/10.1038/s41423-021-00740-6.
55. Ziehr BK, MacDonald JA. Regulation of NLRPs by reactive oxygen species: a story of crosstalk. Biochim Biophys Acta (BBA) Mol Cell Res. 2024;1871(8):119823. https://doi.org/10.1016/j.bbamcr.2024.119823.
56. Maurya R, Sharma A, Naqvi S. Decoding NLRP3 inflammasome activation in Alzheimer's disease: a focus on receptor dynamics. Mol Neurobiol. 2025:1–21. https://doi.org/10.1007/s12035-025-04918-1.
57. Henedak NT, El-Abhar HS, Soubh AA, Abdallah DM. NLRP3 Inflammasome: a central player in renal pathologies and nephropathy. Life Sci. 2024;351:122813. https://doi.org/10.1016/j.lfs.2024.122813.
58. Dadkhah M, Sharifi M. The NLRP3 inflammasome: mechanisms of activation, regulation, and role in diseases. Int Rev Immunol. 2024;44(2):98–111. https://doi.org/10.1080/08830185.2024.2415688.
59. Litwiniuk A, Bik W, Kalisz M, Baranowska-Bik A. Inflammasome NLRP3 potentially links obesity-associated low-grade systemic inflammation and insulin resistance with Alzheimer's disease. Int J Mol Sci. 2021;22(11):5603. https://doi.org/10.3390/ijms22115603.
60. Wendimu MY, Hooks SB. Microglia phenotypes in aging and neurodegenerative diseases. Cells. 2022;11(13):2091. https://doi.org/10.3390/cells11132091.
61. Flores J, Noël A, Fillion ML, et al. Therapeutic potential of Nlrp1 inflammasome, Caspase-1, or Caspase-6 against Alzheimer disease cognitive impairment. Cell Death Differ. 2022;29:657–69. https://doi.org/10.1038/s41418-021-00881-1.
62. Terzioglu G, Young-Pearse TL. Microglial function, INPP5D/SHIP1 signaling, and NLRP3 inflammasome activation: implications for Alzheimer's disease. Mol Neurodegener. 2023;18: 89. https://doi.org/10.1186/s13024-023-00674-9.
63. Wang Q, Yang S, Zhang X, et al. Inflammasomes in neurodegenerative diseases. Transl Neurodegener. 2024;13:65. https://doi.org/10.1186/s40035-024-00459-0.
64. Lee E, Hwang I, Park S, Hong S, Hwang B, Cho Y, Son J, Yu J. MPTP-driven NLRP3 inflammasome activation in microglia plays a central role in dopaminergic neurodegeneration. Cell Death Differ. 2018;26(2):213–28. https://doi.org/10.1038/s41418-018-0124-5.
65. Mishra Y, Kumar A, Kaundal RK. Mitochondrial dysfunction is a crucial immune checkpoint for neuroinflammation and neurodegeneration: mtDAMPs in focus. Mol Neurobiol. 2025;62: 6715–47. https://doi.org/10.1007/s12035-024-04412-0.
66. Olcum M, Tastan B, Kiser C, Genc S, Genc K. Microglial NLRP3 inflammasome activation in multiple sclerosis. Adv Protein Chem Struct Biol. 2019:247–308. https://doi.org/10.1016/bs.apcsb.2019.08.007.

67. Bjelobaba I, Begovic-Kupresanin V, Pekovic S, Lavrnja I. Animal models of multiple sclerosis: focus on experimental autoimmune encephalomyelitis. J Neurosci Res. 2018;96(6):1021–42. https://doi.org/10.1002/jnr.24224.
68. Zhao W, Beers DR, Bell S, Wang J, Wen S, Baloh RH, Appel SH. TDP-43 activates microglia through NF-κB and NLRP3 inflammasome. Exp Neurol. 2015;273:24–35. https://doi.org/10.1016/j.expneurol.2015.07.019.
69. Xu C, Huan X, Luo S, Zhong H, Zhao C, Chen Y, Zou Z, Chen S. Serum cytokines profile changes in amyotrophic lateral sclerosis. Heliyon. 2024;10(7):e28553. https://doi.org/10.1016/j.heliyon.2024.e28553.
70. Takahama M, Akira S, Saitoh T. Autophagy limits activation of the inflammasomes. Immunol Rev. 2017;281(1):62–73. https://doi.org/10.1111/imr.12613.
71. Biasizzo M, Kopitar-Jerala N. Interplay between NLRP3 inflammasome and autophagy. Front Immunol. 2020;11:591803. https://doi.org/10.3389/fimmu.2020.591803.
72. Sasidharakurup H, Diwakar S. Computational modelling of TNFα related pathways regulated by neuroinflammation, oxidative stress and insulin resistance in neurodegeneration. Appl Netw Sci. 2020;5:72. https://doi.org/10.1007/s41109-020-00307-w.
73. Zhang G, Wei H, Zhao A, et al. Mitochondrial DNA leakage: underlying mechanisms and therapeutic implications in neurological disorders. J Neuroinflammation. 2025;22:34. https://doi.org/10.1186/s12974-025-03363-0.
74. Guo F, Liu X, Cai H, Le W. Autophagy in neurodegenerative diseases: pathogenesis and therapy. Brain Pathol. 2017;28(1):3–13. https://doi.org/10.1111/bpa.12545.
75. Niranjan R. The role of inflammatory and oxidative stress mechanisms in the pathogenesis of Parkinson's disease: focus on astrocytes. Mol Neurobiol. 2014;49:28–38. https://doi.org/10.1007/s12035-013-8483-x.
76. Javanmehr N, Saleki K, Alijanizadeh P, et al. Microglia dynamics in aging-related neurobehavioral and neuroinflammatory diseases. J Neuroinflammation. 2022;19:273. https://doi.org/10.1186/s12974-022-02637-1.
77. Yin HZ, Hsu C, Yu S, Rao SD, Sorkin LS, Weiss JH. TNF-α triggers rapid membrane insertion of Ca2+ permeable AMPA receptors into adult motor neurons and enhances their susceptibility to slow excitotoxic injury. Exp Neurol. 2012;238(2):93–102. https://doi.org/10.1016/j.expneurol.2012.08.004.
78. Bellingacci L, Canonichesi J, Mancini A, Parnetti L, Di Filippo M. Cytokines, synaptic plasticity and network dynamics: a matter of balance. Neural Regen Res. 2023;18(12):2569–72. https://doi.org/10.4103/1673-5374.371344.
79. Glass CK, Saijo K, Winner B, Marchetto MC, Gage FH. Mechanisms underlying inflammation in neurodegeneration. Cell. 2010;140(6):918–34. https://doi.org/10.1016/j.cell.2010.02.016.
80. Liang T, Zhang Y, Wu S, Chen Q, Wang L. The role of NLRP3 inflammasome in Alzheimer's disease and potential therapeutic targets. Front Pharmacol. 2022;13:845185. https://doi.org/10.3389/fphar.2022.845185.
81. Li Y, Xu H, Wang H, Yang K, Luan J, Wang S. TREM2: potential therapeutic targeting of microglia for Alzheimer's disease. Biomed Pharmacother. 2023;165:115218. https://doi.org/10.1016/j.biopha.2023.115218.
82. Zhang G, Wang Z, Hu H, Zhao M, Sun L. Microglia in Alzheimer's disease: a target for therapeutic intervention. Front Cell Neurosci. 2021;15:749587. https://doi.org/10.3389/fncel.2021.749587.
83. George J. TREM2 as an evolving therapeutic target in Alzheimer's disease. Neural Regen Res. 2023;18(12):2680–1. https://doi.org/10.4103/1673-5374.371360.
84. Guo S, Wang H, Yin Y. Microglia polarization from M1 to M2 in neurodegenerative diseases. Front Aging Neurosci. 2022;14:815347. https://doi.org/10.3389/fnagi.2022.815347.
85. Lim SW, Su HC, Nyam TTE, et al. Ceftriaxone therapy attenuates brain trauma in rats by affecting glutamate transporters and neuroinflammation and not by its antibacterial effects. BMC Neurosci. 2021;22:54. https://doi.org/10.1186/s12868-021-00659-8.

86. Zhang Y, Khan S, Liu Y, Siddique R, Zhang R, Yong VW, Xue M. Gap junctions and hemichannels composed of connexins and pannexins mediate the secondary brain injury following intracerebral hemorrhage. Biology. 2021;11(1):27. https://doi.org/10.3390/biology11010027.
87. Mendiola AS, Cardona AE. The IL-1β phenomena in neuroinflammatory diseases. J Neural Transm. 2018;125:781–95. https://doi.org/10.1007/s00702-017-1732-9.
88. Xu W, Huang Y, Zhou R. NLRP3 inflammasome in neuroinflammation and central nervous system diseases. Cell Mol Immunol. 2025;22:341–55. https://doi.org/10.1038/s41423-025-01275-w.
89. Zheng Y, Zhang X, Wang Z, Zhang R, Wei H, Yan X, Jiang X, Yang L. MCC950 as a promising candidate for blocking NLRP3 inflammasome activation: a review of preclinical research and future directions. Arch Pharm. 2024;357(11):e2400459. https://doi.org/10.1002/ardp.202400459.
90. Bakhshi S, Shamsi S. MCC950 in the treatment of NLRP3-mediated inflammatory diseases: latest evidence and therapeutic outcomes. Int Immunopharmacol. 2022;106:108595. https://doi.org/10.1016/j.intimp.2022.108595.
91. Zahid A, Li B, Kombe AJK, Jin T, Tao J. Pharmacological inhibitors of the NLRP3 inflammasome. Front Immunol. 2019;10:2538. https://doi.org/10.3389/fimmu.2019.02538.
92. Shao S, Chen C, Shi G, Zhou Y, Wei Y, Fan N, Yang Y, Wu L, Zhang T. Therapeutic potential of the target on NLRP3 inflammasome in multiple sclerosis. Pharmacol Ther. 2021;227:107880. https://doi.org/10.1016/j.pharmthera.2021.107880.
93. Coll RC, Schroder K, Pelegrín P. NLRP3 and pyroptosis blockers for treating inflammatory diseases. Trends Pharmacol Sci. 2022;43(8):653–68. https://doi.org/10.1016/j.tips.2022.04.003.
94. Marcos-Rabal P, González-Fuentes J, Castro-Vázquez L, Lozano MV, Rodríguez-Robledo V, Santander-Ortega MJ, Selva-Clemente J, Villaseca-González N, Del Mar Arroyo-Jiménez M. Neurodegenerative diseases: a multidisciplinary approach. Curr Pharm Des. 2021;27(30):3305–36. https://doi.org/10.2174/1381612827666210608152745.
95. Lu R, Zhang L, Yang X. Interaction between autophagy and the NLRP3 inflammasome in Alzheimer's and Parkinson's disease. Front Aging Neurosci. 2022;14:1018848. https://doi.org/10.3389/fnagi.2022.1018848.
96. Singh K, Sethi P, Datta S, Chaudhary JS, Kumar S, Jain D, Gupta JK, Kumar S, Guru A, Panda SP. Advances in gene therapy approaches targeting neuro-inflammation in neurodegenerative diseases. Ageing Res Rev. 2024;98:102321. https://doi.org/10.1016/j.arr.2024.102321.
97. Lu D, Sun Y, Luan Y, He W. Rational design of siRNA-based delivery systems for effective treatment of brain diseases. Pharm Sci Adv. 2024;2:100041. https://doi.org/10.1016/j.pscia.2024.100041.

Mitochondrial Dysfunction, Endoplasmic Reticulum Stress, and Their Role in Neurodegeneration

4

Akshay Shamnewadi, Vishal S. Patil, and Sunil S. Jalalpure

Abstract

Mitochondrial dysfunction and endoplasmic reticulum stress are emerging as central, interconnected contributors to the pathogenesis of neurodegenerative diseases. This chapter delves into the synergistic disruption of mitochondrial dynamics and endoplasmic reticulum proteostasis, emphasizing their role in promoting oxidative stress, calcium dysregulation, neuroinflammation, and autophagic failure. Special focus is placed on mitochondrial membranes, which facilitate critical crosstalk between organelles and serve as hubs of pathological signaling when disrupted. Key molecular mechanisms such as dynamin-related protein 1-mediated mitochondrial fission, Protein kinase R-like endoplasmic reticulum kinase-C/EBP homologous protein-driven endoplasmic reticulum stress-induced apoptosis, and the breakdown of mitochondrial membrane integrity are highlighted to elucidate the progression of neuronal damage. Additionally, this chapter discusses recent advances in targeted therapies, including mitochondrial reactive oxygen species scavengers and endoplasmic reticulum stress modulators. Clinical relevance is underscored through disease-specific examples such as amyloid β-induced endoplasmic reticulum stress in Alzheimer's

A. Shamnewadi
Department of Pharmaceutics, KLE College of Pharmacy, Hubballi, KLE Academy of Higher Education and Research, Deemed-to-be-University, Belagavi, Karnataka, India

Department of Pharmacology, KLE College of Pharmacy, Belagavi, KLE Academy of Higher Education and Research, Deemed-to-be-University, Belagavi, Karnataka, India

V. S. Patil (✉) · S. S. Jalalpure
Department of Pharmacology, KLE College of Pharmacy, Belagavi, KLE Academy of Higher Education and Research, Deemed-to-be-University, Belagavi, Karnataka, India

Department of Pharmacognosy, KLE College of Pharmacy, Belagavi, KLE Academy of Higher Education and Research, Deemed-to-be-University, Belagavi, Karnataka, India

P. Khanal et al. (eds.), *Neuroinflammation and Autophagy in Neurodegeneration*,
https://doi.org/10.1007/978-981-95-3945-1_4

and superoxide dismutase 1 mutations in amyotrophic lateral sclerosis, offering insight into promising therapeutic interventions aimed at restoring organelle homeostasis in neurodegenerative diseases.

Keywords

Autophagy dysfunction · Mitochondria-associated endoplasmic reticulum membranes · Neurodegeneration · Organelle stress · Unfolded protein response

4.1 Introduction

Over 60% of familial amyotrophic lateral sclerosis (ALS) cases are associated with mutations that impair organelle quality control, driving systemic neurodegeneration through a cascade of cellular dysfunctions [1]. One major contributor is mitochondrial dysfunction, where mutations in genes like superoxide dismutase 1 (SOD1), TAR DNA-binding protein (TARDBP or TDP-43), and fused in sarcoma (FUS) lead to disrupted mitochondrial dynamics, impaired ATP production, and excess reactive oxygen species (ROS), promoting neuronal stress and death [2, 3]. Concurrently, chronic endoplasmic reticulum stress due to the accumulation of misfolded proteins activates the unfolded protein response (UPR). While initially protective, prolonged activation of the Protein kinase R-like endoplasmic reticulum kinase (PERK) to eIF2 alpha to C/EBP homologous protein (CHOP) pathway within the UPR ultimately triggers apoptosis [4]. Disruption of mitochondria and endoplasmic reticulum membranes further impairs calcium signaling, lipid metabolism, and inter-organelle communication, aggravating stress responses [5]. Additionally, mutations interfere with autophagy and mitophagy, hindering the clearance of damaged organelles and reinforcing neurodegeneration [6].

Mitochondrial fission/fusion dynamics and UPR regulation are tightly linked processes that maintain neuronal homeostasis, and their imbalance exacerbates neuronal injury. Furthermore, excessive mitochondrial fission is often driven by dynamin-related protein 1 (DRP1) activation, leading to mitochondria fragmentation and impaired fusion limits organelle recovery, both worsened by chronic UPR activation [7, 8]. Mitochondria-associated endoplasmic reticulum membranes (mitochondrial membranes) act as critical platforms for organelle interaction, coordinating calcium transfer, lipid metabolism, and autophagy. Their disruption results in calcium overload and ROS generation, which further puts stress on mitochondria [5, 9]. Additionally, dysfunctional mitochondria and impaired autophagy release danger-associated molecular patterns, activating glial cells and chronic neuroinflammation. This endoplasmic reticulum mitochondrial stress crosstalk forms a self-reinforcing loop of proteostatic collapse, bioenergetic failure, and immune activation [10, 11].

Therapeutically, targeting organelle stress sensors such as PERK, IRE1, and ATF6 or modulating mitochondrial dynamics and mitochondrial membrane stability may restore inter-organelle communication and break this pathogenic cycle. These approaches aim to reestablish neuronal homeostasis by attenuating endoplasmic reticulum stress, maintaining mitochondrial integrity, and enhancing autophagic clearance, offering promising avenues to slow or halt amyotrophic lateral sclerosis (ALS) and related neurodegenerative conditions [12, 13]. Emerging evidence suggests that pharmacological inhibition of the PERK pathway can reduce neuronal loss and improve motor function in ALS models [14]. Similarly, promoting mitochondrial fusion through the upregulation of mitofusins has been shown to reverse mitochondrial fragmentation and enhance neuronal survival [15, 16]. Furthermore, stabilization of mitochondria-associated endoplasmic reticulum membranes has been linked to improved calcium handling and metabolic balance in neurodegenerative diseases [17].

4.2 Mitochondrial Dysfunction: From Homeostasis to Collapse

4.2.1 Mitochondrial Dynamics in Neuronal Vulnerability

Mitochondrial dynamics are governed by the balance of fission and fusion and are essential for neuronal health and plasticity, yet are often disrupted in neurodegenerative disorders. For example, excessive mitochondrial fission is driven by aberrant activation of DRP1 and has been implicated in Alzheimer's disease (AD) pathogenesis, where amyloid-beta (Aβ) oligomers promote pathological mitochondrial fragmentation, contributing to synaptic loss and cognitive decline [18]. Conversely, impaired mitochondrial fusion is mediated by mitofusin 2 (MFN2) and optic atrophy 1 (OPA1) and is a hallmark of Charcot-Marie-Tooth disease type 2A and ALS, where MFN2 mutations and OPA1 cleavage by presenilin-associated rhomboid-like protein impair mitochondrial elongation and cristae remodeling [19–21]. These structural disturbances compromise mitochondrial trafficking and bioenergetics, specifically in neurons with high metabolic demands. In epilepsy and other neurodevelopmental disorders, sustained hyperexcitability causes ATP depletion, destabilizing mitochondrial integrity and increasing neuronal vulnerability [22]. Together, these disruptions in mitochondrial dynamics converge to exacerbate neuronal dysfunction and degeneration (Fig. 4.1).

4.2.2 Mitophagy Failure and Its Consequences

Mitophagy is the selective degradation of damaged mitochondria, which is essential for maintaining neuronal homeostasis, and its failure contributes significantly to neurodegenerative pathology. Although the PINK1/Parkin pathway is a well-characterized mechanism for mitophagy, alternative Parkin-independent pathways, such as BNIP3 and NIX-mediated mitophagy, are prominent under conditions like hypoxia, regulating mitochondrial clearance and preventing bioenergetic collapse [23, 24]. Interestingly, when mitophagy is impaired, dysfunctional mitochondria accumulate and release ROS, leading to oxidative damage and the spillover of

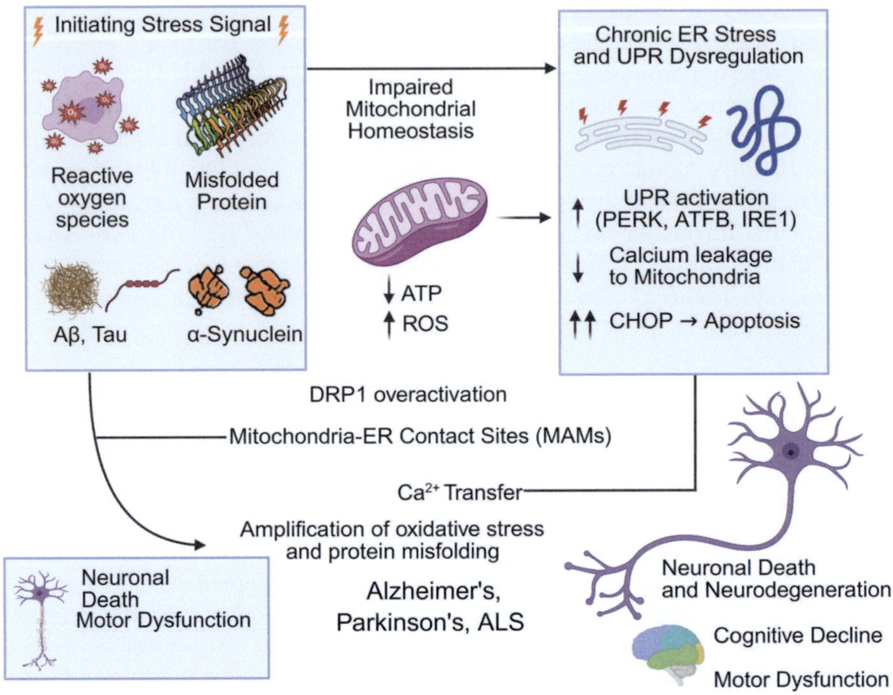

Fig. 4.1 Crosstalk between mitochondrial dysfunction and endoplasmic reticulum stress drives neurodegeneration. Cellular stressors like ROS, calcium overload, and misfolded proteins impair mitochondrial and endoplasmic reticulum function, activating UPR pathways and apoptotic signaling mediated by CHOP. Crosstalk at the mitochondria–endoplasmic reticulum contact sites amplifies stress, leading to neuronal death in AD, Parkinson's disease (PD), and ALS. (Created using BioRender. https://www.biorender.com/)

mitochondrial DNA into the cytosol. This mitochondrial DNA acts as a danger-associated molecular pattern, activating the cyclic GMP–AMP synthase stimulator of the interferon genes pathway and triggering a potent neuroinflammatory response that exacerbates neuronal injury [25, 26]. A well-characterized case of mitophagy failure is observed in Parkinson's disease (PD) patients harboring LRRK2 mutations, which directly impair mitochondrial–lysosomal crosstalk. LRRK2 phosphorylates Rab GTPases (e.g., Rab10 and Rab8), which are crucial for vesicle trafficking. Mutant LRRK2 (e.g., G2019S) hyperphosphorylates these Rab proteins, disrupting their localization and impairing the recruitment of autophagic machinery to damaged mitochondria [27, 28]. Furthermore, LRRK2 delays the degradation of the mitochondrial transport protein Miro1, hindering mitochondrial arrest and engulfment during mitophagy [29, 30]. These disruptions result in dysfunctional mitochondria accumulation, impaired autophagosome–lysosome fusion, and defective clearance, contributing to dopaminergic neuron degeneration in PD. This case exemplifies how a single genetic mutation can derail organelle communication and clearance mechanisms, reinforcing mitochondrial stress and neuroinflammation.

4.3 Endoplasmic Reticulum Stress: Proteostatic Failure and Beyond

4.3.1 UPR Pathways and Neuronal Survival

The UPR is a critical cellular mechanism activated during endoplasmic reticulum stress, balancing adaptation and apoptosis in neurons. One major branch is the IRE1α–XBP1 axis, which facilitates adaptive responses by enhancing protein folding capacity. However, in ALS, chronic IRE1α activation also initiates regulated IRE1-dependent decay, degrading essential mRNAs and exacerbating neurodegeneration [12]. The PERK-p-eIF2α pathway attenuates global protein synthesis to reduce endoplasmic reticulum load, but its prolonged activation leads to ATF4 and CHOP upregulation, which triggers apoptosis, notably implicated in prion disease models where excessive PERK signaling correlates with synaptic failure and neuronal death [31]. In Niemann-Pick type C1-deficient neurons, lipid accumulation induces chronic endoplasmic reticulum stress and disrupts proteostasis, though the specific role of ATF6 in this context remains to be fully elucidated [32]. These divergent UPR outcomes illustrate how neuronal survival or death hinges on the balance between adaptive and maladaptive endoplasmic reticulum stress signaling.

4.3.2 Endoplasmic Reticulum–Mitochondria Crosstalk via Mitochondrial Membranes

Mitochondrial membranes are critical for endoplasmic reticulum mitochondrial communication, facilitating calcium homeostasis, lipid transfer, and immune signaling. The IP3R–GRP75–VDAC1 complex regulates calcium shuttling between the endoplasmic reticulum and mitochondria, and its dysregulation in AD leads to mitochondrial calcium overload and neuronal dysfunction [33]. Additionally, phosphatidylserine synthesis at mitochondrial membranes is essential for mitochondrial membrane biogenesis, and defects in this lipid exchange impair mitochondrial structure and function [34]. Endoplasmic reticulum stress can activate the NLRP3 inflammasome through induction and recruitment of thioredoxin-interacting protein, an effect well demonstrated in endoplasmic reticulum stress models and has been implicated in neuroinflammation and demyelination in multiple sclerosis models [35, 36]. Disruption of mitochondria-associated endoplasmic reticulum membranes has been shown to impair insulin signaling and contribute to metabolic dysregulation in neurodegenerative diseases [37]. Mitofusin 2, a key tethering protein at mitochondrial membranes, not only regulates mitochondrial dynamics but also controls endoplasmic reticulum mitochondria juxtaposition and calcium flux; its loss results in endoplasmic reticulum stress and mitochondrial fragmentation [38]. Moreover, altered mitochondrial membrane function is increasingly recognized as a convergent mechanism underlying both amyloid and tau pathologies in AD [39].

4.4 Neurodegenerative Amplification Loops

4.4.1 Mitochondria–Endoplasmic Reticulum Stress Synergy

Mitochondria and the endoplasmic reticulum are key organelles orchestrating cellular stress responses, and their crosstalk becomes pathologically amplified in neurodegenerative conditions. One central mechanism involves the release of cytochrome c from dysfunctional mitochondria into the cytoplasm, which triggers apoptosome formation and caspase-9 activation, initiating an intrinsic apoptotic pathway. Meanwhile, persistent endoplasmic reticulum stress leads to caspase-12 activation, which further enhances apoptotic cascades either directly or via interaction with mitochondrial apoptotic signals, creating a dual-organelle amplification of cell death [40–42]. Furthermore, this synergy is compounded by dysregulated cellular metabolism through overactivation of mechanistic target of rapamycin complex 1 (mTORC1), which suppresses autophagy by inhibiting the ULK1 complex, impairing the cell's ability to clear dysfunctional organelles and misfolded proteins, and accelerating proteostatic collapse [43]. Additionally, mTORC1 promotes anabolic biosynthetic processes, including protein synthesis, thereby increasing the folding load on the endoplasmic reticulum and further exacerbating endoplasmic reticulum dysfunction under stress. Under endoplasmic reticulum stress, mTORC1 signaling also enhances apoptotic signaling via the IRE1–JNK axis, compounding organelle stress and cell death [44, 45]. These interconnected mechanisms create a self-perpetuating cycle in which endoplasmic reticulum stress and mitochondrial dysfunction intensify bioenergetic collapse, apoptosis, and disrupted proteostasis, hallmarks of neuronal decline in disorders such as AD, PD, and ALS (Fig. 4.2).

4.4.2 Neuroinflammatory Triggers in Organelle Stress and Neurodegeneration

Mitochondrial DNA released from damaged mitochondria acts as a damage-associated molecular pattern that activates endosomal TLR9 and cytosolic DNA sensors in microglia, triggering cytokine release. Such mitochondrial DNA-driven innate immune activation contributes to neuroinflammation and has been linked to exacerbation of tau pathology in AD [46, 47]. Endoplasmic reticulum stress-induced CHOP activation exacerbates neuroinflammation by promoting apoptotic pathways through B-cell lymphoma 2 suppression and Bcl-2-associated X protein/Bcl-2-antagonist/killer activation [48, 49] and upregulating inflammatory mediators such as IL-23 in glial cells, which further drive Th17 cell differentiation. The UPR–IL-23/Th17 axis is further implicated in neuroimmune dysregulation in pathogenesis, such

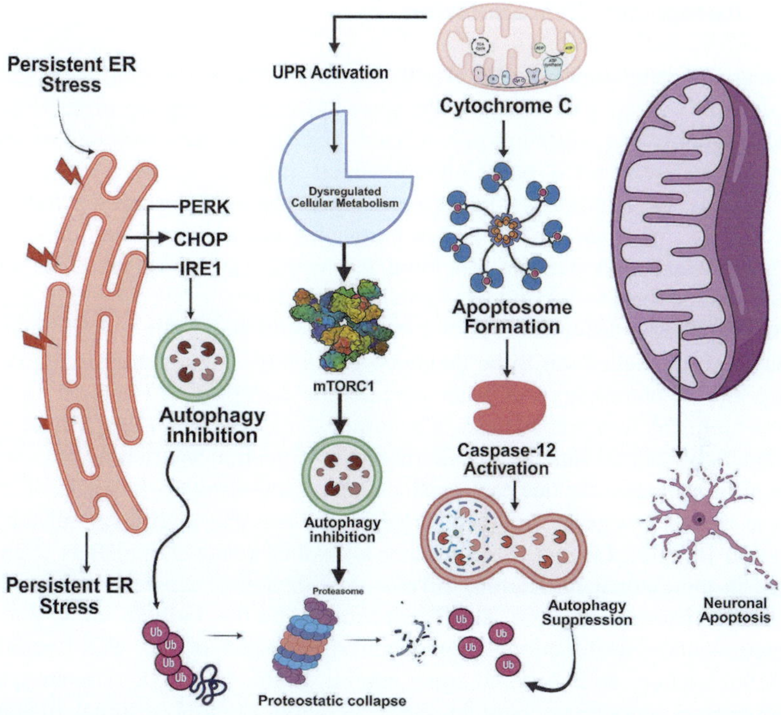

Fig. 4.2 Mechanisms of endoplasmic reticulum and mitochondrial stress leading to neuronal apoptosis. Persistent stress inhibits autophagy, disrupts proteostasis, and activates apoptotic signaling, leading to neuronal death. (Created using BioRender. https://www.biorender.com/)

as multiple sclerosis, where endoplasmic reticulum–mitochondria crosstalk at mitochondrial membranes facilitates NLRP3 inflammasome activation and pro-inflammatory cytokine secretion and UPR-mediated IL-23 induction promotes Th17 cell differentiation [50, 51]. Chronic endoplasmic reticulum stress amplifies neuroinflammation through microglial activation [52, 53] and is functionally linked to mitochondrial dysfunction, which together establish a self-perpetuating cycle that contributes to neurodegeneration [49]. Additionally, mitochondrial ROS generated during endoplasmic reticulum stress contributes to NF-κB activation, which upregulates pro-inflammatory cytokine expression and drives microglial activation in neurodegenerative models [11, 54]. Disrupted mitophagy has also been shown to contribute to the accumulation of dysfunctional mitochondria, intensifying inflammatory responses in neurodegenerative disease models [55]. In experimental autoimmune encephalomyelitis, inhibition of endoplasmic reticulum stress was found to significantly reduce Th17 cell infiltration and demyelination, highlighting its therapeutic potential [56, 57].

4.4.3 Autophagy Dysregulation

Autophagy dysregulation is a central pathological mechanism in neurodegeneration, as neurons rely heavily on efficient clearance systems to maintain proteostasis and organelle quality. In Huntington's disease (HD), the accumulation of mutant huntingtin with expanded polyglutamine tracts impairs selective autophagy by overwhelming the aggrephagy machinery. These aggregates sequester autophagic receptors like p62 and NBR1, disrupting the normal turnover of damaged mitochondria and endoplasmic reticulum fragments, leading to the accumulation of dysfunctional organelles and exacerbating cellular stress [58, 59]. This competition for autophagic resources creates a substrate bottleneck, where both misfolded proteins and organelles fail to be degraded efficiently, thereby amplifying cellular toxicity. In PD, autophagy–lysosome signaling is disrupted by impaired endoplasmic reticulum calcium homeostasis. Calcium leakage from the endoplasmic reticulum impairs lysosomal function by altering its acidification and membrane integrity, which in turn blocks the nuclear translocation of transcription factor EB (TFEB), which is a master regulator of lysosomal biogenesis and autophagy-related gene expression [60–62]. Instead of promoting lysosomal renewal, TFEB is aberrantly retained in the cytoplasm, leading to lysosomal depletion and reduced autophagic flux. This mislocalization of TFEB contributes to the buildup of α-synuclein aggregates and dysfunctional mitochondria observed in PD, promoting neuroinflammation and dopaminergic neuron death [60, 63]. Together, these impairments in autophagy, both in cargo recognition and lysosomal biogenesis, form a maladaptive loop that intensifies neurodegenerative cascades in both HD and PD.

4.5 Multimodal Therapeutic Interventions

4.5.1 Mitochondrial Rescue Strategies

Restoring mitochondrial health has emerged as a key therapeutic goal in neurodegenerative disorders, where mitochondrial fragmentation, oxidative stress, and energy deficits are prominent. In HD, excessive mitochondrial fission driven by DRP1 contributes to neuronal vulnerability. In patient-derived neuronal and fibroblast models, treatment with the DRP1 inhibitor P110-TAT corrected mitochondrial fragmentation, restored membrane potential, and reduced apoptosis, highlighting its therapeutic potential to counteract fission-induced dysfunction and supporting the role of DRP [64, 65]. Alongside dynamics, mitochondrial oxidative stress is a central pathological feature. In Friedreich's ataxia, impaired frataxin function leads to iron accumulation and elevated ROS, damaging mitochondrial DNA and membranes. In vitro work has shown that mitochondria-targeted antioxidants protect Friedreich's ataxia fibroblasts more effectively than untargeted antioxidants [66]. Moreover, early clinical trials in other neurodegenerative conditions, such as PD, have demonstrated that mitochondria-targeted antioxidants can reduce oxidative biomarkers and improve mitochondrial function [67]. Beyond structural and redox interventions, stimulating mitochondrial biogenesis offers another therapeutic axis. Activation of the PPARγ/PGC-1α pathway enhances mitochondrial mass and energy output. In AD models, the PPARγ agonist pioglitazone promotes amyloid clearance, increases mitochondrial gene expression, improves synaptic function, and attenuates cognitive decline [68]. By addressing mitochondrial fission, oxidative stress, and impaired biogenesis, this combinatorial strategy offers a robust approach to restoring mitochondrial integrity and enhancing neuronal resilience in neurodegenerative disorders.

4.5.2 Endoplasmic Reticulum Stress Alleviation

Chronic endoplasmic reticulum stress and maladaptive UPR signaling are key drivers of neurodegeneration across diseases such as AD, ALS, and prion disorders. Therapeutically targeting this stress axis has shown promise. PERK inhibitor GSK2606414 blocks excessive eIF2α phosphorylation and helps to restore normal protein synthesis, which is otherwise globally repressed during prolonged endoplasmic reticulum stress. In prion disease models, GSK2606414 treatment preserved synaptic integrity, reduced neuronal loss, and improved behavioral outcomes, highlighting the neuroprotective potential of early PERK pathway intervention [69]. However, long-term systemic PERK inhibition can cause pancreatic toxicity, prompting the development of next-generation compounds with improved selectivity and safety profiles [70]. In parallel, restoration of endoplasmic reticulum calcium homeostasis is another viable strategy to alleviate neurodegeneration. The endoplasmic reticulum resident SERCA pump maintains luminal Ca^{2+} levels, which are defective in several neurodegenerative states. SERCA-activating small molecule CDN1163 has demonstrated efficacy in presenilin-1 mutant AD models by increasing endoplasmic reticulum calcium uptake, reducing stress markers, and improving cognitive performance, linking calcium homeostasis with proteostasis restoration [71, 72]. Additionally, UPR reprogramming via gene therapy is gaining traction. For example, viral delivery of the spliced, active form of XBP1 in SOD1-ALS mouse models not only reduced endoplasmic reticulum stress but also enhanced proteostasis, delayed disease onset, and extended survival [73, 74]. In addition to these strategies, researchers are also investigating the modulation of ATF6, the third branch of the UPR. ATF6 activation increases the expression of endoplasmic reticulum chaperones and folding enzymes, and novel ATF6 agonists such as AA147 have shown protective effects in ischemia-reperfusion and PD models by reducing oxidative and endoplasmic reticulum stress [75]. These pharmacological and genetic approaches highlight the therapeutic potential of precisely modulating endoplasmic reticulum stress pathways to halt or slow neurodegeneration.

4.5.3 Combinatorial Approaches

Neurodegenerative diseases often involve concurrent disruption of mitochondrial dynamics, endoplasmic reticulum stress signaling, and autophagy. To address this complexity, combinatorial strategies must target multiple organelle dysfunctions simultaneously. Sigma-1 receptor agonists such as pridopidine have shown promise in HD models. Pridopidine restores mitochondrial endoplasmic reticulum contacts, improves membrane potential and respiration, reduces oxidative stress, and attenuates endoplasmic reticulum stress via the Sigma-1 receptor, complementing its known effects on calcium signaling and neuronal survival [76, 77]. Rapamycin demonstrates complementary mechanisms in ameliorating AD pathology in APP/PS1 mice. For example, rapamycin enhances mitophagy by promoting Parkin-mediated mitochondrial clearance and lysosomal fusion, which reduces oxidative stress, improves synaptic plasticity, and rescues cognitive deficits [78, 79]. Similarly, in 5×FAD mice, rapamycin activates PINK1–Parkin-mediated mitophagy, alleviating neuronal loss and mitochondrial dysfunction [80]. Meanwhile, tauroursodeoxycholic acid attenuates Aβ deposition by modulating amyloid precursor protein processing, reducing glial activation, and suppressing glycogen synthase kinase-3β-driven tau hyperphosphorylation. Together, these agents target distinct organelle systems. For example, rapamycin enhances mitochondrial quality control via autophagy, and tauroursodeoxycholic acid alleviates endoplasmic reticulum stress and Aβ toxicity [81, 82]. Nanoparticle-based delivery systems targeting mitochondria or the endoplasmic reticulum have shown significant promise in enhancing the therapeutic efficacy of curcumin derivatives for reducing pathological plaque burden in neurodegenerative diseases. Mitochondria-targeted nanoparticles encapsulating curcumin can directly improve mitochondrial function, restore mitochondrial dynamics and biogenesis, and decrease oxidative stress and cell apoptosis, at lower therapeutic concentrations compared to non-targeted formulations, thereby minimizing side effects [83]. Interestingly, endoplasmic reticulum-targeted curcumin-loaded nanoparticles have demonstrated an efficacy in modulating calcium signaling, protecting against mitochondrial damage, and preventing apoptosis, further contributing to cellular protection and potentially reducing plaque formation [84]. These targeted delivery strategies not only enhance the bioavailability and stability of curcumin but also allow precise accumulation at subcellular sites and are most implicated in disease processes, offering a potent approach to mitigate plaque burden and improve clinical outcomes [85, 86].

4.6 Case Study: ALS—A Perfect Storm of Organelle Pathologies

ALS exemplifies a convergence of organelle pathologies, with mutant SOD1 protein aggregates playing a central role in cellular dysfunction. Mutant SOD1 misfolds and accumulates on the outer mitochondrial membrane, where it disrupts mitochondrial function by interacting with voltage-dependent anion channel 1, impairing its function and thus altering mitochondrial permeability [87–89]. This leads to decreased mitochondrial membrane potential, increased oxidative stress, and, ultimately, cytochrome c release and neuronal apoptosis. Additionally, SOD1 aggregates interfere with endoplasmic reticulum-associated degradation pathways, further exacerbating proteostasis disturbances and mitochondrial damage. Another major genetic contributor to ALS is C9orf72 repeat expansions, which produce dipeptide repeat proteins that disrupt autophagic processes, notably reticulophagy and mitophagy. These dipeptides repeat proteins impair the clearance of dysfunctional mitochondria and endoplasmic reticulum, leading to damaged organelles and cellular stress [87, 90]. The combined effects of impaired mitophagy and reticulophagy compromise neuronal survival by promoting the persistence of toxic aggregates and dysfunctional organelles, which are hallmarks of ALS pathogenesis. Arimoclomol is a small-molecule co-inducer of heat shock proteins and has emerged as a clinical candidate for modulating endoplasmic reticulum stress and enhancing proteostasis in neurodegenerative diseases [91]. By amplifying HSP70 expression and other chaperones under stress conditions, arimoclomol supports protein folding, reduces aggregation, and stabilizes misfolded proteins in the endoplasmic reticulum. Additionally, arimoclomol demonstrated a favorable safety profile and delayed disease progression in a subset of patients carrying SOD1 mutations, which correlates with biomarker signatures of reduced protein aggregation and oxidative stress [92, 93]. Similarly, in Niemann-Pick Disease type C, arimoclomol treatment was associated with stabilized neurological function and biomarker improvements in lysosomal lipid storage and heat shock protein expression [94]. Despite modest primary endpoint outcomes, this evidence suggests arimoclomol efficacy may depend on patient stratification and disease stage, emphasizing the need for precision-based biomarker selection in future studies.

4.7 Conclusion and Emerging Frontiers

This chapter highlighted how organelle stress, involving mitochondria and the endoplasmic reticulum, underlies the converging pathologies of protein aggregation, neuroinflammation, and metabolic failure in neurodegenerative diseases. Disruptions in mitochondrial dynamics, mitophagy, endoplasmic reticulum proteostasis, and inter-organelle communication mediated by mitochondrial membranes reinforce each other to create self-sustaining cycles of cellular dysfunction. These organelle-level perturbations not only impair neuronal viability but also serve as upstream triggers of apoptotic signaling and immune activation, as observed in neurodegenerative disorders such as ALS, AD, and PD. This chapter also emphasized how these dysfunctions form the basis for neurodegenerative amplification loops and maladaptive feedback mechanisms involving autophagy collapse, inflammasome activation, and failed adaptive stress responses. Looking forward, several emerging strategies hold promise for disrupting these pathological circuits. High-throughput CRISPR-Cas9 screening is poised to uncover novel stress-buffering factors, such as PDIA6 and HSPA5, which may offer organelle-specific therapeutic leverage. In parallel, advances in biosensor technologies enable dynamic monitoring of endoplasmic reticulum–mitochondria crosstalk, providing valuable tools for drug screening and mechanistic insight. Furthermore, integrating molecular stratification into clinical trial design, such as subtyping AD patients based on APOE4 status and mitochondrial membrane dysfunction profiles, may enhance treatment responsiveness and precision. These next-generation approaches, such as real-time subcellular imaging and patient-specific biomarkers, may pave the way for more targeted and effective interventions against organelle-driven neurodegeneration.

References

1. Wang H, Guan L, Deng M. Recent progress of the genetics of amyotrophic lateral sclerosis and challenges of gene therapy. Front Neurosci. 2023;17:1170996. https://doi.org/10.3389/fnins.2023.1170996.
2. Magrané J, Hervias I, Henning MS, Damiano M, Kawamata H, Manfredi G. Mutant SOD1 in neuronal mitochondria causes toxicity and mitochondrial dynamics abnormalities. Hum Mol Genet. 2009;18(23):4552–64. https://doi.org/10.1093/hmg/ddp421.
3. Wang W, Li L, Lin WL, et al. The ALS disease-associated mutant TDP-43 impairs mitochondrial dynamics and function in motor neurons. Hum Mol Genet. 2013;22(23):4706–19. https://doi.org/10.1093/hmg/ddt319.
4. Hetz C, Mollereau B. Disturbance of endoplasmic reticulum proteostasis in neurodegenerative diseases. Nat Rev Neurosci. 2014;15:233–49. https://doi.org/10.1038/nrn3689.
5. Paillusson S, Stoica R, Gomez-Suaga P, et al. There's something wrong with my MAM; the ER-mitochondria axis and neurodegenerative diseases. Trends Neurosci. 2016;39(3):146–57. https://doi.org/10.1016/j.tins.2016.01.008.
6. Wong YC, Holzbaur EL. Autophagosome dynamics in neurodegeneration at a glance. J Cell Sci. 2015;128(7):1259–67. https://doi.org/10.1242/jcs.161216.
7. Wang X, Su B, Lee HG, et al. Impaired balance of mitochondrial fission and fusion in Alzheimer's disease. J Neurosci. 2009;29(28):9090–103. https://doi.org/10.1523/JNEUROSCI.1357-09.2009.
8. Iqbal S, Hood DA. Oxidative stress-induced mitochondrial fragmentation and movement in skeletal muscle myoblasts. Am J Physiol Cell Physiol. 2014;306(12):C1176–83. https://doi.org/10.1152/ajpcell.00017.2014.
9. Rowland AA, Voeltz GK. Endoplasmic reticulum-mitochondria contacts: function of the junction. Nat Rev Mol Cell Biol. 2012;13(10):607–25. https://doi.org/10.1038/nrm3440.
10. Lin MM, Liu N, Qin ZH, Wang Y. Mitochondrial-derived damage-associated molecular patterns amplify neuroinflammation in neurodegenerative diseases. Acta Pharmacol Sin. 2022;43(10):2439–47. https://doi.org/10.1038/s41401-022-00879-6.
11. Zhou R, Yazdi A, Menu P, et al. A role for mitochondria in NLRP3 inflammasome activation. Nature. 2011;469:221–5. https://doi.org/10.1038/nature09663.
12. Hetz C, Saxena S. ER stress and the unfolded protein response in neurodegeneration. Nat Rev Neurol. 2017;13(8):477–91. https://doi.org/10.1038/nrneurol.2017.99.
13. Bernard-Marissal N, Chrast R, Schneider BL. Endoplasmic reticulum and mitochondria in diseases of motor and sensory neurons: a broken relationship? Cell Death Dis. 2018;9:333. https://doi.org/10.1038/s41419-017-0125-1.
14. Wang L, Popko B, Roos RP. The unfolded protein response in familial amyotrophic lateral sclerosis. Hum Mol Genet. 2011;20(5):1008–15. https://doi.org/10.1093/hmg/ddq546.

15. Liesa M, Shirihai OS. Mitochondrial dynamics in the regulation of nutrient utilization and energy expenditure. Cell Metab. 2013;17(4):491–506. https://doi.org/10.1016/j.cmet.2013.03.002.
16. Dang X, Zhang L, Franco A, Dorn Ii GW. Activating mitofusins interrupts mitochondrial degeneration and delays disease progression in SOD1 mutant amyotrophic lateral sclerosis. Hum Mol Genet. 2023;32(7):1208–22. https://doi.org/10.1093/hmg/ddac287.
17. Area-Gomez E, Del Carmen Lara Castillo M, Tambini MD, et al. Upregulated function of mitochondria-associated ER membranes in Alzheimer disease. EMBO J. 2012;31(21):4106–23. https://doi.org/10.1038/emboj.2012.202.
18. Manczak M, Calkins MJ, Reddy PH. Impaired mitochondrial dynamics and abnormal interaction of amyloid beta with mitochondrial protein Drp1 in neurons from patients with Alzheimer's disease: implications for neuronal damage. Hum Mol Genet. 2011;20(13):2495–509. https://doi.org/10.1093/hmg/ddr139.
19. Züchner S, Mersiyanova IV, Muglia M, et al. Mutations in the mitochondrial GTPase mitofusin 2 cause Charcot-Marie-Tooth neuropathy type 2A. Nat Genet. 2004;36(6):660.
20. Greene AW, Grenier K, Aguileta MA, et al. Mitochondrial processing peptidase regulates PINK1 processing, import and Parkin recruitment. EMBO Rep. 2012;13(4):378–85. https://doi.org/10.1038/embor.2012.14.
21. Cipolat S, Rudka T, Hartmann D, et al. Mitochondrial rhomboid PARL regulates cytochrome c release during apoptosis via OPA1-dependent cristae remodeling. Cell. 2006;126(1):163–75. https://doi.org/10.1016/j.cell.2006.06.021.
22. Waldbaum S, Patel M. Mitochondrial dysfunction and oxidative stress: a contributing link to acquired epilepsy? J Bioenerg Biomembr. 2010;42:449–55. https://doi.org/10.1007/s10863-010-9320-9.
23. Li Y, Zheng W, Lu Y, et al. BNIP3L/NIX-mediated mitophagy: molecular mechanisms and implications for human disease. Cell Death Dis. 2021;13(1):14. https://doi.org/10.1038/s41419-021-04469-y.
24. Pickrell AM, Youle RJ. The roles of PINK1, parkin, and mitochondrial fidelity in Parkinson's disease. Neuron. 2015;85(2):257–73. https://doi.org/10.1016/j.neuron.2014.12.007.
25. Sliter DA, Martinez J, Hao L, et al. Parkin and PINK1 mitigate STING-induced inflammation. Nature. 2018;561:258–62. https://doi.org/10.1038/s41586-018-0448-9.
26. Zorov DB, Juhaszova M, Sollott SJ. Mitochondrial reactive oxygen species (ROS) and ROS-induced ROS release. Physiol Rev. 2014;94(3):909–50. https://doi.org/10.1152/physrev.00026.2013.
27. Steger M, Tonelli F, Ito G, et al. Phosphoproteomics reveals that Parkinson's disease kinase LRRK2 regulates a subset of Rab GTPases. Elife. 2016;5:e12813. https://doi.org/10.7554/eLife.12813.
28. Wauters F, Cornelissen T, Imberechts D, et al. LRRK2 mutations impair depolarization-induced mitophagy through inhibition of mitochondrial accumulation of RAB10. Autophagy. 2020;16(2):203–22. https://doi.org/10.1080/15548627.2019.1603548.
29. Wang X, Winter D, Ashrafi G, et al. PINK1 and Parkin target Miro for phosphorylation and degradation to arrest mitochondrial motility. Cell. 2011;147(4):893–906. https://doi.org/10.1016/j.cell.2011.10.018.
30. Hsieh CH, Shaltouki A, Gonzalez AE, et al. Functional impairment in miro degradation and mitophagy is a shared feature in familial and sporadic Parkinson's disease. Cell Stem Cell. 2016;19(6):709–24. https://doi.org/10.1016/j.stem.2016.08.002.
31. Moreno JA, Radford H, Peretti D, et al. Sustained translational repression by eIF2α-P mediates prion neurodegeneration. Nature. 2014;511(7509):370.
32. Cabrera-Reyes F, Parra-Ruiz C, Yuseff MI, Zanlungo S. Alterations in lysosome homeostasis in lipid-related disorders: impact on metabolic tissues and immune cells. Front Cell Dev Biol. 2021;9:790568. https://doi.org/10.3389/fcell.2021.790568.

33. Atakpa-Adaji P, Ivanova A. IP3R at ER-mitochondrial contact sites: beyond the IP3R-GRP75-VDAC1 Ca2+ funnel. Contact. 2023;6:25152564231181020. https://doi.org/10.1177/25152564231181020.
34. Yang X, Liang J, Ding L, Li X, Lam S-M, Shui G, et al. Phosphatidylserine synthase regulates cellular homeostasis through distinct metabolic mechanisms. PLoS Genet. 2019;15(12):e1008548. https://doi.org/10.1371/journal.pgen.1008548.
35. Cui Y, Yu H, Bu Z, Wen L, Yan L, Feng J. Focus on the role of the NLRP3 inflammasome in multiple sclerosis: pathogenesis, diagnosis, and therapeutics. Front Mol Neurosci. 2022;15:894298. https://doi.org/10.3389/fnmol.2022.894298.
36. Lerner AG, Upton JP, Praveen PV, et al. IRE1α induces thioredoxin-interacting protein to activate the NLRP3 inflammasome and promote programmed cell death under irremediable ER stress. Cell Metab. 2012;16(2):250–64. https://doi.org/10.1016/j.cmet.2012.07.007.
37. Tubbs E, Rieusset J. Metabolic signaling functions of ER-mitochondria contact sites: role in metabolic diseases. J Mol Endocrinol. 2017;58(2):R87–R106. https://doi.org/10.1530/JME-16-0189.
38. de Brito OM, Scorrano L. Mitofusin 2 tethers endoplasmic reticulum to mitochondria [published correction appears in Nature. 2014;513(7517):266]. Nature. 2008;456(7222):605–10. https://doi.org/10.1038/nature07534.
39. Area-Gomez E, Schon EA. Mitochondria-associated ER membranes and Alzheimer disease. Curr Opin Genet Dev. 2016;38:90–6. https://doi.org/10.1016/j.gde.2016.04.006.
40. Lamkanfi M, Kalai M, Vandenabeele P. Caspase-12: an overview. Cell Death Differ. 2004;11(4):365–8. https://doi.org/10.1038/sj.cdd.4401364.
41. Szegezdi E, Logue SE, Gorman AM, Samali A. Mediators of endoplasmic reticulum stress-induced apoptosis. EMBO Rep. 2006;7(9):880–5. https://doi.org/10.1038/sj.embor.7400779.
42. Liu X, Kim CN, Yang J, Jemmerson R, Wang X. Induction of apoptotic program in cell-free extracts: requirement for dATP and cytochrome c. Cell. 1996;86(1):147–57. https://doi.org/10.1016/s0092-8674(00)80085-9.
43. Kim J, Kundu M, Viollet B, et al. AMPK and mTOR regulate autophagy through direct phosphorylation of Ulk1. Nat Cell Biol. 2011;13:132–41. https://doi.org/10.1038/ncb2152.
44. Kato H, Nakajima S, Saito Y, Takahashi S, Katoh R, Kitamura M. mTORC1 serves ER stress-triggered apoptosis via selective activation of the IRE1-JNK pathway. Cell Death Differ. 2012;19(2):310–20. https://doi.org/10.1038/cdd.2011.98.
45. Rabanal-Ruiz Y, Korolchuk VI. mTORC1 and nutrient homeostasis: the central role of the lysosome. Int J Mol Sci. 2018;19(3):818. Published 2018 Mar 12. https://doi.org/10.3390/ijms19030818.
46. Oka T, Hikoso S, Yamaguchi O, et al. Mitochondrial DNA that escapes from autophagy causes inflammation and heart failure. Nature. 2012;485(7397):251–5. https://doi.org/10.1038/nature10992.
47. Udeochu JC, Amin S, Huang Y, et al. Tau activation of microglial cGAS-IFN reduces MEF2C-mediated cognitive resilience. Nat Neurosci. 2023;26(5):737–50. https://doi.org/10.1038/s41593-023-01315-6.
48. Malhotra JD, Kaufman RJ. ER stress and its functional link to mitochondria: role in cell survival and death. Cold Spring Harb Perspect Biol. 2011;3(9):a004424. https://doi.org/10.1101/cshperspect.a004424.
49. Chen X, Shi C, He M, et al. Endoplasmic reticulum stress: molecular mechanism and therapeutic targets. Signal Transduct Target Ther. 2023;8:352. https://doi.org/10.1038/s41392-023-01570-w.
50. Pereira AC, De Pascale J, Resende R, et al. ER-mitochondria communication is involved in NLRP3 inflammasome activation under stress conditions in the innate immune system. Cell Mol Life Sci. 2022;79(4):213. https://doi.org/10.1007/s00018-022-04211-7.
51. DeLay ML, Turner MJ, Klenk EI, Smith JA, Sowders DP, Colbert RA. HLA-B27 misfolding and the unfolded protein response augment interleukin-23 production and are associated with

Th17 activation in transgenic rats. Arthritis Rheum. 2009;60(9):2633–43. https://doi.org/10.1002/art.24763.
52. Sprenkle NT, Sims SG, Sánchez CL, Meares GP. Endoplasmic reticulum stress and inflammation in the central nervous system. Mol Neurodegener. 2017;12(1):42. https://doi.org/10.1186/s13024-017-0183-y.
53. Yang Z, Shao Y, Zhao Y, et al. Endoplasmic reticulum stress-related neuroinflammation and neural stem cells decrease in mice exposure to paraquat. Sci Rep. 2020;10:17757. https://doi.org/10.1038/s41598-020-74916-x.
54. Stilgenbauer L, Chen Q, Pungi D, James N, Jayarathne H, Koshko L, Scofield S, Zhang K, Sadagurski M. Microglial ER stress response via IRE1α regulates diet-induced metabolic imbalance and obesity in mice. Mol Metab. 2025;95:102128.
55. Fang EF, Hou Y, Palikaras K, et al. Mitophagy inhibits amyloid-β and tau pathology and reverses cognitive deficits in models of Alzheimer's disease. Nat Neurosci. 2019;22(3):401–12. https://doi.org/10.1038/s41593-018-0332-9.
56. Stone S, Lin W. The unfolded protein response in multiple sclerosis. Front Neurosci. 2015;9:264. https://doi.org/10.3389/fnins.2015.00264.
57. Brucklacher-Waldert V, Ferreira C, Stebegg M, et al. Cellular stress in the context of an inflammatory environment supports TGF-β-independent T helper-17 differentiation. Cell Rep. 2017;19(11):2357–70. https://doi.org/10.1016/j.celrep.2017.05.052.
58. Martinez-Vicente M, Talloczy Z, Wong E, et al. Cargo recognition failure is responsible for inefficient autophagy in Huntington's disease. Nat Neurosci. 2010;13(5):567–76. https://doi.org/10.1038/nn.2528.
59. Wong E, Cuervo AM. Autophagy gone awry in neurodegenerative diseases. Nat Neurosci. 2010;13(7):805–11. https://doi.org/10.1038/nn.2575.
60. Bonam SR, Tranchant C, Muller S. Autophagy-lysosomal pathway as potential therapeutic target in Parkinson's disease. Cells. 2021;10(12):3547. https://doi.org/10.3390/cells10123547.
61. Sepúlveda D, Cisternas-Olmedo M, Arcos J, Nassif M, Vidal RL. Contribution of autophagy-lysosomal pathway in the exosomal secretion of alpha-synuclein and its impact in the progression of Parkinson's disease. Front Mol Neurosci. 2022;15:805087. https://doi.org/10.3389/fnmol.2022.805087.
62. Medina DL, Di Paola S, Peluso I, et al. Lysosomal calcium signalling regulates autophagy through calcineurin and TFEB. Nat Cell Biol. 2015;17(3):288–99. https://doi.org/10.1038/ncb3114.
63. Jiao F, Meng L, Du K, Li X. The autophagy-lysosome pathway: a potential target in the chemical and gene therapeutic strategies for Parkinson's disease. Neural Regen Res. 2025;20(1):139–58. https://doi.org/10.4103/NRR.NRR-D-23-01195.
64. Song W, Chen J, Petrilli A, et al. Mutant huntingtin binds the mitochondrial fission GTPase dynamin-related protein-1 and increases its enzymatic activity. Nat Med. 2011;17(3):377–82. https://doi.org/10.1038/nm.2313.
65. Guo X, Disatnik MH, Monbureau M, Shamloo M, Mochly-Rosen D, Qi X. Inhibition of mitochondrial fragmentation diminishes Huntington's disease-associated neurodegeneration. J Clin Invest. 2013;123(12):5371–88. https://doi.org/10.1172/JCI70911.
66. Jauslin ML, Meier T, Smith RA, Murphy MP. Mitochondria-targeted antioxidants protect Friedreich Ataxia fibroblasts from endogenous oxidative stress more effectively than untargeted antioxidants. FASEB J. 2003;17(13):1972–4. https://doi.org/10.1096/fj.03-0240fje.
67. Snow BJ, Rolfe FL, Lockhart MM, et al. A double-blind, placebo-controlled study to assess the mitochondria-targeted antioxidant MitoQ as a disease-modifying therapy in Parkinson's disease. Mov Disord. 2010;25(11):1670–4. https://doi.org/10.1002/mds.23148.
68. Mandrekar-Colucci S, Karlo JC, Landreth GE. Mechanisms underlying the rapid peroxisome proliferator-activated receptor-γ-mediated amyloid clearance and reversal of cognitive deficits in a murine model of Alzheimer's disease. J Neurosci. 2012;32(30):10117–28. https://doi.org/10.1523/JNEUROSCI.5268-11.2012.

69. Moreno JA, Halliday M, Molloy C, et al. Oral treatment targeting the unfolded protein response prevents neurodegeneration and clinical disease in prion-infected mice. Sci Transl Med. 2013;5 (206):206ra138. https://doi.org/10.1126/scitranslmed.3006767.
70. Radford H, Moreno JA, Verity N, Halliday M, Mallucci GR. PERK inhibition prevents tau-mediated neurodegeneration in a mouse model of frontotemporal dementia. Acta Neuropathol. 2015;130(5):633–42. https://doi.org/10.1007/s00401-015-1487-z.
71. Klocke B, Moore C, Ott H, Pitychoutis PM. Chronic pharmacological activation of SERCA with CDN1163 affects spatial cognitive flexibility but not attention and impulsivity in mice. Behav Pharmacol. 2023;34(8):477–87. https://doi.org/10.1097/FBP.0000000000000756.
72. Krajnak K, Dahl R. A new target for Alzheimer's disease: a small molecule SERCA activator is neuroprotective in vitro and improves memory and cognition in APP/PS1 mice. Bioorg Med Chem Lett. 2018;28(9):1591–4. https://doi.org/10.1016/j.bmcl.2018.03.052.
73. Hetz C, Thielen P, Matus S, et al. XBP-1 deficiency in the nervous system protects against amyotrophic lateral sclerosis by increasing autophagy [published correction appears in Genes Dev. 2024;38(15–16):785]. https://doi.org/10.1101/gad.352249.124.
74. Valenzuela V, Becerra D, Astorga JI, et al. Artificial enforcement of the unfolded protein response reduces disease features in multiple preclinical models of ALS/FTD. Mol Ther. 2025;33(3):1226–45. https://doi.org/10.1016/j.ymthe.2025.01.004.
75. Blackwood EA, Azizi K, Thuerauf DJ, et al. Pharmacologic ATF6 activation confers global protection in widespread disease models by reprograming cellular proteostasis. Nat Commun. 2019;10(1):187. https://doi.org/10.1038/s41467-018-08129-2.
76. Ryskamp DA, Korban S, Zhemkov V, Kraskovskaya N, Bezprozvanny I. Neuronal Sigma-1 receptors: signaling functions and protective roles in neurodegenerative diseases. Front Neurosci. 2019;13:862. https://doi.org/10.3389/fnins.2019.00862.
77. Naia L, Ly P, Mota SI, et al. The sigma-1 receptor mediates pridopidine rescue of mitochondrial function in Huntington disease models. Neurotherapeutics. 2021;18(2):1017–38. https://doi.org/10.1007/s13311-021-01022-9.
78. Wang H, Fu J, Xu X, Yang Z, Zhang T. Rapamycin activates mitophagy and alleviates cognitive and synaptic plasticity deficits in a mouse model of Alzheimer's disease. J Gerontol A Biol Sci Med Sci. 2021;76(10):1707–13. https://doi.org/10.1093/gerona/glab142.
79. Fernandes SM, Mayer J, Nilsson P, Shimozawa M. How close is autophagytargeting therapy for Alzheimer's disease to clinical use? A summary of autophagy modulators in clinical studies. Front Cell Dev Biol. 2025;12:1520949. https://doi.org/10.3389/fcell.2024.1520949.
80. Zheng W, Li K, Zhong M, et al. Mitophagy activation by rapamycin enhances mitochondrial function and cognition in 5×FAD mice. Behav Brain Res. 2024;463:114889. https://doi.org/10.1016/j.bbr.2024.114889.
81. Dionísio PA, Amaral JD, Ribeiro MF, Lo AC, D'Hooge R, Rodrigues CM. Amyloid-β pathology is attenuated by tauroursodeoxycholic acid treatment in APP/PS1 mice after disease onset. Neurobiol Aging. 2015;36(1):228–40. https://doi.org/10.1016/j.neurobiolaging.2014.08.034.
82. Nunes AF, Amaral JD, Lo AC, et al. TUDCA, a bile acid, attenuates amyloid precursor protein processing and amyloid-β deposition in APP/PS1 mice. Mol Neurobiol. 2012;45(3):440–54. https://doi.org/10.1007/s12035-012-8256-y.
83. Buck AC, Maarman GJ, Dube A, Bardien S. Mitochondria targeted nanoparticles for the treatment of mitochondrial dysfunction-associated brain disorders. Front Bioeng Biotechnol. 2025;13:1563701. https://doi.org/10.3389/fbioe.2025.1563701.
84. Yen CL, Liao YC, Chen RF, et al. Targeted delivery of curcumin rescues endoplasmic reticulum-retained mutant NOX2 protein and avoids leukocyte apoptosis. J Immunol. 2019;202(12):3394–403. https://doi.org/10.4049/jimmunol.1801599.
85. Li Y, Li X, Wei L, Ye J. Advancements in mitochondrial-targeted nanotherapeutics: overcoming biological obstacles and optimizing drug delivery. Front Immunol. 2024;15: 1451989. https://doi.org/10.3389/fimmu.2024.1451989.

86. Jang D. Mitochondria-directed glycated oat protein nanoparticles for improved curcumin delivery and antioxidant efficacy. J Nanomed Nanotechnol. 2024;15:727.
87. Genin EC, Abou-Ali M, Paquis-Flucklinger V. Mitochondria, a key target in amyotrophic lateral sclerosis pathogenesis. Genes (Basel). 2023;14(11):1981. Published 2023 Oct 24. https://doi.org/10.3390/genes14111981.
88. Tafuri F, Ronchi D, Magri F, Comi GP, Corti S. SOD1 misplacing and mitochondrial dysfunction in amyotrophic lateral sclerosis pathogenesis. Front Cell Neurosci. 2015;9:336. https://doi.org/10.3389/fncel.2015.00336.
89. Günther R, Pal A, Williams C, et al. Alteration of mitochondrial integrity as upstream event in the pathophysiology of SOD1-ALS. Cells. 2022;11(7):1246. Published 2022 Apr 6. https://doi.org/10.3390/cells11071246.
90. Jiang S, Xu R. The current potential pathogenesis of amyotrophic lateral sclerosis. Mol Neurobiol. 2025;62(1):221–32. https://doi.org/10.1007/s12035-024-04269-3.
91. Kalmar B, Lu CH, Greensmith L. The role of heat shock proteins in amyotrophic lateral sclerosis: the therapeutic potential of arimoclomol. Pharmacol Ther. 2014;141(1):40–54. https://doi.org/10.1016/j.pharmthera.2013.08.003.
92. Benatar M, Wuu J, Andersen PM, et al. Randomized, double-blind, placebo-controlled trial of arimoclomol in rapidly progressive SOD1 ALS. Neurology. 2018;90(7):e565–74. https://doi.org/10.1212/WNL.0000000000004960.
93. Kalmar B, Edet-Amana E, Greensmith L. Treatment with a coinducer of the heat shock response delays muscle denervation in the SOD1-G93A mouse model of amyotrophic lateral sclerosis. Amyotroph Lateral Scler. 2012;13(4):378–92. https://doi.org/10.3109/17482968.2012.660953.
94. Kirkegaard T, Gray J, Priestman DA, et al. Heat shock protein-based therapy as a potential candidate for treating the sphingolipidoses. Sci Transl Med. 2016;8(355):355ra118. https://doi.org/10.1126/scitranslmed.aad9823.

Protein Aggregation, Oxidative Stress, and Autophagic Dysfunction in Neurodegenerative Diseases

5

Prarambh S. R. Dwivedi, Pavithra Pradeep Prabhu, and Abhishek Balmik

Abstract

Neurodegenerative diseases, such as Alzheimer's and Parkinson's, are likely to be the leading cause of disability in the upcoming decades. The pathological development in these diseases is driven by a triad of protein aggregation, oxidative stress, and autophagic dysfunction. This chapter examines how these factors, accelerate neuronal degeneration. Misfolded proteins like amyloid-β, tau, and α-synuclein directly impair autophagic clearance. Reactive oxygen species inactivate redox-sensitive autophagy regulators and proteostasis machinery. Altogether, these disruptions create toxic feedback loops that trigger neuroinflammation. An ideal therapeutic strategy should be focused on targeting multiple pathways of the triad rather than a single target. Gene editing, redox-modulating prodrugs, and biomarker-guided timing are some of the promising tools which takes into account the multifaceted pathological aspects of neurodegenerative diseases. This chapter discuss the role and inter-relation of aforementioned pathological mechanisms and multi-target therapeutic approach to combat neurodegeneration in an effective manner.

P. S. R. Dwivedi
Department of Pharmacology, NGSM Institute of Pharmaceutical Sciences (NGSMIPS), Nitte (Deemed to be University), Mangalore, India

P. P. Prabhu
Department of Pharmacognosy, NGSM Institute of Pharmaceutical Sciences (NGSMIPS), Nitte (Deemed to be University), Mangalore, India
e-mail: pavithra.pradeep@nitte.edu.in

A. Balmik (✉)
Department of Pharmacology and Chemical Biology, School of Medicine, Emory University, Atlanta, GA, USA
e-mail: abalmik@emory.edu

P. Khanal et al. (eds.), *Neuroinflammation and Autophagy in Neurodegeneration*,
https://doi.org/10.1007/978-981-95-3945-1_5

Keywords

Autophagic dysfunction · Protein aggregation · Oxidative stress · Neurodegeneration · Multimodal therapy

5.1 Introduction

Neurodegenerative disorders, driven primarily by proteostasis collapse, are projected to become the leading cause of disability by the year 2040 [1–3]. Despite advances in the management of neurodegenerative diseases like Alzheimer's and Parkinson's, they continue to devastate global health, mainly because of an incomplete understanding of the molecular mechanisms that promote their progression. Among the multiple reasons, protein aggregation and oxidative stress are studied to be the prime reasons that not only damage neurons independently but also affect the most critical mechanism of the cellular recycling system, autophagy [4]. Proteotoxicity [5], redox imbalance [6], and autophagic dysfunction [7] form a triad of self-amplifying systems that irreversibly damage the neurons. This chapter provides a mechanistic overview of how protein aggregation and oxidative stress work together to disrupt autophagy in neurodegenerative diseases. Misfolded proteins, in addition to being just deleterious to cells, also undergo chemical and structural modifications to protect themselves from being degraded [8]. Reactive oxygen species (ROS) also attack redox-sensitive proteins, crucial for forming autophagosomes and maintaining lysosome function, thereby disrupting the cellular balance, making it more vulnerable to metabolic stress [9]. This further triggers inflammation in the brain. At the heart of this triad is autophagic dysfunction, which not only triggers the damage to neurons but also eventually becomes the root cause of neuronal death. This chapter describes in depth how misfolded proteins and ROS affect autophagy and whether combination therapy can restore balance without further complications.

5.2 Protein Aggregation: Direct Sabotage of Autophagy

5.2.1 Aggregate-Specific Interference Mechanisms

Misfolded proteins, one of the major hallmarks in neurodegenerative diseases, have evolved in astonishingly precise ways to affect the autophagy–lysosomal pathway [10, 11]. In Alzheimer's disease (AD), soluble amyloid-β (Aβ) oligomers target the Beclin-1-VPS34 initiation complex in the endoplasmic reticulum membrane. It attacks the BH3 domain of Beclin-1 and strips off VPS34 [12], reducing the supply of nascent phagophores, instantly lowering autophagosome biogenesis throughout the neuron. The more hydrophobic Aβ-42 species can migrate to lysosomes and neutralize luminal protons, raising the pH by nearly one log unit. This inactivates cathepsin B/L and hinders proteolysis [13, 14]. In contrast, hyperphosphorylated tau disrupts the transportation phase rather than the initiation phase. Polyubiquitinated Tau tangles bind to the C-terminal zinc-binding domain of histone deacetylase-6 (HDAC6), which acts as the scaffold that attaches dynein motors to their cargo. Without HDAC6, autophagosomes accumulate in distal axons, unable to travel toward lysosomes clustered in the soma [15–17]. α-Synucleinopathies are involved in a completely different mechanism: elongated α-synuclein fibrils blocks the lysosome's chaperone-mediated autophagy (CMA) translocon by crowding the luminal side of lysosomal-associated membrane protein-2A (LAMP2A) [18, 19]. This saturation hinders the turnover of not only α-synuclein but also of hundreds of other single-peptide CMA substrates, resulting in a system-wide accumulation of proteins that rely on the same pathway. Even smaller α-synuclein oligomers affect this process by binding to cytosolic Hsc70 and diverting it away from productive substrate delivery [20]. Collectively, Aβ, tau, and α-synuclein affect initiation, transport, and CMA import, illustrating why their coexistence can be so devastating.

5.2.2 Lysosomal Compromise

Once protein aggregates disrupt autophagy, they can further affect the lysosomal pathway. Studies on transgenic AD mice show that Aβ-laden lysosomes undergo lipid-peroxidation-driven fragilization; when oxidized sphingomyelin and bis (monoacylglycerol)-phosphate cause nanoscale pores on lysosomes. This results in cathepsins D and L leaking into the cytosol, where they activate caspase-dependent apoptosis [21–23]. Concomitantly, Aβ interacts with Rab7 and HOPS, blocking the fusion of autophagosome and lysosome. This results in the buildup of waste-filled vacuoles in the brain, as studied in post-mortem cortex [24–27]. Tau adds to the complication by breaking down proton-pumping vacuolar ATPase [28]. α-Synuclein also affects the function of lysosomal enzymes. α-Synuclein fibrils bind to the enzyme glucocerebrosidase in the endoplasmic reticulum–Golgi region and block its movement to the lysosome. This lowers glucocerebrosidase activity, slows down fat breakdown, and in turn accelerates α-synuclein clumping [29–32]. Meanwhile, damaged lysosomes release potassium and ROS. This results in activation of TBK1 and cGAS-STING inflammatory pathways that further weaken autophagy gene expression [33–35]. This results in lysosomal dysfunction and add to cellular toxicity.

5.2.3 Therapeutic Strategies

As discussed, the accumulation of harmful protein aggregates disrupts multiple stages of autophagy. Hence, it becomes important to ensure that the treatment approaches are also multitargeted. Currently, the focus is on developing the next-generation antibodies that can bind to two distinct proteinaceuos species – extracellular Aβ and intracellular phosphorylated tau. This facilitates the immune cells, microglia, to clear these protein aggregates without causing inflammation [36]. Similarly, special nanobodies are designed to attach α-synuclein fibers to the immune system, directing them to lysosomal breakdown [37]. Gene therapy is also progressing at a remarkable pace; for example, using viral vectors to increase levels of LAMP2A in brain cells can help restore CMA [38]. Efforts are also being made to protect the lysosomes from undergoing degradation. This approach focus on stabilizing the lysosomal membrane and preventing oxidative damage, ensuring that the acidic environment of the lysosomes needed for digestion is not affected [39]. Moreover, small molecules are under investigation that can promote and improve autophagy, such as those that improve the function of CMA proteins or activate macroautophagy pathways more safely than conventional drugs [40, 41]. Looking forward, integrating such treatment strategies, like immunotherapy with gene editing, is likely to slow down disease progression and restore the brain's natural cleaning systems (Table 5.1), thus offering a new hope in treating neurodegenerative diseases [42].

5.3 Oxidative Stress: Disabling Autophagy's Redox Sensors

5.3.1 ROS Generation Hotspots

In neurodegenerative diseases, ROS arises from specific "hotspots" within the cell, in addition to being scattered everywhere. For example, in Parkinson's disease (PD), the major source of ROS production is dysfunctional mitochondria. In PD, mutations to PINK1 and Parkin impair mitophagy, which is an important process to remove dysfunctional mitochondria [62, 63]. These dysfunctional mitochondria tend to accumulate in dopaminergic neurons, continuously leaking superoxide and hydrogen peroxide, leading to chronic oxidative stress on both mitochondrial and cytosolic targets. Moreover, astrocytes, which are normally neuroprotective, can also become a source of toxicity in neurodegenerative diseases. Increased inflammatory cytokines and glutamate levels activate NOX4, an isoform of NADPH oxidase, predominantly expressed in astrocytes. This activation persistently releases ROS into the extracellular matrix, increasing oxidative stress [64]. In AD, the mismanagement of redox-active metals, particularly iron and copper, becomes a source of ROS. Impaired ferritinophagy and transferrin receptor upregulation are the root cause of the increase in labile Fe^{2+} pools. When Fe^{2+} interacts with hydrogen peroxide, specifically in the presence of Aβ plaques, it catalyzes hydroxyl radical formation via Fenton chemistry. This is further intensified by Aβ's direct ability to bind metal ions [65]. Altogether, these ROS hotspots damage DNA, lipids, and proteins, and serve as molecular inhibitors of autophagy. This links redox imbalance with progressive proteostatic collapse in neurodegenerative diseases.

Table 5.1 Effect of protein aggregation on autophagy in neurodegenerative diseases

Step into autophagy	Key proteins	Effect of aggregates	Aggregates	References
Initiation and phagophore formation	Beclin-1 (BECN1)	Sequestration, inhibition of autophagy initiation	Mutant huntingtin, α-synuclein	[43–45]
	ATG proteins (ATG5, ATG7, ATG12, ATG16L1)	Sequestration or reduced availability, impaired phagophore elongation	Mutant huntingtin, tau	[46, 47]
Cargo recognition and delivery	p62/SQSTM1	Trapped in aggregates, reduced cargo targeting	α-synuclein, TDP-43, tau	[44, 48]
	NBR1	Similar to p62, impaired cargo recognition	α-Synuclein	[49]
	LC3	Sequestration reduces autophagosome maturation	Tau, huntingtin	[50]
Autophagosome–lysosome fusion	SNARE proteins (syntaxin 17, VAMP8, SNAP29)	Disrupted vesicle fusion machinery	α-Synuclein, tau	[51]
	Rab7	Mislocalization or dysfunction impairs fusion	α-Synuclein	[52, 53]
	HOPS complex (VPS33A, VPS16)	Impaired tethering and fusion	α-Synuclein	[27]
Lysosomal function	LAMP-1, LAMP-2	Lysosomal membrane instability, decreased fusion efficiency	α-Synuclein, Aβ	[54, 55]
	Cathepsins (e.g., cathepsin D)	Reduced protease activity, lysosomal dysfunction	Tau, Aβ	[56]
	TFEB (transcription factor EB)	Dysregulated lysosomal biogenesis, impaired clearance	Huntingtin, α-synuclein	[57, 58]
Upstream regulation	mTOR	Hyperactivation inhibits autophagy initiation	Tau, Aβ	[59]
	HDAC6	Sequestration impairs aggresome–autophagy pathway	PolyQ proteins, α-synuclein	[60, 61]

5.3.2 Oxidation of Autophagy Machinery

Besides causing cellular damage in general, ROS disrupts the function of key autophagy proteins. One key target is ATG4, a cysteine protease responsible for the priming (C-terminal cleavage and conjugation with phosphatidylethanolamine) and recycling of LC3 by enabling its deconjugation. LC3 is a protein further crucial for autophagosome formation and turnover. Under physiological conditions, ATG4 cycles LC3 between its cytosolic (LC3-I) and membrane-bound (LC3-II) forms to maintain the autophagic cycle [66]. However, under oxidative stress, ATG4's catalytic cysteine undergoes sulfonation. This oxidative modification is stable and irreversible, and results in functional inactivation, leading to the accumulation of undegraded LC3-II. Similarly, ULK1, the kinase that starts autophagy, becomes functionally paralyzed when S-glutathionylation occurs on its cysteine residues. This redox modification inhibits its ATP-binding ability, further resulting in the shutdown of the signaling cascade required for autophagy induction [67]. A third vulnerable node is TFEB (transcription factor EB), the master transcriptional regulator of lysosomal and autophagy genes. Normally, dephosphorylated TFEB moves into the nucleus when the cell needs more lysosomal activity. However, under chronic oxidative stress, elevated ROS levels interfere with this process [68, 69]. Consequently, the expression of lysosomal enzymes, transporters, and membrane proteins will be suppressed, greatly affecting autophagy.

5.3.3 Therapeutic Strategies

Since autophagy proteins are sensitive to ROS, treatments must be aimed at lowering the harmful ROS without blocking the normal signaling they provide. One promising approach involves the use of redox-activated prodrugs, such as gamma-glutamylcysteine ethyl ester. It is a cell-permeable precursor to glutathione. Upon entering the neurons, it bypasses the rate-limiting γ-glutamylcysteine synthetase step, rapidly increasing the levels of glutathione even under oxidative stress [70, 71]. Elevated glutathione scavenges excess ROS and protects autophagy components like ATG4 and ULK1 from undergoing irreversible oxidative damage. This helps restore autophagic flux, promote proper lysosomal function, and reduce proteostatic burden in neurons. Parallelly, chelation therapy has also shown considerable promise, particularly in metal-induced redox cycling seen in AD. Deferiprone, a clinically approved iron chelator, crosses the blood–brain barrier, binds free Fe^{2+}, and blocks Fenton reaction mediated by Aβ plaques. In animal models, deferiprone treatment is shown to lower both oxidative damage markers and amyloid load, while simultaneously boosting autophagic clearance of tau and Aβ [72]. Importantly, such interventions can be utilized to avoid depletion of essential iron stores needed for neuronal metabolism [73]. Additional strategies include improving endogenous antioxidant defenses via Nrf2 activators or by delivering targeted SOD and catalase mimetics. By lowering ROS and protecting autophagy proteins at the same time, these combinatorial therapies may aid in breaking the vicious cycle that links oxidative stress, autophagy failure, and neurodegeneration [74, 75].

5.4 Feed-Forward Loops: Synergistic Toxicity

5.4.1 Aggregate-Driven Oxidative Stress

Protein aggregates are cytotoxic, that is, they stimulate the cell's production of harmful oxidants [76, 77]. In AD, soluble Aβ oligomer clusters along with cellular prion protein in cholesterol-rich microdomains, bring in Fyn kinase, making NMDA receptors overly sensitive [78]. This excessive NMDA activity allows abnormally high influx of calcium into mitochondria, disrupting mitochondrial membrane potential, and turns the electron-transport chain into a massive source of superoxide [79, 80]. Simultaneously, Ca^{2+}-activated nitric-oxide synthase and phospholipase A_2 start lipid-peroxidation cascades. This generates 4-hydroxynonenal and malondialdehyde, which are cross-linkers of both tau and α-synuclein [81]. In PD, long α-synuclein stick where endoplasmic reticulum and mitochondria contact, uncoupling the VAPB–PTPIP51 link. This normally controls the transfer of Ca^{2+} for oxidative phosphorylation [78]. This uncoupling deprives mitochondria of timely Ca^{2+} buffering, precipitating metabolic mismatch, complex-I "electron leak," and massive ROS efflux. ROS an oxidize α-synuclein itself, making the aggregate fibers even more likely to clump, creating a vicious cycle of increased aggregate burden, and subsequently more ROS [82, 83].

5.4.2 ROS-Accelerated Proteostasis Collapse

The neuronal proteostasis network, which includes molecular chaperones, the ubiquitin–proteasome system, and autophagy, relies on redox-sensitive enzymes that are highly susceptible to oxidative damage by reactive oxygen species (ROS). Long-lasting ROS damage slows the 20S proteasome core and the 19S cap, cutting peptide breakdown by about half [84, 85]. As the proteasomal activity declines, ubiquitinated substrates accumulates and p62/SQSTM1 bodies aggregates, which in turn signals to promote transport of excess cargo toward macroautophagy [86, 87]. Under oxidative stress, HSP70 loses activity when its Cys306 forms disulfides. Without a functional HSP70 cycle, nascent or partially folded proteins also collapse into β-rich oligomers that seed further aggregation. ROS also affects ER disulfide-isomerases, blocking proper folding and starting the unfolded-protein response [88, 89]. PERK, an ER-resident kinase, then phosphorylates eIF2α to curb translation, but chronic PERK activation diverts ATG gene transcription toward pro-apoptotic CHOP, further shrinking autophagic capacity [90, 91]. Together, the failing proteasome, chaperones, and autophagy create a complete loss of protein control that speeds neuronal death [92, 93] (Fig. 5.1).

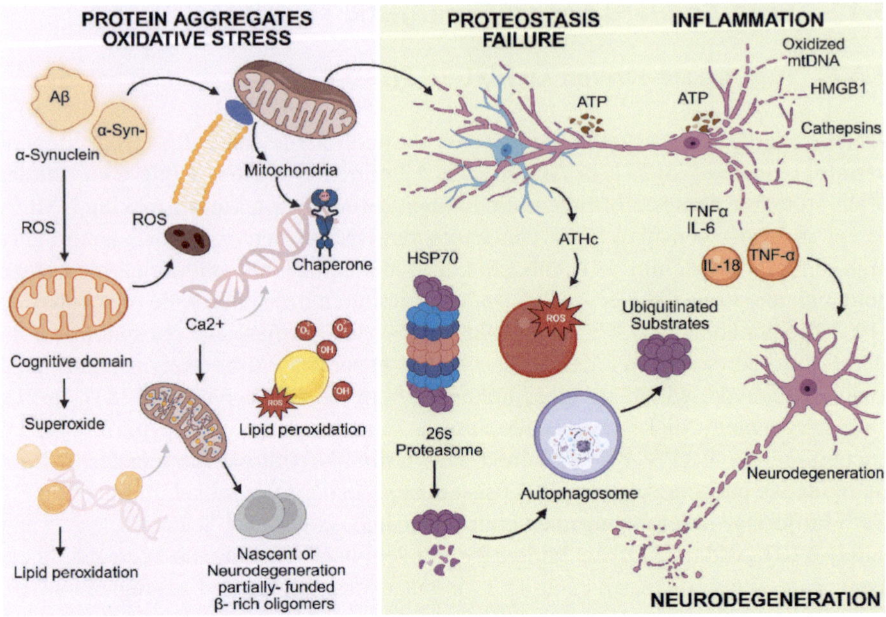

Fig. 5.1 Mechanisms of oxidative stress, proteostasis failure, and inflammation-driven neurodegeneration. Protein aggregates, oxidative stress, proteostasis failure, and inflammation drive neurodegeneration through mitochondrial dysfunction, impaired autophagy, and pro-inflammatory signaling. (Created using BioRender. https://www.biorender.com/)

5.4.3 Neuroinflammatory Crosstalk

When protein control collapses and ROS rises, neurons transmit danger signals outside the cell in the form of ruptured lysosomes and depolarized mitochondria that leak cathepsins, oxidized mtDNA, and HMGB1. This results in damage-associated molecular patterns [94, 95] that bind toll-like receptors (TLRs) and the NLRP3 inflammasome in microglia, further prompting ASC speck formation, caspase-1 activation, and cytokines IL-1β and IL-18 secretion. Secreted cytokines activate NF-κB in both microglia and astrocytes, and this further induces a feed-forward wave of TNF-α and IL-6 [96]. TNF-α, in turn, binds to TNFR1, stimulating IKKβ and mTORC1, which keep TFEB out of the nucleus and block lysosomal biogenesis [97, 98]. Simultaneously, NF-κB also boosts iNOS and NOX2, adding more ROS into the environment [99]. The heightened oxidative environment further promotes oxidation in autophagy regulators (ATG4, ULK1) and proteasome subunits, tightening the vice of proteostasis failure. Microglia caught in this loop switch to a phagoptotic phenotype. This results in engulfment of stressed but viable neurons, while astrocytes adopt the A1 neurotoxic state, secreting complement proteins (C3b) that opsonize synapses for removal [96]. Thus, inflammation, oxidation, and aggregate clump together and promote neurodegeneration even faster.

5.5 Multimodal Therapeutics: Targeting Shared Nodes

5.5.1 Pharmacological Combinations

One promising therapeutic approach is to develop drugs that target both mTORC1 signaling and the Keap1-Nrf2 axis [100]. A well-known mTORC1 inhibitor, Sirolimus (rapamycin), boosts autophagy and reduces α-synuclein accumulation in PD models [101, 102]. Bardoxolone methyl (CDDO-Me), a synthetic triterpenoid, acts as a potent Keap1-Nrf2 activator. It promotes the expression of endogenous antioxidant genes and has shown neuroprotective effects in models of ischemic brain injury and optic nerve degeneration [103, 104]. Combining sirolimus with bardoxolone is now gaining interest in neurodegenerative contexts where enhancing autophagic clearance and improving redox resilience could prove especially beneficial. Additionally, lysosomotropic agents such as hydroxychloroquine and its analogues have also been explored for their ability to accumulate in lysosomes and stabilize their membranes. At therapeutic doses, they can limit lysosomal membrane permeabilization induced by oxidative stress and are useful in the later disease stages when lysosomal integrity is compromised [105].

5.5.2 Gene-Editing Approaches

Advances in gene therapy and genome editing provide tools to directly repair or enhance the function of autophagy-related proteins. For example, CRISPR-Cas9-mediated correction of LAMP2A mutations restores CMA, thereby reducing protein aggregation [106]. In parallel, researchers have developed oxidation-resistant mutants of ATG4B, such as C292S/C361S variants. They can retain their enzymatic function even in oxidative environments. These mutants preserve LC3 processing under redox stress and can be delivered to neurons via adeno-associated viral vectors optimized for brain drug delivery [107]. Such highly specific targeted intervention could give long-term protection by keeping autophagy active and stopping toxic buildup.

5.5.3 Clinical Challenges

Despite these advances, significant challenges remain in translating multimodal therapies to clinical practice. One main concern is the timing of intervention. In early stages, boosting autophagy and antioxidant defenses is most helpful, while in later stages, marked by lysosomal damage and high oxidative stress, stabilizing lysosomes and supporting the remaining autophagy become more important. Another vital challenge is the stratification of patients based on biomarkers that reflect the underlying cellular pathology. For example, LC3B-II/p62 ratios in cerebrospinal fluid or blood-derived extracellular vesicles can serve as markers of autophagic flux, while levels of 8-hydroxy-2′-deoxyguanosine in plasma or urine reflect oxidative DNA damage and can be used to monitor the effectiveness of Nrf2-targeted therapies [108, 109]. These biomarkers are essential for tailoring therapies to individual patients and optimizing treatment windows.

5.6 Emerging Mechanistic Insights and Unanswered Questions

5.6.1 Novel Pathways in Autophagy Regulation

Recent work reveals that cells can start autophagy through a non-canonical AMPK-ULK1 route, bypassing the classic mTOR switch. Under oxidative stress, AMPK directly stabilizes the ULK1 complex and keeps it from caspase cleavage. This allows autophagy to begin even when energy is low [110, 111]. A second breakthrough concern is liquid–liquid phase separation. Certain autophagy proteins and misfolded proteins condense into liquid-like droplets on the endoplasmic reticulum surface. These droplets reshape local membranes and mark where autophagosomes will form. This will link the physical state of aggregates to autophagosome biogenesis [111, 112]. Together, these findings broaden our view of how oxidative signals and phase-separated condensates fine-tune the early autophagy phases.

5.6.2 Cross-Disease Commonalities

Evidence shows that α-synuclein spreads between neurons like a prion-like manner. When autophagy is weak, stressed neurons pack the proteins into exosomes, which are then taken up by neighboring cells and potentially seed new aggregates [113, 114]. A parallel theme is the presence of shared redox weak points across sporadic and familial disorders. For example, mutant or oxidized SOD1 in amyotrophic lateral sclerosis and APP/Aβ in AD both amplify ROS and overwhelm antioxidant defenses, despite arising from different genes [115, 116]. These overlaps suggest that therapies aimed at limiting exosomal spread or boosting redox buffering might help multiple neurodegenerative diseases.

5.6.3 Future Direction

In recent times, two therapeutic strategies have garnered increased attention. First, researchers are engineering autophagy receptors such as OPTN and NBR1 so that their LC3-interacting and ubiquitin-binding domains preferentially associate with disease-specific aggregates. This will accelerate their clearance without disturbing normal turnover [117, 118]. Second, attention is turning to the glymphatic system. It is the brain-wide fluid network that flushes waste during sleep. Enhancing perivascular flow or repairing aquaporin-4 polarity has been proposed as a way to drain extracellular aggregates and lower oxidative load. This will complement intracellular autophagy-based therapies [119, 120]. These novel mechanisms and open questions underscore the need for integrative approaches that span intracellular autophagy, intercellular propagation, and whole-brain clearance pathways.

5.7 Conclusion

Protein aggregation and oxidative stress are not mechanisms that work in isolation but essentially work together to impact autophagy, governing the progression of neurodegenerative diseases. This chapter has briefly illustrated how autophagy, proteotoxicity, and neuronal imbalance as a triad impact neurodegenerative diseases. Breaking this triad requires a multimodal therapeutic approach that simultaneously targets them, rather than treating them as separate entities. Therapeutic strategies are also focused on the immunological aspect of neurodegenerative diseases, specifically how microglial autophagy and cytokine-driven lysosomal dysfunction contribute to the neuroinflammatory amplification of this cycle.

References

1. Gammon K. Neurodegenerative disease: brain windfall. Nature. 2014;515(7526):299–300.
2. Labbadia J, Morimoto RI. The biology of Proteostasis in aging and disease. Annu Rev Biochem. 2015;2015(84):435–64.
3. Kumar P, Choudhary A, Kinger S, Jagtap YA, Dubey AR, Gutti RK, et al. Proteostasis defects: medicinal challenges of imperfect aging & neurodegeneration. Transl Med Aging. 2023;7(6968):87–97.
4. Tabner BJ, El-Agnaf OMA, German MJ, Fullwood NJ, Allsop D. Protein aggregation, metals and oxidative stress in neurodegenerative diseases. Biochem Soc Trans. 2005;33(5):1082–6.
5. Ruz C, Alcantud JL, Montero FV, Duran R, Bandres-Ciga S. Proteotoxicity and neurodegenerative diseases. Int J Mol Sci. 2020;21(16):5646.
6. Sbodio JI, Snyder SH, Paul BD. Redox mechanisms in neurodegeneration: from disease outcomes to therapeutic opportunities. Antioxid Redox Signal. 2019;30(11):1450.
7. Guo F, Liu X, Cai H, Le W. Autophagy in neurodegenerative diseases: pathogenesis and therapy. Brain Pathol. 2017;28(1):3–13.
8. Koszła O, Sołek P. Misfolding and aggregation in neurodegenerative diseases: protein quality control machinery as potential therapeutic clearance pathways. Cell Commun Signal. 2024;22(1):1–23.
9. Redza-Dutordoir M, Averill-Bates DA. Interactions between reactive oxygen species and autophagy: special issue: death mechanisms in cellular homeostasis. Biochim Biophys Acta, Mol Cell Res. 2021;1868(8):119041.
10. Finkbeiner S. The autophagy lysosomal pathway and neurodegeneration. Cold Spring Harb Perspect Biol. 2020;12(3):a033993.
11. Metcalf DJ, García-Arencibia M, Hochfeld WE, Rubinsztein DC. Autophagy and misfolded proteins in neurodegeneration. Exp Neurol. 2012;238(1):22–8.
12. Yuan M, Wang Y, Huang Z, Jing F, Qiao P, Zou Q, et al. Impaired autophagy in amyloid-beta pathology: a traditional review of recent Alzheimer's research. J Biomed Res. 2022;37(1):30–46.
13. Wolfe DM, Lee J, Hyun KA, Lee S, Orenstein SJ, Nixon RA. Autophagy failure in Alzheimer's disease and the role of defective lysosomal acidification. Eur J Neurosci. 2013;37(12):1949–61.

14. Liu RQ, Zhou QH, Ji SR, Zhou Q, Feng D, Wu Y, et al. Membrane localization of β-Amyloid 1–42 in lysosomes: a possible mechanism for lysosome labilization. J Biol Chem. 2010;285 (26):19986–96.
15. Onishi T, Maeda R, Terada M, Sato S, Fujii T, Ito M, et al. A novel orally active HDAC6 inhibitor T-518 shows a therapeutic potential for Alzheimer's disease and tauopathy in mice. Sci Rep. 2021;11(1):1–11.
16. Zhang L, Sheng S, Qin C. The role of HDAC6 in Alzheimer's disease. J Alzheimer's Dis. 2013;33(2):283–95.
17. Leyk J, Goldbaum O, Noack M, Richter-Landsberg C. Inhibition of HDAC6 modifies Tau inclusion body formation and impairs Autophagic clearance. J Mol Neurosci. 2015;55(4): 1031–46.
18. Xilouri M, Brekk OR, Kirik D, Stefanis L. LAMP2A as a therapeutic target in Parkinson disease. Autophagy. 2013;9(12):2166–8.
19. Stefanis L. α-Synuclein in Parkinson's disease. Cold Spring Harb Perspect Med. 2012;2(2): a009399.
20. Pemberton S, Melki R. The interaction of Hsc70 protein with fibrillar α-Synuclein and its therapeutic potential in Parkinson's disease. Commun Integr Biol. 2012;5(1):94–5.
21. Oberle C, Huai J, Reinheckel T, Tacke M, Rassner M, Ekert PG, et al. Lysosomal membrane permeabilization and cathepsin release is a Bax/Bak-dependent, amplifying event of apoptosis in fibroblasts and monocytes. Cell Death Differ. 2010;17(7):1167–78.
22. Zaretsky DV, Zaretskaia MV, Molkov YI. Membrane channel hypothesis of lysosomal permeabilization by beta-amyloid. Neurosci Lett. 2022;2022(770):136338.
23. Grabner GF, Fawzy N, Schreiber R, Pusch LM, Bulfon D, Koefeler H, et al. Metabolic regulation of the lysosomal cofactor bis(monoacylglycero)phosphate in mice. J Lipid Res. 2020;61(7):995–1003.
24. Hyttinen JMT, Niittykoski M, Salminen A, Kaarniranta K. Maturation of autophagosomes and endosomes: a key role for Rab7. Biochim Biophys Acta, Mol Cell Res. 2013;1833(3):503–10.
25. Ba L, Chen XH, Chen YL, Nie Q, Li ZJ, Ding FF, Zhang M. Distinct Rab7-related endosomal-Autophagic-lysosomal dysregulation observed in Cortex and Hippocampus in APPswe/ PSEN1dE9 Mouse Model of Alzheimer's disease. Chin Med J. 2017;130(24):2941–50.
26. Cai CZ, Yang C, Zhuang XX, Yuan NN, Wu MY, Tan JQ, et al. NRBF2 is a RAB7 effector required for autophagosome maturation and mediates the association of APP-CTFs with active form of RAB7 for degradation. Autophagy. 2021;17(5):1112–30.
27. McEwan DG, Popovic D, Gubas A, Terawaki S, Suzuki H, Stadel D, et al. PLEKHM1 regulates autophagosome-lysosome fusion through HOPS complex and LC3/GABARAP proteins. Mol Cell. 2015;57(1):39–54.
28. Jiang P, Nishimura T, Sakamaki Y, Itakura E, Hatta T, Natsume T, Mizushima N. The HOPS complex mediates autophagosome-lysosome fusion through interaction with syntaxin 17. Mol Biol Cell. 2014;25(8):1327–37.
29. Mazzulli JR, Xu YH, Sun Y, Knight AL, McLean PJ, Caldwell GA, et al. Gaucher disease Glucocerebrosidase and α-Synuclein form a bidirectional pathogenic loop in Synucleinopathies. Cell. 2011;146(1):37–52.
30. Stojkovska I, Krainc D, Mazzulli JR. Molecular mechanisms of α-synuclein and GBA1 in Parkinson's disease. Cell Tissue Res. 2018;373(1):51–60.
31. Bo RX, Li YY, Zhou TT, Chen NH, Yuan YH. The neuroinflammatory role of glucocerebrosidase in Parkinson's disease. Neuropharmacology. 2022;2022(207):108964.
32. Gegg ME, Verona G, Schapira AHV. Glucocerebrosidase deficiency promotes release of α-synuclein fibrils from cultured neurons. Hum Mol Genet. 2020;29(10):1716–28.
33. Lv J, Zhu X, Xing C, Chen Y, Bian H, Yin H, et al. Stimulator of interferon genes (STING): key therapeutic targets in ischemia/reperfusion injury. Biomed Pharmacother. 2023;2023 (167):115458.

34. Tang Z, Xing C, Araszkiewicz A, Yang K, Huai W, Jeltema D, et al. STING mediates lysosomal quality control and recovery through its proton channel function and TFEB activation in lysosomal storage disorders. Mol Cell. 2025;85(8):1624–1639.e5.
35. Park NY, Jo DS, Yang JY, Bae JE, Kim JB, Kim YH, et al. Activation of lysophagy by a TBK1-SCFFBXO3-TMEM192-TAX1BP1 axis in response to lysosomal damage. Nat Commun. 2025;16(1):1109.
36. Kedage V, Ellerman D, Chen Y, Liang WC, Borneo J, Wu Y, et al. Harnessing MerTK agonism for targeted therapeutics. MAbs. 2020;12(1):1685832.
37. Messer A, Butler DC. Optimizing intracellular antibodies (intrabodies/nanobodies) to treat neurodegenerative disorders. Neurobiol Dis. 2020;2020(134):104619.
38. Grochowska KM, Sperveslage M, Raman R, Failla AV, Głów D, Schulze C, et al. Chaperone-mediated autophagy in neuronal dendrites utilizes activity-dependent lysosomal exocytosis for protein disposal. Cell Rep. 2023;42(8):112998.
39. Kuzu OF, Toprak M, Noory MA, Robertson GP. Effect of lysosomotropic molecules on cellular homeostasis. Pharmacol Res. 2017;2017(117):177–84.
40. Egan DF, Chun MGH, Vamos M, Zou H, Rong J, Miller CJ, et al. Small molecule inhibition of the autophagy Kinase ULK1 and identification of ULK1 substrates. Mol Cell. 2015;59(2): 285–97.
41. Choi YJ, Yun SH, Yu J, Mun Y, Lee W, Park CJ, et al. Chaperone-mediated autophagy dysregulation during aging impairs hepatic fatty acid oxidation via accumulation of NCoR1. Mol Metab. 2023;2023(76):101784.
42. Wang Y, Liu J, Wang H, Jiang P, Cao L, Lu S, et al. Multiple regulatory mechanisms, functions and therapeutic potential of chaperone-mediated autophagy. Theranostics. 2025;15 (7):2778–93.
43. Shibata M, Lu T, Furuya T, Degterev A, Mizushima N, Yoshimori T, et al. Regulation of intracellular accumulation of mutant Huntingtin by Beclin 1. J Biol Chem. 2006;281(20): 14474–85.
44. Wong E, Cuervo AM. Autophagy gone awry in neurodegenerative diseases. Nat Neurosci. 2010;13(7):805–11.
45. Rui YN, Xu Z, Patel B, Chen Z, Chen D, Tito A, et al. Huntingtin functions as a scaffold for selective macroautophagy. Nat Cell Biol. 2015;17(3):262–75.
46. Martinez-Vicente M, Talloczy Z, Wong E, Tang G, Koga H, Kaushik S, et al. Cargo recognition failure is responsible for inefficient autophagy in Huntington's disease. Nat Neurosci. 2010;13(5):567–76.
47. Nixon RA. The role of autophagy in neurodegenerative disease. Nat Med. 2013;19(8):983–97.
48. Komatsu M, Wang QJ, Holstein GR, Friedrich VL, Iwata J, Ichi KE, et al. Essential role for autophagy protein Atg7 in the maintenance of axonal homeostasis and the prevention of axonal degeneration. Proc Natl Acad Sci. 2007;104(36):14489–94.
49. Kirkin V, Lamark T, Sou YS, Bjørkøy G, Nunn JL, Bruun JA, et al. A role for NBR1 in Autophagosomal degradation of Ubiquitinated substrates. Mol Cell. 2009;33(4):505–16.
50. Menzies FM, Fleming A, Caricasole A, Bento CF, Andrews SP, Ashkenazi A, et al. Autophagy and neurodegeneration: pathogenic mechanisms and therapeutic opportunities. Neuron. 2017;93(5):1015–34.
51. Tang FL, Liu W, Hu JX, Erion JR, Ye J, Mei L, et al. VPS35 deficiency or mutation causes dopaminergic neuronal loss by impairing mitochondrial fusion and function. Cell Rep. 2015;12(10):1631–43.
52. Winslow AR, Chen CW, Corrochano S, Acevedo-Arozena A, Gordon DE, Peden AA, et al. α-Synuclein impairs macroautophagy: implications for Parkinson's disease. J Cell Biol. 2010;190(6):1023–37.
53. Malik BR, Maddison DC, Smith GA, Peters OM. Autophagic and endo-lysosomal dysfunction in neurodegenerative disease. Mol Brain. 2019;12(1):1–21.
54. Cuervo AM, Stefanis L, Fredenburg R, Lansbury PT, Sulzer D. Impaired degradation of mutant α-Synuclein by Chaperone-mediated autophagy. Science. 2004;305(5688):1292–5.

55. Alvarez-Erviti L, Rodriguez-Oroz MC, Cooper JM, Caballero C, Ferrer I, Obeso JA, et al. Chaperone-mediated autophagy markers in Parkinson disease brains. Arch Neurol. 2010;67 (12):1464–72.
56. Kenessey A, Nacharaju P, Ko L, Yen S. Degradation of Tau by lysosomal enzyme Cathepsin D: implication for Alzheimer neurofibrillary degeneration. J Neurochem. 1997;69 (5):2026–38.
57. Settembre C, Di Malta C, Polito VA, Arencibia MG, Vetrini F, Erdin S, et al. TFEB links autophagy to lysosomal biogenesis. Science. 2011;332(6036):1429–33.
58. Pastore N, Brady OA, Diab HI, Martina JA, Sun L, Huynh T, et al. TFEB and TFE3 cooperate in the regulation of the innate immune response in activated macrophages. Autophagy. 2016;12(8):1240–58.
59. Caccamo A, Magrì A, Medina DX, Wisely EV, López-Aranda MF, Silva AJ, et al. mTOR regulates tau phosphorylation and degradation: implications for Alzheimer's disease and other tauopathies. Aging Cell. 2013;12(3):370–80.
60. Kawaguchi Y, Kovacs JJ, McLaurin A, Vance JM, Ito A, Yao TP. The deacetylase HDAC6 regulates Aggresome formation and cell viability in response to misfolded protein stress. Cell. 2003;115(6):727–38.
61. Pandey UB, Nie Z, Batlevi Y, McCray BA, Ritson GP, Nedelsky NB, et al. HDAC6 rescues neurodegeneration and provides an essential link between autophagy and the UPS. Nature. 2007;447(7146):860–4.
62. Bus C, Zizmare L, Feldkaemper M, Geisler S, Zarani M, Schaedler A, et al. Human dopaminergic neurons lacking PINK1 exhibit disrupted dopamine metabolism related to vitamin B6 co-factors. iScience. 2020;23(12):101797.
63. O'Callaghan B, Hardy J, Plun-Favreau H. PINK1: from Parkinson's disease to mitophagy and back again. PLoS Biol. 2023;21(6):e3002196.
64. Park MW, Cha HW, Kim J, Kim JH, Yang H, Yoon S, et al. NOX4 promotes ferroptosis of astrocytes by oxidative stress-induced lipid peroxidation via the impairment of mitochondrial metabolism in Alzheimer's diseases. Redox Biol. 2021;2021(41):101947.
65. Chen Z, Zheng N, Wang F, Zhou Q, Chen Z, Xie L, et al. The role of ferritinophagy and ferroptosis in Alzheimer's disease. Brain Res. 2025;2025(1850):149340.
66. Kataura T, Hattori N, Saiki S. Chapter 3: the autophagy pathway and its key regulators. In: Autophagy Dysfunct Alzheimer's Dis Dement; 2022. p. 47–69.
67. Wang B, Pareek G, Kundu M. ULK/Atg1: phasing in and out of autophagy. Trends Biochem Sci. 2024;49(6):494–505.
68. El-Houjeiri L, Possik E, Vijayaraghavan T, Paquette M, Martina JA, Kazan JM, et al. The transcription factors TFEB and TFE3 link the FLCN-AMPK signaling axis to innate immune response and pathogen resistance. Cell Rep. 2019;26(13):3613–3628.e6.
69. Chen H, Gong S, Zhang H, Chen Y, Liu Y, Hao J, et al. From the regulatory mechanism of TFEB to its therapeutic implications. Cell Death Discov. 2024;10(1):1–11.
70. Allen NJ, Barres BA. Signaling between glia and neurons: focus on synaptic plasticity. Curr Opin Neurobiol. 2005;15(5):542–8.
71. Azad MB, Chen Y, Gibson SB. Regulation of autophagy by reactive oxygen species (ROS): implications for cancer progression and treatment. Antioxid Redox Signal. 2009;11(4): 777–90.
72. Ayton S, Bush AI. investigators 3D trial. A trial of the iron-chelator deferiprone in Alzheimer's disease. Alzheimers Dement. 2025;20(Suppl 6):e086330.
73. Ayton S, Barton D, Brew B, Brodtmann A, Clarnette R, Desmond P, et al. Deferiprone in Alzheimer disease: a randomized clinical trial. JAMA Neurol. 2024;82(1):11–8.
74. Liu M, Liu S, Lin Z, Chen X, Jiao Q, Du X, et al. Targeting the interplay between autophagy and the Nrf2 pathway in Parkinson's disease with potential therapeutic implications. Biomolecules. 2025;15(1):149.

75. Armeli F, Mengoni B, Laskin DL, Businaro R. Interplay among oxidative stress, autophagy, and the endocannabinoid system in neurodegenerative diseases: role of the Nrf2-p62/SQSTM1 pathway and nutraceutical activation. Curr Issues Mol Biol. 2024;46(7):6868.
76. Ali J, Choe K, Park JS, Park HY, Kang H, Park TJ, et al. The interplay of protein aggregation, genetics, and oxidative stress in Alzheimer's disease: role for natural antioxidants and immunotherapeutics. Antioxidants. 2024;13(7):862.
77. Texidó L, Martín-Satué M, Alberdi E, Solsona C, Matute C. Amyloid β peptide oligomers directly activate NMDA receptors. Cell Calcium. 2011;49(3):184–90.
78. Um JW, Nygaard HB, Heiss JK, Kostylev MA, Stagi M, Vortmeyer A, Wisniewski T, Gunther EC and Strittmatter SM, 2012. Alzheimer amyloid-β oligomer bound to postsynaptic prion protein activates Fyn to impair neurons. Nature neuroscience, 15(9), pp.1227-1235.
79. Lemasters JJ, Theruvath TP, Zhong Z, Nieminen AL. Mitochondrial calcium and the permeability transition in cell death. Biochim Biophys Acta Bioenerg. 2009;1787(11):1395–401.
80. Bernardi P, Gerle C, Halestrap AP, Jonas EA, Karch J, Mnatsakanyan N, et al. Identity, structure, and function of the mitochondrial permeability transition pore: controversies, consensus, recent advances, and future directions. Cell Death Differ. 2023;30(8):1869–85.
81. MacAinsh M, Muhammedkutty FNK, Zhou HX. BPS2025 - membrane-assisted fibril formation of α-synuclein. Biophys J. 2025;124(3):408a.
82. Zeng H, Liu Y, Liu X, Li J, Lu L, Xue C, et al. Interplay of α-Synuclein oligomers and endoplasmic reticulum stress in Parkinson'S disease: insights into cellular dysfunctions. Inflammation. 2024;2024:1–17.
83. Bernal-Conde LD, Ramos-Acevedo R, Reyes-Hernández MA, Balbuena-Olvera AJ, Morales-Moreno ID, Argüero-Sánchez R, et al. Alpha-Synuclein physiology and pathology: a perspective on cellular structures and organelles. Front Neurosci. 2020;2020(13):502007.
84. Xu J, Wang S, Wu Y, Song P, Zou MH. Tyrosine nitration of PA700 activates the 26S proteasome to induce endothelial dysfunction in mice with angiotensin II-induced hypertension. Hypertension. 2009;54(3):625–32.
85. Aiken CT, Kaake RM, Wang X, Huang L. Oxidative stress-mediated regulation of proteasome complexes. Mol Cell Proteomics. 2011;10(5):R110.006924.
86. Kumar AV, Mills J, Lapierre LR. Selective autophagy receptor p62/SQSTM1, a pivotal player in stress and aging. Front Cell Dev Biol. 2022;2022(10):793328.
87. Sánchez-Martın P, Komatsu M. p62/SQSTM1 – steering the cell through health and disease. J Cell Sci. 2018;131(21):jcs222836.
88. Bhandary B, Marahatta A, Kim HR, Chae HJ. An involvement of oxidative stress in endoplasmic reticulum stress and its associated diseases. Int J Mol Sci. 2012;14(1):434–56.
89. Zhang Z, Zhang L, Zhou L, Lei Y, Zhang Y, Huang C. Redox signaling and unfolded protein response coordinate cell fate decisions under ER stress. Redox Biol. 2019;2019(25):101047.
90. Wang X, Chemmama IE, Yu C, Huszagh A, Xu Y, Viner R, et al. The proteasome-interacting Ecm29 protein disassembles the 26S proteasome in response to oxidative stress. J Biol Chem. 2017;292(39):16310–20.
91. Wang MG, Fan RF, Li WH, Zhang D, Yang DB, Wang ZY, et al. Activation of PERK-eIF2α-ATF4-CHOP axis triggered by excessive ER stress contributes to lead-induced nephrotoxicity. Biochim Biophys Acta, Mol Cell Res. 2019;1866(4):713–26.
92. Shang F, Taylor A. Ubiquitin–proteasome pathway and cellular responses to oxidative stress. Free Radic Biol Med. 2011;51(1):5–16.
93. Livnat-Levanon N, Kevei É, Kleifeld O, Krutauz D, Segref A, Rinaldi T, et al. Reversible 26S proteasome disassembly upon mitochondrial stress. Cell Rep. 2014;7(5):1371–80.
94. Ma M, Jiang W, Zhou R. DAMPs and DAMP-sensing receptors in inflammation and diseases. Immunity. 2024;57(4):752–71.
95. Kaur J, Singh H, Naqvi S. Intracellular DAMPs in neurodegeneration and their role in clinical therapeutics. Mol Neurobiol. 2023;60(7):3600–16.

96. Wang S, Pan Y, Zhang C, Zhao Y, Wang H, Ma H, et al. Transcriptome analysis reveals dynamic microglial-induced A1 astrocyte reactivity via C3/C3aR/NF-κB signaling after Ischemic stroke. Mol Neurobiol. 2024;61(12):10246–70.
97. Battaglioni S, Benjamin D, Wälchli M, Maier T, Hall MN. mTOR substrate phosphorylation in growth control. Cell. 2022;185(11):1814–36.
98. Adli M, Merkhofer E, Cogswell P, Baldwin AS. IKKα and IKKβ each function to regulate NF-κB activation in the TNF-induced/canonical pathway. PLoS One. 2010;5(2):e9428.
99. Davizon-Castillo P, McMahon B, Aguila S, Bark D, Ashworth K, Allawzi A, et al. TNF-α–driven inflammation and mitochondrial dysfunction define the platelet hyperreactivity of aging. Blood. 2019;134(9):727–40.
100. Pucha KA, Ma T, York W, Kang UJ, Kaufmann H, Kapogiannis D, et al. Neuron-derived extracellular vesicles to examine brain mTOR target engagement with sirolimus in patients with multiple system atrophy. Parkinsonism Relat Disord. 2023;2023(115):105821.
101. Pan T, Kondo S, Le W, Jankovic J. The role of autophagy-lysosome pathway in neurodegeneration associated with Parkinson's disease. Brain. 2008;131(8):1969–78.
102. Bové J, Martínez-Vicente M, Vila M. Fighting neurodegeneration with rapamycin: mechanistic insights. Nat Rev Neurosci. 2011;12(8):437–52.
103. Chien JY, Chou YY, Ciou JW, Liu FY, Huang SP. The effects of two Nrf2 activators, Bardoxolone methyl and Omaveloxolone, on retinal Ganglion cell survival during ischemic optic neuropathy. Antioxidants. 2021;10(9):1466.
104. Takagi T, Kitashoji A, Iwawaki T, Tsuruma K, Shimazawa M, Yoshimura S, et al. Temporal activation of Nrf2 in the penumbra and Nrf2 activator-mediated neuroprotection in ischemia–reperfusion injury. Free Radic Biol Med. 2014;2014(72):124–33.
105. Mauthe M, Orhon I, Rocchi C, Zhou X, Luhr M, Hijlkema KJ, et al. Chloroquine inhibits autophagic flux by decreasing autophagosome-lysosome fusion. Autophagy. 2018;14(8):1435–55.
106. Lescat L, Véeron V, Mourot B, Péron S, Chenais N, Dias K, et al. Chaperone-mediated autophagy in the light of evolution: insight from fish. Mol Biol Evol. 2020;37(10):2887–99.
107. Xu L, Yao S, Ding YE, Xie M, Feng D, Sha P, et al. Designing and optimizing AAV-mediated gene therapy for neurodegenerative diseases: from bench to bedside. J Transl Med. 2024;22(1):1–17.
108. Jumaah SH, Ali DAF. Exploring 8-hydroxy-2-deoxyguanosine as a biomarker of oxidative DNA damage in various diseases and pathological conditions: review. Int J Clin Biol Biochem. 2024;6(1):126–39.
109. Choi YJ, Lee JU, Sim SJ. Label-free and highly sensitive nanoplasmonic biosensor-based autophagy flux sensing for clinical application. Sensors Actuators B Chem. 2022;2022(350):130880.
110. Park JM, Lee DH, Kim DH. Redefining the role of AMPK in autophagy and the energy stress response. Nat Commun. 2023;14(1):2994.
111. Zhu L, Li Z, Wang H, Cheng Z, Zhang L. Unraveling the mysterious veil of ULK1: from non-canonical functions to therapeutic applications. Int J Biol Macromol. 2025;2025(318):145177.
112. Wang Z, Zhang H. Phase-separated condensates in autophagosome formation and autophagy regulation. J Mol Biol. 2025;437:168964.
113. Fan RZ, Guo M, Luo S, Cui M, Tieu K. Exosome release and neuropathology induced by α-synuclein: new insights into protective mechanisms of Drp1 inhibition. Acta Neuropathol Commun. 2019;7(1):1–18.
114. Bayati A, McPherson PS. Alpha-synuclein, autophagy-lysosomal pathway, and Lewy bodies: mutations, propagation, aggregation, and the formation of inclusions. J Biol Chem. 2024;300(10):107742.
115. Dhapola R, Beura SK, Sharma P, Singh SK, HariKrishnaReddy D. Oxidative stress in Alzheimer's disease: current knowledge of signaling pathways and therapeutics. Mol Biol Rep. 2024;51(1):1–18.

116. Barber SC, Mead RJ, Shaw PJ. Oxidative stress in ALS: a mechanism of neurodegeneration and a therapeutic target. Biochim Biophys Acta Mol basis Dis. 2006;1762(11–12):1051–67.
117. Rasmussen NL, Kournoutis A, Lamark T, Johansen T. NBR1: the archetypal selective autophagy receptor. J Cell Biol. 2022;221(11):e202208092.
118. Deng Z, Purtell K, Lachance V, Wold MS, Chen S, Yue Z. Autophagy receptors and neurodegenerative diseases. Trends Cell Biol. 2017;27(7):491–504.
119. Jiang-Xie LF, Drieu A, Kipnis J. Waste clearance shapes aging brain health. Neuron. 2025;113(1):71–81.
120. Sun Y, Rui LQK, Liu JY, Wang F, Liu CF. New perspectives on the glymphatic system and the relationship between glymphatic system and neurodegenerative diseases. Neurobiol Dis. 2025;2025(205):106791.

Blood–Brain Barrier Dysfunction in Neuroinflammation and Autophagy

6

Mousumi Paul and Yadu Nandan Dey

Abstract

Blood–brain barrier dysfunction plays a key role in the development and progression of neurodegenerative diseases like Alzheimer's disease, Parkinson's disease, multiple sclerosis, stroke, and brain injuries from trauma. The structural and functional integrity of the blood-brain barrier is essential for maintaining balance in the central nervous system. It allows selective permeability and protects neurons from changes in the environment. The link between blood–brain barrier dysfunction and a failed autophagy process involves several mechanisms, including lysosomal trapping, mammalian target of rapamycin 1 (mTORC1) overactivation, and endoplasmic reticulum stress. New treatment approaches can aim to restore blood–brain barrier function and support autophagy with pharmacological inhibitors, antioxidants, gene-editing techniques, senolytics, and interventions that regulate the microbiota. Challenges with drug delivery pose significant issues for targeting the blood–brain barrier. Improvements in organoids from patients, computational screenings for new drugs, and microfluidic models of the blood–brain barrier offer promising new options for better neuroprotection.

Keywords

Autophagy · Blood · brain barrier · Mitochondrial dysfunction · Neurodegenerative diseases · Neuroinflammation · Oxidative stress

M. Paul · Y. N. Dey (✉)
Pharmacology Division, Dr. B.C. Roy College of Pharmacy and Allied Health Sciences, Durgapur, West Bengal, India
e-mail: yadunandan.dey@bcrcp.org

P. Khanal et al. (eds.), *Neuroinflammation and Autophagy in Neurodegeneration*,
https://doi.org/10.1007/978-981-95-3945-1_6

6.1 Introduction

Neurodegenerative diseases are one of the fastest-growing global health burdens of twenty-first century. With aging populations, the prevalence of these disorders is on a sharp incline, leading to significant personal, societal, and economic impacts. Neurodegenerative diseases such as including Alzheimer's disease (AD), Parkinson's disease (PD), Huntington's disease (HD), and amyotrophic lateral sclerosis (ALS) are manifested by progressive loss of structure and function of neurons [1]. However, even with extensive research on neurodegenerative diseases, the cause remains elusive, thus limiting the treatment approach. As a consequence, this growing crisis demands a greater emphasis on investigation to unravel underlying mechanisms, detect early biomarkers, and identify targets for drug development. In majority of PD cases, blood–brain barrier (BBB) leakage precedes motor symptoms, implicating barrier failure as a silent instigator of neurodegeneration [2]. It is believed that neurodegeneration is a result of several factors that consequently lead to neurovascular unit dysfunction, which leads to neuronal dysfunction and death. This is popularly referred to as the vascular model of neurodegeneration. Two principal factors form the pillar of this model, including hypoperfusion and BBB disruption. In PD, vascular remodeling, altered vasculature, and abnormal angiogenesis make a significant contribution to disease progression and pathogenesis [3]. In PD, BBB leakage often occurs before the appearance of motor symptoms. This indicates that BBB dysfunction may serve as an early and silent driver of neurodegeneration. This early BBB compromise permits peripheral immune cells and neurotoxic molecules to enter the brain, triggering inflammation, microglial activation, and neuronal damage that contribute to PD progression [4]. Additionally, it has been proven that the selective and gradual loss of dopaminergic neurons in substantia nigra pars compacta and their interactions with other basal ganglia play a central role in BBB disruption, contributing to neurodegeneration [5].

Recent studies show that BBB disruption not only acts as a downstream effector of neurodegeneration but also as a bidirectional inducer of neuroinflammation and dysregulation in cell autophagy, creating a self-sustaining cycle that leads to neuronal loss [6, 7]. In BBB disruption, peripheral immune cells and toxic macromolecules infiltrate, increasing inflammatory cascades, whereas neuronal homeostasis and clearance processes like autophagy are essential for α-synuclein turnover and mitochondrial quality control [6, 8]. Disruption in communication between pericytes, microglia, neurons, astrocytes, and other brain endothelial cells causes BBB dysfunction [9, 10]. Additionally, astrocyte abnormal function and central nervous system (CNS) protein leakage into plasma result in increased S100β levels [11]. Next, an increase in plasma matrix metalloproteinases-9 (MMP-9) results from tight junction and basement membrane degradation in ischemia or inflammatory states [12–14]. Interestingly, this occurs before visible edema or bleeding, suggesting a critical window for early treatment [15].

Certain physiological ailments like hypertension have been studied to disrupt the BBB due to disturbed endothelial dysfunction, increasing vasoconstriction, oxidative stress, as well as reduced nitric oxide level and adhesion molecules expression

[16, 17]. Next, diabetes leads to advanced glycation end-products accumulation in basement membranes, thickening vessel walls, increasing oxidative stress, and pericyte loss, which are the hallmarks of microangiopathy [9]. Thus, an in-depth study of BBB failure in aging and physiological ailments can pave the way to understand the convergence of evolutionary adaptive mechanisms, altered brain cell function, and hallmarks like S100β and MMP-9 as measurable early signs.

6.2 Structural and Functional Foundations of the BBB

The BBB is a specialized vascular structure that protects CNS by regulating ions, molecules, and cells between blood and brain to maintain neuronal function, defending against toxins and pathogens [16]. The BBB is composed of continuous non-fenestrated endothelial cells that form the wall of blood vessels and render them with unique properties to allow trafficking between the brain and blood [18]. These cells are modified versions of mesodermally derived simple squamous epithelial cells. There are two primary types of transporters found in CNS endothelial cells. The first type comprises efflux transporters, which are localized to the luminal surface and transport a diverse range of lipophilic substances that would otherwise passively diffuse across the cell membrane toward the bloodstream. The second type is highly specialized nutrient transporters that enable specific nutrients to be transported across the BBB into CNS, in addition to facilitating the waste products from CNS into the bloodstream. CNS endothelial cells also exhibit expression of leukocyte adhesion molecules, which restricts immune cells into CNS [19]. The BBB also contains mural cells, which encompass vascular smooth muscle cells surrounding larger blood vessels and pericytes that wrap endothelial surfaces of microvasculature (Fig. 6.1). Pericytes play a role in regulating angiogenesis, extracellular matrix deposition, wound healing, immune cell infiltration, and blood flow, and they are crucial for the formation and maintenance of BBB [9]. The next structural component for BBB integrity is the basal membrane, which can roughly be divided into the inner vascular and outer parenchymal basal membrane, which is also referred to as vascular glia limitans perivascularis. These membranes consist of type IV collagens, heparin sulfate proteoglycans, nidogen, laminin, including glycoproteins [20, 21]. When metalloproteinase disrupts the basal membrane, it leads to BBB disruption and leukocyte infiltration [22]. Next, astrocytes form a major part of BBB vasculature and mediate BBB function by regulating blood flow in response to neuronal activity [23]. Astrocytes represent a primary type of glial cell, which are characterized by polarized cellular processes that enclose either neuronal structures or blood vessels. Astrocytic basal-process endfeet almost entirely envelop the cerebral microvasculature and contain specific proteins, including dystroglycan, dystrophin, and aquaporin-4 [24]. The dystroglycan–dystrophin complex plays a crucial role in connecting the endfeet cytoskeleton to the basement membrane by interacting with Agrin [25]. This facilitates the arrangement of aquaporin 4 into orthogonal particle arrays, which is essential for maintaining water homeostasis in the CNS [26]. Moreover, astrocytes serve as a cellular bridge

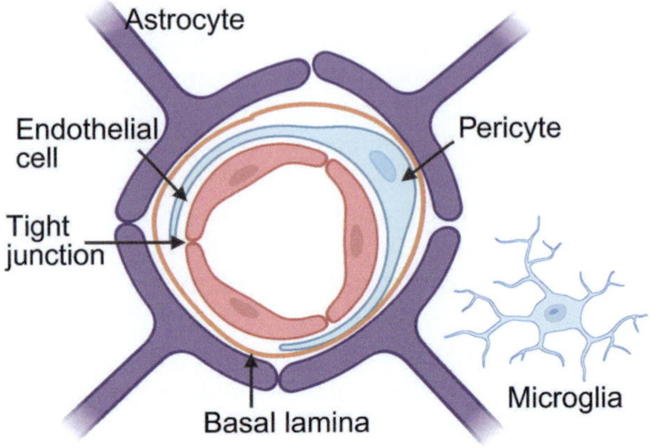

Fig. 6.1 Cellular components of the neurovascular unit and blood–brain barrier. The neurovascular unit is composed of endothelial cells, pericytes, astrocytes, and the basal lamina, working together to maintain blood–brain barrier integrity. Microglia interacts with these structures to regulate immune responses and support neural health. (Created using BioRender, https://www.biorender.com/)

between neuronal networks and blood vessels, and this neurovascular coupling allows astrocytes to transmit signals that modulate blood flow in response to the activity of neurons. This includes the contraction and relaxation of vascular smooth muscle cells that encircle arterioles, including pericytes that are found around capillaries [27]. The BBB vasculature also contains certain immune cells, like perivascular macrophages and microglial cells that serve as primary immune cell populations in CNS [28]. Immune cells from bloodstream can engage with CNS blood vessels and modulate the properties of BBB in response to infection, injury, and disease [29]. Apart from this, the BBB also functions in modulating other permeability and transport factors that are dependent on the time of day, for example, Bmal1 coordination circadian rhythms with endothelial functions [30]. These transcription factors regulate tight junction proteins (e.g., claudin-5 and occludin), transporter systems expression, and immune response regulation [31]. It has also been studied that cerebral endothelial cells are constantly subjected to hemodynamic forces, and to overcome that, ion channels containing Piezo1 have been identified as crucial sensors for shear stress [32]. If activated by increased blood flow, these Piezo1 channels convert mechanical signals into an influx of intracellular calcium stabilizing adherens and tight junctions through downstream effectors such as calmodulin and small GTPases like Rac1 [33].

6.3 Mechanisms of BBB Disruption in Neuroinflammation

From the above discussion, we can conclude that the structure and integrity of the BBB are vital to maintain CNS homeostasis and that it acts as a hallmark feature in neurodegenerative diseases. Next, the pathways that contribute to BBB disruption during neuroinflammation have to be well discussed. The most common among them include the role of matrix metalloproteinases (MMPs), pro-inflammatory cytokines, and oxidative stress [22]. The MMPs have been studied to digest tight junction proteins like claudin-1, claudin-5, occludin, and zonula occludens-1, along with the basement membrane proteins, which lead to increased BBB permeability. Studies have also shown that MMP-2/9 levels rise significantly during conditions like stroke, particularly during reperfusion. MMPs like MMP-1/2/9 have been identified in microglial nodules and microglial-like cells in lesional MS tissue, proving their role in neuroinflammation and degrading BBB integrity [34]. MMPs also assist leukocyte movement through the brain's endothelium by activating adhesion molecules and breaking down the basement membrane. Additionally, certain pro-inflammatory cytokines like tumor necrosis factor- α (TNF-α), interleukin-1β (IL-1β), and IL-6 are released by activated microglia, that have been studied to cause BBB disruptions. Oxidative stress markers like reactive oxygen species have been studied to affect BBB integrity by transitory activating the PI3K/Akt pathway employing factors like RhoA, further leading to tight junction leakage. The NADPH oxidase system present in microglia has also been observed to produce superoxide ions, which further compromise BBB integrity. The excessive accumulation of free radicals not only damages cells but also causes neurodegenerative diseases [35]. Moreover, oxidative stress also leads to the peroxidation of membrane polyunsaturated fatty acids, increasing BBB permeability to toxins. Recent findings have identified multiple mechanisms such as mitochondrial dysfunction, inflammasome activation, metabolites dysregulation, and imbalance in signaling pathways, thereby broadening our understanding of dysregulation in the neurovascular unit [36]. Below, some of the factors are discussed briefly.

Mitochondrial fission needs the employment of a major group of GTPases called DRP1, which further regulates the electron transport chain and reduces ATP synthesis. Increased DRP1activity is linked to a decrease in electron transport chain efficiency and ATP reduction, which further affects cell survival and endothelial junctions, contributing to vascular dysfunction [37, 38]. Next, NLRP3 inflammasome activation has been studied to activate pro-inflammatory cytokines. The NLRP3 activation in cells like pericytes and endothelial cells stimulates the production and release of IL-1β and IL-18 in vascular injury. Both of these compromise the endothelial barrier integrity by disrupting the tight junctions between endothelial cells, which in turn leads to vascular leakage and barrier dysfunction [39, 40]. Furthermore, HDAC3 has been reported to suppress gene transcription by histone deacetylation in the gene promoter site [41]. Claudin-5 is an essential tight junction protein in endothelial cells and is susceptible to this form of epigenetic silencing, leading to reduced expression and barrier dysfunction [42]. Additionally, the alternative complement pathway also produces certain anaphylatoxins like C3a and C5a, which help in recognizing and enhancing paracellular permeability in endothelial cells by activating RhoA [43]. This activation leads to a reorganization of junctions and heightened vascular leakage, compromising BBB integrity.

6.4 BBB Dysfunction and Autophagy Failure: Mechanistic Crosstalk

The disruption of BBB integrity not only permits peripheral toxins and proteins to infiltrate the brain parenchyma but also hinders lysosomal flow in glial and endothelial cells, which leads to autophagy [1]. This, in turn, weakens the brain proteostasis and worsens the state of neuroinflammation and neurodegeneration. Under normal physiological conditions, amyloid-β (Aβ) and α-synuclein efflux out from the brain through efflux transporters such as LRP1 and P-gp transporters in the BBB, and autophagy-lysosomal degradation in microglia and astrocytes [44, 45]. When there is an imbalance or compromise in BBB function, it can result in Aβ and harmful protein accumulation, creating a substantial load on the neural autophagy cascade [46]. Impairment or a halt in the autophagy cascade in neural cells prohibits the intracellular aggregates' clearance (Fig. 6.2), which are composed of misfolded proteins (Aβ, α-synuclein), worsening their accumulation and toxicity [47].

6.5 Emerging Synergies into the Cascade of Cellular Stress

Once the BBB breaks down, fibrinogen circulating in the bloodstream enters the CNS, where it gets phagocytosed by the microglia. Due to increased load, an excessive accumulation of non-degradable fibrinogen is entrapped in lysosomes, blocking the degradation of neuronal byproducts or protein aggregates [48]. This blockage of lysosomes (autophagolysosomes) results in a disrupted autophagic process, increases lysosomal stress, and increases microglial activation, producing inflammatory cytokines, and establishing a cycle of neuroinflammation and cellular stress [49].

The BBB in normal physiological conditions blocks the entry of albumin, but is compromised in a leaky compartment. Once infiltrated, albumin stimulates mammalian target of rapamycin 1 (mTORC1) signaling in both glial cells and neurons through metabolic sensing mechanisms, and this in turn inhibits the AMPK activity. In this series of alterations follows such as dysregulation of macroautophagy, leading to reduced LC3 activity, which leads to irregular or compromised autophagosome formation [50]. Under normal physiological conditions, a brain network ensures that proteins are folded correctly. This mechanism focuses on the removal of superfluously folded proteins either by autophagy or by degradation mediated via the proteasome, lysosome, and other pathways [51]. Thus, proteostasis is a crucial mechanism to preserve proper cell function and overall cellular function. The major element to identify the misfolded proteins and thereby start their degradation process is through chaperones that reside in the cytosol and endoplasmic

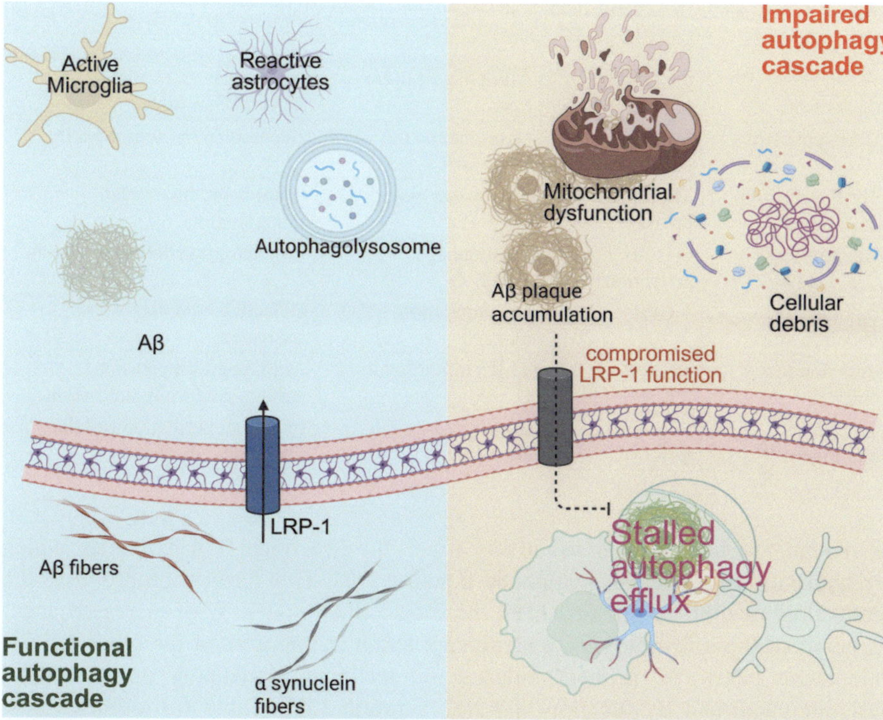

Fig. 6.2 Role of LRP-1 in regulating autophagy and Aβ clearance in neural cells. A functional autophagy cascade is supported by LRP-1, which facilitates the clearance of Aβ and α-synuclein fibers, maintaining cellular homeostasis. In contrast, compromised LRP-1 function leads to stalled autophagy efflux, mitochondrial dysfunction, Aβ plaque accumulation, and buildup of cellular debris, contributing to neurodegenerative processes. (Created using BioRender, https://www.biorender.com/)

reticulum. In neurodegenerative diseases, this mechanism of sustaining cellular proteostasis is under threat, followed by the complexities arising due to the accumulation of superfluous proteins. The endoplasmic reticulum stress in turn triggers an unfolded protein response, which is one of the principal consequences of accumulating misfolded protein [52]. Additionally, oxidative stress or metabolic disruptions caused due to dysfunctional BBB lead to compromised endothelial mitochondria, which are not efficiently removed through mitophagy [53]. Furthermore, the mitochondrial DNA released into the brain tissue acts as a damage-associated molecular pattern, which activates TLR9 on glial and endothelial cells, leading to an increase in pro-inflammatory signaling, obstructing autophagic clearance even more [54, 55]. Additionally, the glymphatic system plays a crucial role in the removal and clearance of the interstitial waste products, like Aβ, by leveraging perivascular spaces and by employing aquaporin-4-mediated water flow [56]. This is most active, especially during sleep. However, when the BBB integrity and vasculature are compromised, the resulting leakage also leads to the disruption of the

Table 6.1 Mechanisms of BBB dysfunction and autophagy failure

Mechanism	BBB dysfunction effect	Autophagy failure link	Downstream impact
Aβ/α-synuclein	Impaired efflux	Overwhelms the lysosomal system	Accumulation, neurotoxicity
Fibrinogen influx	Microglial uptake	Lysosomal clogging	Autophagic flux arrest, inflammation
Albumin leak	mTORC1 activation	↓ Autophagy initiation	Protein/organelle accumulation
Endoplasmic reticulum stress	↑ UPR	LC3 lipidation blocked	Defective autophagosome formation
Mitochondrial injury	mtDNA release	TLR9 inflammation	Autophagy suppression, sustained neuroinflammation
Glymphatic impairment	↓ Interstitial flow	Impaired waste clearance	Increased burden on cellular autophagic machinery

perivascular architecture and negatively affects the directional flow dynamics, which are both vital for effective glymphatic function [57]. This further compromises the removal of neurotoxic proteins from the extracellular space. With increased accumulation of these neurotoxins, a significant strain is imposed on the other cellular autophagic pathways, further hindering intracellular degradation processes and hastening neuronal dysfunction and degeneration [58]. Table 6.1 outlines how BBB dysfunction leads to autophagy failure and impacts brain function.

6.6 Multimodal Therapeutics in Repairing the BBB and Restoring Autophagy

Disruption in BBB integrity and autophagy failure are closely interlinked in the pathogenesis of neurodegenerative diseases such as AD, PD, stroke, MS, and traumatic brain injury [1, 44]. Currently, many novel therapeutics that address the underlying cause of BBB disruption can be developed and are in the research pipeline. The role of MMP-2/9 in regulating neuroinflammatory response has already been discussed above. Additionally, we explored the role of oxidative stress contributing to mitochondrial dysfunction and neuroinflammation, resulting in lipid peroxidation and damage to tight junctions [59]. Reactive oxygen species also inhibit the function of essential autophagy regulators, such as AMPK [60]. Small molecule inhibitors like minocycline, SB-3CT, and BB-94, including broad-spectrum matrix metalloproteinase inhibitors, can reduce tight junction degradation, thereby maintaining the BBB function [61, 62]. SB-3CT is a selective MMP-2/9 inhibitor, whereas BB-94 (batimastat) is a broad-spectrum matrix metalloproteinase inhibitor. The major idea behind inhibiting matrix metalloproteinases is to mitigate inflammatory matrix remodeling, which may further lead to a decrease in secondary oxidative stress and cell death signaling molecules that typically hinder autophagic activity. The inhibition of M matrix metalloproteinases was observed to maintain the integrity of the BBB, reduce edema, and reduce the neurological implications of multiple sclerosis [63, 64]. Antioxidant compounds like N-acetylcysteine, edaravone, resveratrol, and curcumin have been widely known to scavenge reactive oxygen species, stabilize tight junctions, and promote the upregulation of other endogenous antioxidant enzymes, such as those of superoxide dismutase and catalase. They have been studied to promote autophagy indirectly by alleviating reactive oxygen species-induced inhibition and restoring cellular redox balance [65].

Since, CRISPR/Cas9 system can be used in gene activation and editing, it can be utilized as a strategy to boost claudin-5 expression specifically in brain endothelial cells. This can be done by linking a catalytically inactive Cas9 to activators like VP64 or p300, and using endothelial-specific promoters to direct the construct to the claudin-5 promoter; this method can enhance claudin-5 in the BBB [66]. This focused activation can promote increased integrity of tight junctions with minimal risk of influencing gene expression in non-endothelial tissues. Preclinical research has shown that CRISPR/Cas9-mediated claudin-5 upregulation restores BBB function after injury and effectively inhibits Aβ infiltration, with studies indicating reduced cognitive impairment following neurovascular damage [67]. Additionally, the Exosome-encapsulated miRNA technique shows great promise as a miRNA-focused therapeutic approach for restoring the BBB as well as the autophagy cascade [68]. This method involves loading exosomes derived from stem cells, such as bone marrow or induced pluripotent stem cell-derived mesenchymal stem cells, allowing for their passage through the damaged BBB to deliver miRNA to brain pericytes and endothelial cells [69]. Once within the target cells, miRNA inhibits PTEN, which in turn activates the PI3K/Akt-mTOR signaling. This provides an opportunity for cell survival by increasing the expression of tight junction proteins and boosting autophagy by influencing the mTOR pathways. In early preclinical tests on mice, treatment with extracellular vesicles significantly improved neurological recovery, preserved blood–brain barrier integrity, and reduced brain infarct size highlighting their therapeutic promise [70].

Moreover, senescent endothelial cells are marked by pro-inflammatory secretions, cell death resistance, and a disrupted autophagy cascade. A combination of drugs like dasatinib and quercetin triggers apoptosis in such senescent cells [71]. This eliminates senescent cells, which lowers inflammation, enhances vascular reactivity, and boosts the production of non-senescent endothelial cells, thereby restoring proteostatic equilibrium. In a study conducted in aged mice for the study of neurodegeneration with disrupted BBB, senolytic therapy restores the structure and function of the barrier and boosts autophagy markers in both endothelial and glial cells [72]. Currently development of peptides that employs tight junction repair sequences along with motifs that enhance autophagy can be engineered for their dual action [73]. For example, facilitating the assembly of tight junctions and using Beclin-1 to activate the formation of autophagosomes through class III PI3K activation can be used to target both the cell junction and proteostasis [74]. Additionally, nanocarriers encapsulating MMP-9 siRNA and rapamycin have also been developed. The MMP-9 siRNA suppresses endothelial-expressed MMP-9, and rapamycin acts as a potent mTOR inhibitor that activates the autophagy cascade. Thus, this method can be utilized to target the brain microvasculature directly [75].

In addition, gut dysbiosis is characterized by a decrease in butyrate-producing bacteria such as *Faecalibacterium* and *Eubacterium*, leading to increased permeability and neuroinflammatory responses [76, 77]. Fecal microbiota transplantation reinstates a balanced microbiome, which boosts the depleted butyrate production, enhancing tight junction expression, diminishing stimulation of TLR4-mediated inflammatory pathways, thereby strengthening endothelial cells of the BBB vasculature. It also acts by promoting the expression of autophagy-related genes like BECN1 and LC3 [78].

6.7 Translational Challenges and Next-Gen Solutions

Treatment of AD and PD still requires vast medical research and patient acceptance, even after decades of the burden of its prevalence in the senile population [79]. Therapies for BBB repair and autophagy restoration encounter major obstacles from successful preclinical results to effective clinical trials due to multidimensional problems such as variations in BBB structure and regulation between species, along with the absolute difficulty in administering and targeting drugs across the human BBB [80]. A range of innovative strategies is now emerging to address these challenges. In studies conducted using patient-derived organoids created from induced pluripotent stem cells, it was seen to permit scope for in vitro investigation of genetic risk factors like APOE4 and support personalized drug testing [81]. In silico studies involving computational drug screening using artificial intelligence easily predict and enhance candidates from small molecule libraries, or synthetic libraries, which are not only capable of penetrating the BBB but also modulating autophagy [82]. Dynamic microfluidic devices that mimic physiological shear stress employing pulsatile flow provide a more precise representation of the BBB response to treatments and injuries [83]. Considering gender-specific vulnerabilities, compounds such as estrogen receptor-β agonists can be assessed for their benefits on the BBB in women who have undergone menopause [84]. Moreover, non-drug interventions like photobiomodulation utilizing near-infrared light exhibit potential in promoting mitochondrial repair and improving the function of endothelial cells [85].

6.8 Conclusion

BBB dysfunction can be a crucial link to peripheral inflammation, disrupted autophagy, and the spread of proteinopathies in the CNS, further resulting in neurodegenerative diseases. This multidimensional integrative function indicates that a compromised BBB not only enables neuroinflammatory and neurodegenerative processes but also hinders the clearance of harmful proteins and disturbs neuronal homeostasis. Thus, future research regarding the BBB integrity, including promising strategies for precision neuroprotection, will involve multimodal approaches that address the systemic factors that overall affect brain health. Combining targeted treatments such as senolytics to eliminate senescent vascular cells with systemic approaches like fecal microbiota transplant to modulate peripheral immune responses, as well as aligning the circadian rhythm to enhance innate neuroprotective cascades, can be employed to protect the brain from BBB damage and neurodegeneration.

References

1. Sweeney MD, Sagare AP, Zlokovic BV. Blood–brain barrier breakdown in Alzheimer disease and other neurodegenerative disorders. Nat Rev Neurol. 2018;14(3):133–50.
2. Takata F, Nakagawa S, Matsumoto J, Dohgu S. Blood-brain barrier dysfunction amplifies the development of neuroinflammation: understanding of cellular events in brain microvascular endothelial cells for prevention and treatment of BBB dysfunction. Front Cell Neurosci. 2021;15:661838.
3. Brandl S, Reindl M. Blood–brain barrier breakdown in neuroinflammation: current in vitro models. Int J Mol Sci. 2023;24(16):12699.
4. Zlokovic BV. Neurovascular pathways to neurodegeneration in Alzheimer's disease and other disorders. Nat Rev Neurosci. 2011;12(12):723–38.
5. Al-Bachari S, Naish JH, Parker GJM, Emsley HCA, Parkes LM. Blood-brain barrier leakage is increased in Parkinson's disease. Front Physiol. 2020;11:593026.
6. Yang Z, Huang C, Wu Y, Chen B, Zhang W, Zhang J. Autophagy protects the blood-brain barrier through regulating the dynamic of claudin-5 in short-term starvation. Front Physiol. 2019;10:2.
7. Nixon RA. The role of autophagy in neurodegenerative disease. Nat Med. 2013;19(8):983–97.
8. Tohidpour A, Morgun AV, Boitsova EB, Malinovskaya NA, Martynova GP, Khilazheva ED, et al. Neuroinflammation and infection: molecular mechanisms associated with dysfunction of neurovascular unit. Front Cell Infect Microbiol. 2017;7:276.
9. Armulik A, Genové G, Mäe M, Nisancioglu MH, Wallgard E, Niaudet C, et al. Pericytes regulate the blood–brain barrier. Nature. 2010;468(7323):557–61.
10. Abbott JN. Astrocyte-endothelial interactions at the blood-brain barrier. Nat Rev Neurosci. 2006;7:41–53.
11. Marchi N, Rasmussen P, Kapural M, Fazio V, Kight K, Mayberg MR, et al. Peripheral markers of brain damage and blood-brain barrier dysfunction. Restor Neurol Neurosci. 2003;21(3–4): 109–21.
12. Lopez-Navarro ER, Gutierrez J. Metalloproteinases and their inhibitors in neurological disease. Naunyn Schmiedeberg's Arch Pharmacol. 2022;395(1):27–38.
13. Rosenberg GA. Matrix metalloproteinases and their multiple roles in neurodegenerative diseases. Lancet Neurol. 2009;8(2):205–16.
14. Ronaldson PT, Davis TP. Regulation of blood–brain barrier integrity by microglia in health and disease: a therapeutic opportunity. J Cereb Blood Flow Metab. 2020;40(1_suppl):S6–S24.
15. Candelario-Jalil E, Dijkhuizen RM, Magnus T. Neuroinflammation, stroke, blood-brain barrier dysfunction, and imaging modalities. Stroke. 2022;53(5):1473–86.
16. Abbott NJ, Patabendige AA, Dolman DE, Yusof SR, Begley DJ. Structure and function of the blood–brain barrier. Neurobiol Dis. 2010;37(1):13–25.
17. Keep RF, Andjelkovic AV, Xiang J, Stamatovic SM, Antonetti DA, Hua Y, et al. Brain endothelial cell junctions after cerebral hemorrhage: changes, mechanisms and therapeutic targets. J Cereb Blood Flow Metab. 2018;38(8):1255–75.

18. Pardridge WM. Drug transport across the blood–brain barrier. J Cereb Blood Flow Metab. 2012;32(11):1959–72.
19. Engelhardt B, Ransohoff RM. Capture, crawl, cross: the T cell code to breach the blood–brain barriers. Trends Immunol. 2012;33(12):579–89.
20. Daneman R, Prat A. The blood–brain barrier. Cold Spring Harb Perspect Biol. 2015;7(1): a020412.
21. Castro Dias M, Mapunda JA, Vladymyrov M, Engelhardt B. Structure and junctional complexes of endothelial, epithelial and Glial brain barriers. Int J Mol Sci. 2019;20(21):5372.
22. Rempe RG, Hartz AM, Bauer B. Matrix metalloproteinases in the brain and blood–brain barrier: versatile breakers and makers. J Cereb Blood Flow Metab. 2016;36(9):1481–507.
23. Zonta M, Angulo MC, Gobbo S, Rosengarten B, Hossmann K-A, Pozzan T, et al. Neuron-to-astrocyte signaling is central to the dynamic control of brain microcirculation. Nat Neurosci. 2003;6(1):43–50.
24. Nagelhus EA, Ottersen OP. Physiological roles of aquaporin-4 in brain. Physiol Rev. 2013;93 (4):1543–62.
25. Satz JS, Ostendorf AP, Hou S, Turner A, Kusano H, Lee JC, et al. Distinct functions of glial and neuronal dystroglycan in the developing and adult mouse brain. J Neurosci. 2010;30(43): 14560–72.
26. Nielsen S, Nagelhus EA, Amiry-Moghaddam M, Bourque C, Agre P, Ottersen OP. Specialized membrane domains for water transport in glial cells: high-resolution immunogold cytochemistry of aquaporin-4 in rat brain. J Neurosci. 1997;17(1):171–80.
27. Iadecola C. The neurovascular unit coming of age: a journey through neurovascular coupling in health and disease. Neuron. 2017;96(1):17–42.
28. Prinz M, Priller J. Microglia and brain macrophages in the molecular age: from origin to neuropsychiatric disease. Nat Rev Neurosci. 2014;15(5):300–12.
29. Ransohoff RM, Engelhardt B. The anatomical and cellular basis of immune surveillance in the central nervous system. Nat Rev Immunol. 2012;12(9):623–35.
30. Zhang SL, Lahens NF, Yue Z, Arnold DM, Pakstis PP, Schwarz JE, Sehgal A. A circadian clock regulates efflux by the blood-brain barrier in mice and human cells. Nat Commun. 2021;12(1):617.
31. Cazalla E, Cuadrado A, García-Yagüe ÁJ. Role of the transcription factor NRF2 in maintaining the integrity of the Blood-Brain Barrier. Fluids Barrier CNS. 2024;21(1):93.
32. Li J, Hou B, Tumova S, Muraki K, Bruns A, Ludlow MJ, et al. Piezo1 integration of vascular architecture with physiological force. Nature. 2014;515(7526):279–82.
33. Bae C, Sachs F, Gottlieb PA. The mechanosensitive ion channel Piezo1 is inhibited by the peptide GsMTx4. Biochemistry. 2011;50(29):6295–300.
34. Turner RJ, Sharp FR. Implications of MMP9 for blood brain barrier disruption and hemorrhagic transformation following ischemic stroke. Front Cell Neurosci. 2016;10:56.
35. Lakhan SE, Kirchgessner A, Tepper D, Leonard A. Matrix metalloproteinases and blood-brain barrier disruption in acute ischemic stroke. Front Neurol. 2013;4:32.
36. Kelley N, Jeltema D, Duan Y, He Y. The NLRP3 Inflammasome: an overview of mechanisms of activation and regulation. Int J Mol Sci. 2019;20(13):3328.
37. Wang X, Su B, Lee H-g, Li X, Perry G, Smith MA, et al. Impaired balance of mitochondrial fission and fusion in Alzheimer's disease. J Neurosci. 2009;29(28):9090–103.
38. Joshi AU, Saw NL, Shamloo M, Mochly-Rosen D. Drp1/Fis1 interaction mediates mitochondrial dysfunction, bioenergetic failure and cognitive decline in Alzheimer's disease. Oncotarget. 2018;9(5):6128–43.
39. Zhou R, Yazdi AS, Menu P, Tschopp J. A role for mitochondria in NLRP3 inflammasome activation. Nature. 2011;469(7329):221–5.
40. Zhang Y, Yang H, Hou S, Xia Y, Wang Y-Q. Influence of the brain-gut axis on neuroinflammation in cerebral ischemia-reperfusion injury. Int J Mol Med. 2024;53(3):30.

41. Nott A, Nitarska J, Veenvliet JV, Schacke S, Derijck AA, Sirko P, et al. S-nitrosylation of HDAC2 regulates the expression of the chromatin-remodeling factor Brm during radial neuron migration. Proc Natl Acad Sci USA. 2013;110(8):3113–8.
42. Hashimoto Y, Greene C, Munnich A, Campbell M. The CLDN5 gene at the blood-brain barrier in health and disease. Fluids Barrier CNS. 2023;20(1):22.
43. Alawieh A, Elvington A, Tomlinson S. Complement in the homeostatic and ischemic brain. Front Immunol. 2015;6:417.
44. Erickson MA, Banks WA. Blood–brain barrier dysfunction as a cause and consequence of Alzheimer's disease. J Cereb Blood Flow Metab. 2013;33(10):1500–13.
45. Tarasoff-Conway JM, Carare RO, Osorio RS, Glodzik L, Butler T, Fieremans E, et al. Clearance systems in the brain—implications for Alzheimer disease. Nat Rev Neurol. 2015;11(8): 457–70.
46. Lee J-H, Yu WH, Kumar A, Lee S, Mohan PS, Peterhoff CM, et al. Lysosomal proteolysis and autophagy require presenilin 1 and are disrupted by Alzheimer-related PS1 mutations. Cell. 2010;141(7):1146–58.
47. Guo F, Liu X, Cai H, Le W. Autophagy in neurodegenerative diseases: pathogenesis and therapy. Brain Pathol. 2018;28(1):3–13.
48. Petersen MA, Ryu JK, Akassoglou K. Fibrinogen in neurological diseases: mechanisms, imaging and therapeutics. Nat Rev Neurosci. 2018;19(5):283–301.
49. Winkler EA, Sagare AP, Zlokovic BV. The Pericyte: a forgotten cell type with important implications for Alzheimer's disease? Brain Pathol. 2014;24(4):371–86.
50. Saxton RA, Sabatini DM. mTOR signaling in growth, metabolism, and disease. Cell. 2017;168 (6):960–76.
51. Hetz C, Saxena S. ER stress and the unfolded protein response in neurodegeneration. Nat Rev Neurol. 2017;13(8):477–91.
52. Forman MS, Lee VM, Trojanowski JQ. 'Unfolding' pathways in neurodegenerative disease. Trends Neurosci. 2003;26(8):407–10.
53. Fang EF, Hou Y, Palikaras K, Adriaanse BA, Kerr JS, Yang B, et al. Mitophagy inhibits amyloid-β and tau pathology and reverses cognitive deficits in models of Alzheimer's disease. Nat Neurosci. 2019;22(3):401–12.
54. Zhang G, Wei H, Zhao A, Yan X, Zhang X, Gan J, Guo M, Wang J, Zhang F, Jiang Y, Liu X, Yang Z, Jiang X. Mitochondrial DNA leakage: underlying mechanisms and therapeutic implications in neurological disorders. J Neuroinflammation. 2025;22(1):34.
55. Pilard M, Ollivier EL, Gourdou-Latyszenok V, Couturaud F, Lemarié CA. Endothelial cell phenotype, a major determinant of venous Thrombo-inflammation. Front Cardiovasc Med. 2022;9:864735.
56. Jessen NA, Munk ASF, Lundgaard I, Nedergaard M. The glymphatic system: a beginner's guide. Neurochem Res. 2015;40(12):2583–99.
57. Gaberel T, Gakuba C, Goulay R, De Lizarrondo SM, Hanouz J-L, Emery E, et al. Impaired glymphatic perfusion after strokes revealed by contrast-enhanced MRI: a new target for fibrinolysis? Stroke. 2014;45(10):3092–6.
58. Nedergaard M, Goldman SA. Glymphatic failure as a final common pathway to dementia. Science. 2020;370(6512):50–6.
59. Ahmed SMU, Luo L, Namani A, Wang XJ, Tang X. Nrf2 signaling pathway: pivotal roles in inflammation. Biochim et Biophys Acta. 2017;1863(2):585–97.
60. Paquette M, El-Houjeiri L, Pause A. mTOR pathways in cancer and autophagy. Cancers (Basel). 2018;10(1):18.
61. Zhao B-Q, Wang S, Kim H-Y, Storrie H, Rosen BR, Mooney DJ, et al. Role of matrix metalloproteinases in delayed cortical responses after stroke. Nat Med. 2006;12(4):441–5.
62. Dong X, Song YN, Liu WG, Guo XL. Mmp-9, a potential target for cerebral ischemic treatment. Curr Neuropharmacol. 2009;7(4):269–75.
63. Yang Y, Estrada EY, Thompson JF, Liu W, Rosenberg GA. Matrix metalloproteinase-mediated disruption of tight junction proteins in cerebral vessels is reversed by synthetic matrix

metalloproteinase inhibitor in focal ischemia in rat. J Cereb Blood Flow Metab. 2007;27(4): 697–709.
64. Onwuha-Ekpete LC, Tokmina-Roszyk D, Fields GB. Selective inhibition of matrix metalloproteinase-9 reduces clinical severity in a murine model of Multiple Sclerosis via a reduced immune response. J Immunol. 2016;196(1_Supplement):139.12.
65. Halliwell B. Free radicals and antioxidants: updating a personal view. Nutr Rev. 2012;70(5): 257–65.
66. Ryu JY, Cerecedo-Lopez C, Yang H, Ryu I, Du R. Brain-targeted intranasal delivery of protein-based gene therapy for treatment of ischemic stroke. Theranostics. 2024;14(12):4773–86.
67. Yin H, Xue W, Chen S, Bogorad RL, Benedetti E, Grompe M, et al. Genome editing with Cas9 in adult mice corrects a disease mutation and phenotype. Nat Biotechnol. 2014;32(6):551–3.
68. Tastan B, Tarakcioglu E, Birinci Y, Park Y, Genc S. Role of exosomal microRNAs in cell-to-cell communication. Methods Mol Biol. 2022;2257:269–92. https://doi.org/10.1007/978-1-0716-1170-8_14. PMID: 34432284
69. Pan Q, Kuang X, Cai S, Wang X, Du D, Wang J, et al. miR-132-3p priming enhances the effects of mesenchymal stromal cell-derived exosomes on ameliorating brain ischemic injury. Stem Cell Res Ther. 2020;11(1):260.
70. Sun Y, Wan G, Bao X. Extracellular vesicles as a potential therapy for stroke. Int J Mol Sci. 2025;26(7):3130.
71. Lelarge V, Capelle R, Oger F, Mathieu T, Le Calvé B. Senolytics: from pharmacological inhibitors to immunotherapies, a promising future for patients' treatment. NPJ Aging. 2024;10 (1):12.
72. Xu M, Pirtskhalava T, Farr JN, Weigand BM, Palmer AK, Weivoda MM, et al. Senolytics improve physical function and increase lifespan in old age. Nat Med. 2018;24(8):1246–56.
73. Amin A, Perera ND, Tomas D, Cuic B, Radwan M, Hatters DM, et al. Systemic administration of a novel Beclin 1-derived peptide significantly upregulates autophagy in the spinal motor neurons of autophagy reporter mice. Int J Pharm. 2024;659:124198.
74. Guo Q, Xu S, Yang P, Wang P, Lu S, Sheng D, et al. A dual-ligand fusion peptide improves the brain-neuron targeting of nanocarriers in Alzheimer's disease mice. J Control Release. 2020;320:347–62.
75. Li C, Sun T, Jiang C. Recent advances in nanomedicines for the treatment of ischemic stroke. Acta Pharm Sin B. 2021;11(7):1767–88.
76. Seo D-o, Holtzman DM. Current understanding of the Alzheimer's disease-associated microbiome and therapeutic strategies. Exp Mol Med. 2024;56(1):86–94.
77. Jung YH, Chae CW, Han HJ. The potential role of gut microbiota-derived metabolites as regulators of metabolic syndrome-associated mitochondrial and endolysosomal dysfunction in Alzheimer's disease. Exp Mol Med. 2024;56(8):1691–702.
78. Sampson TR, Debelius JW, Thron T, Janssen S, Shastri GG, Ilhan ZE, et al. Gut microbiota regulate motor deficits and neuroinflammation in a model of Parkinson's disease. Cell. 2016;167(6):1469–80. e12.
79. Jiang Q, Liu J, Huang S, Wang XY, Chen X, Liu GH, Ye K, Song W, Masters CL, Wang J, Wang YJ. Antiageing strategy for neurodegenerative diseases: from mechanisms to clinical advances. Signal Transduct Target Ther. 2025;10(1):76.
80. Cruz Hernández JC, Bracko O, Kersbergen CJ, Muse V, Haft-Javaherian M, Berg M, et al. Neutrophil adhesion in brain capillaries reduces cortical blood flow and impairs memory function in Alzheimer's disease mouse models. Nat Neurosci. 2019;22(3):413–20.
81. Katt ME, Linville RM, Mayo LN, Xu ZS, Searson PC. Functional brain-specific microvessels from iPSC-derived human brain microvascular endothelial cells: the role of matrix composition on monolayer formation. Fluids Barriers CNS. 2018;15(1):7.
82. Mohanan P. Artificial intelligence and biological sciences. CRC Press/Taylor & Francis Group; 2025.

83. Vatine GD, Barrile R, Workman MJ, Sances S, Barriga BK, Rahnama M, et al. Human iPSC-derived blood-brain barrier chips enable disease modeling and personalized medicine applications. Cell Stem Cell. 2019;24(6):995–1005.e6.
84. Arevalo M-A, Azcoitia I, Garcia-Segura LM. The neuroprotective actions of oestradiol and oestrogen receptors. Nat Rev Neurosci. 2015;16(1):17–29.
85. Wong-Riley MT, Liang HL, Eells JT, Chance B, Henry MM, Buchmann E, et al. Photobiomodulation directly benefits primary neurons functionally inactivated by toxins: role of cytochrome c oxidase. J Biol Chem. 2005;280(6):4761–71.

Epigenetic Regulation of Neuroinflammatory and Autophagic Pathways

7

Yadu Nandan Dey, Siddhartha Das Pramanik, Partha Roy, Atreyee Ganguly, and Samir Kumar Samanta

Abstract

This chapter summarizes how autophagy is regulated by epigenetic mechanisms and how neuroinflammation plays a role in neurodegenerative diseases. This could be the secret to creating fresh approaches to the diagnosis and management of age-related brain disorders. According to recent studies, epigenetic instability can cause neuroinflammation by triggering pathways such as the microglial NLRP3 inflammasome. It can alter genes like ATG5 and BECN1 that promote autophagy. A vicious cycle that damages neurons and accelerates cognitive decline is produced by this effect. We review how chromatin structure, DNA alterations, and regulatory RNAs impact brain processes like microglial behavior, protein balance, neuron survival, and inflammation regulation. We also go over promising treatments that target these epigenetic changes, such as RNA-based therapies, CRISPR/dCas9-based gene editing, and modulators of histone deacetylases and histone acetyltransferases. We may be able to slow or even reverse neurodegeneration in ageing brains by comprehending and reprogramming the epigenetic connections between autophagy and neuroinflammation. This chapter provides a road map for creating multimodal interventions by connecting mechanistic insights with cutting-edge therapeutic approaches.

Y. N. Dey (✉) · A. Ganguly · S. K. Samanta
Pharmacology Division, Dr. B.C. Roy College of Pharmacy and Allied Health Sciences, Durgapur, West Bengal, India
e-mail: yadunandan.dey@bcrcp.org; atreyee.ganguly@bcrcp.org; samir.samanta@bcrcp.org

S. Das Pramanik · P. Roy
Molecular Endocrinology Laboratory, Department of Biosciences and Bioengineering, Indian Institute of Technology Roorkee, Roorkee, Uttarakhand, India
e-mail: siddhartha_p@bt.iit.ac.in; partha.roy@bt.iitr.ac.in

P. Khanal et al. (eds.), *Neuroinflammation and Autophagy in Neurodegeneration*,
https://doi.org/10.1007/978-981-95-3945-1_7

Keywords

Autophagy; DNA Methylation; Epigenetic Regulation; Histone Modification; Inflammasome; Neuroinflammation; Neurodegeneration; Non-coding RNA

7.1 Introduction

Many biological processes are regulated by DNA methylation through suppression of gene expression [1]. Further, methylation of cytosine–phosphate–guanine dinucleotides leads to the modification of various transcription factor binding and chromatin structure. DNA methylation is very crucial to maintain homeostasis, which is controlled by methyltransferases (DNMT1/3A/3B) and demethylases (TET1/2/3). Mutations in enzymes like DNMT1 can result in neurological or neurodegenerative disorders due to impaired methylation and incomplete gene silencing [2, 3]. One class of enzymes with therapeutic potential in anticancer applications is represented by HDAC inhibitors. These substances prevent the removal of acetyl groups from protein lysines by blocking the active sites of zinc-dependent class I, II, and IV HDAC. This action can induce cell cycle arrest, differentiation, and cell death in various preclinical cancer models. The Food and Drug Administration (FDA) has approved three HDAC inhibitors for use as anticancer medications thus far [4]. The sequence homology, domain organization, and functional characteristics, including substrate specificity and regulatory interactions, are used to classify histone acetyltransferases into several major families, including HAT1, GNAT (e.g., GCN5/PCAF), p300/CBP, and MYST [5]. They are important in ageing, neurological, and cardiovascular disorders because they take part in signal transduction and molecular processes like DNA repair, cancer cell transformation, and cell fate decisions [6]. Epigenetic disorders and chromatin structural instability can result from aberrant gene expression caused by an imbalance between the activities of histone acetyltransferases and deacetylases. Therefore, control over the activity of histone acetyltransferases and deacetylases is crucial for gene expression in signal transduction, cell division, and cell death [7]. HDAC inhibitors are thought to increase global chromatin acetylation [8], but it is still unknown how exactly they stop cell growth. Historically, chromatin relaxation and improved accessibility of transcription factors have been linked to increased histone tail acetylation [9]. Chromatin modifiers also target signaling pathways, such as oxidative stress, energy metabolism, and programmed cell death, that are linked to the pathophysiology of myocardial infarction. Sodium caprylate has been shown to increase H3K9ac at the HO1 and NQO1 gene promoters in rats by stimulating the histone acetyltransferases KAT2A, improving defense against oxidative stress, and inhibiting apoptosis [10]. Histidine decarboxylase may modulate chromatin accessibility through SWI/SNF complexes to regulate PRMT1 expression, according to emerging evidence, which connects histamine metabolism to epigenetic regulation in immune responses [11].

Next, microRNAs affect cell division, proliferation, and apoptosis by controlling post-transcriptional gene expression and protein synthesis [12]. Numerous microRNAs, including miR-21, miR-124, miR-146a, and miR-132, are important in neurological disorders and may be targets for treatment [13]. MiR-124 makes up 5–48% of all miRNA in the cerebral cortex, making it extremely common in the mammalian nervous system [14]. By encouraging cell differentiation and preventing proliferation in both mature neurons and during brain development, it controls nerve differentiation [15]. MiR-124 is highly expressed in microglia, controls neuronal differentiation, and has anti-inflammatory properties [16]. Additionally, MiR-21 is highly expressed in immune cells and plays a critical role in central nervous system (CNS) cell types such as neurons, astrocytes, and microglia [17, 18]. By targeting PDCD4 to downregulate nuclear factor kappa B (NF-κB) and induce IL-10, it functions as an anti-inflammatory regulator via toll-like receptor signaling [19].

Long non-coding RNAs, which make up a significant portion of the non-coding genome, primarily control gene expression by acting either cis- (on the same chromosome) or trans- (on different chromosomes) [20]. For example, by disrupting the equilibrium of the NLRP3/NLRP6 inflammasome, the retinas of ischemia/reperfusion mice with elevated levels of long non-coding RNA H19 control neuroinflammation [21]. In the ischemic cascade, H19 facilitates the production of PDCD4 by promoting inflammation and neuronal damage brought on by ischemia/reperfusion. On the other hand, retinal ganglion cell survival against ischemia is improved by blocking brain-derived neurotrophic factor antisense RNA [22]. MALAT1 expression is increased in cerebral ischemia/reperfusion damage, which triggers inflammatory responses in microglia via MyD88/IRAK1/TRAF6 signaling [23]. MicroRNAs, autophagy, monocyte activation, ion channels, T cell activation, and epigenetic factors are some of the many components that contribute to neuroinflammation and orchestrate intricate neuroimmune responses in both health and disease (see Fig. 7.1). Figure 7.1 illustrates the multifaceted contributors to neuroinflammation, including microRNAs, autophagy, monocyte activation, ion channels, T cell activation, and epigenetic factors, all of which orchestrate complex neuroimmune responses in health and disease. The various mechanisms contributing to neuroinflammation are depicted in Fig. 7.1.

7.2 Epigenetic Orchestration of Neuroinflammatory Signaling

By identifying host-derived danger signals and pathogen-associated molecular patterns, the NLRP3 inflammasome plays a crucial role in innate immunity and the advancement of disease [24]. When NLRP3 is activated, it sets off caspase-1, which encourages the release of tumor necrosis factor-alpha (TNF-α) and interleukin-1β (IL-1β). NLRP3 deficiency improves memory outcomes in Alzheimer's disease (AD) by increasing amyloid-β (Aβ) clearance and decreasing caspase-1 activity [25]. AD patients show higher levels of pro-inflammatory cytokines, aggravating neuroinflammation [26]. In a model induced by sevoflurane, autophagy deficiency and NLRP3 activation are facilitated by histone acetylation dysregulation.

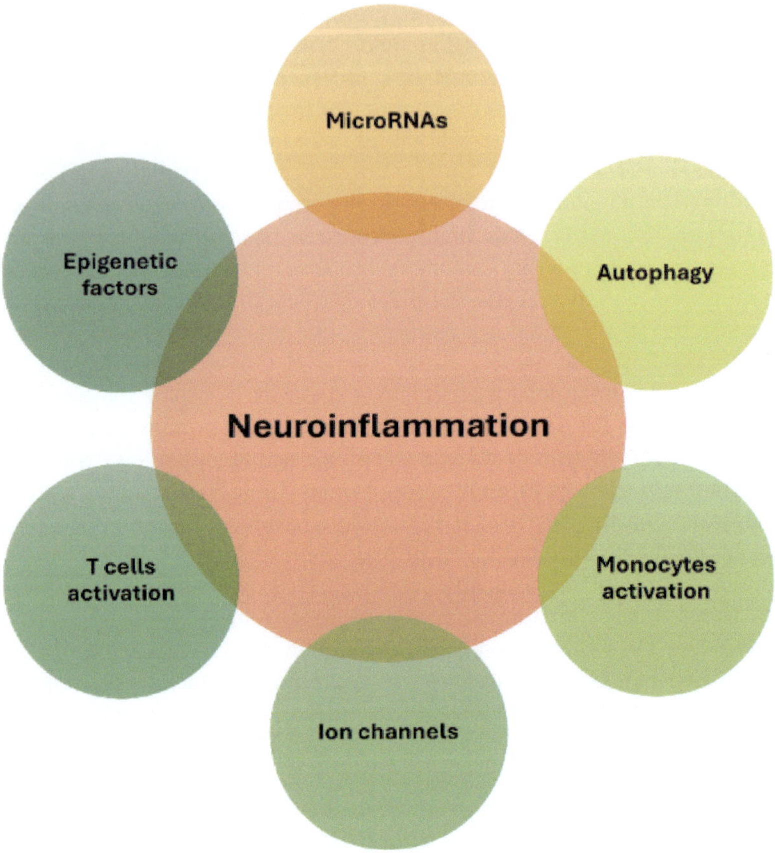

Fig. 7.1 Key regulators of neuroinflammation. Neuroinflammation is stimulated by multiple factors, including microRNAs, autophagy, monocyte and T cell activation, ion channels, and epigenetic factors

Suberoylanilide hydroxamic acid, an inhibitor of HDAC, enhanced cognition by restoring H3 and H4 acetylation, reactivating autophagy, and reducing inflammation [27]. NLRP3 is stimulated by Aβ-induced lysosomal rupture and release of cathepsin B, and cytosolic Aβ directly activates inflammasome activity by the production of reactive oxygen species during phagocytosis [28, 29]. Cytosolic Aβ directly activates inflammasome activity by producing reactive oxygen species during phagocytosis, whereas Aβ-induced lysosomal rupture and cathepsin B release stimulate NLRP3 [28, 29]. Parkinson's disease (PD) patients' substantia nigra dopaminergic neurodegeneration is influenced by NLRP3 [30]. By inhibiting NLRP3 and SNCA, miR-7 enhances microglial activity and safeguards neurons [31]. Likewise, miR-190 reduces neuronal damage and microglial activation by upregulating IL-10 and inhibiting inflammatory mediators [32]. In the pathophysiology of PD, NLRP3 also interacts with the JAK/STAT pathway [33]. Through caspase-1, NLRP3 inflammasome activation stimulates the release of IL-1β and IL-18 [34]. The SNHG1 acts as a competing endogenous RNA to miR-7, influencing NLRP3 expression. Decreased miR-7 expression is linked with α-synuclein accumulation and dopaminergic neuron death in PD [35]. Age-related decline in TET2 expression in astrocytes results in reduced DNA demethylation contributes to cognitive deficits. TET2, as a part of the Tet family (TET1, TET2, TET3) is essential for astrocytic lineage differentiation. Overexpression promotes astrocyte development, while knockdown impairs it [36, 37].

Targeting HDAC Class I/II has shown anti-inflammatory potential. siRNA silencing of HDAC1/2 inhibits lipopolysaccharide-induced TNF-α and IL-6 in microglia [38]. HDAC3 inhibition reverses inflammatory morphology and improves outcomes post-brain injury [39]. However, the exact role in regulating inflammation-related gene expression remains unclear [40]. Histone mark H3K4me3 at NF-κB binding sites modulates gene expression in chronic neuroinflammation. NF-κB drives EZH2 transcription, enhancing H3K27me3 methylation, which is a key change observed in glioblastoma [41, 42]. Non-coding RNAs such as miR-155 play a dual role in inflammation. miR-155 promotes TNF-α, IL-6, and iNOS by targeting anti-inflammatory SOCS1 [43], whereas feedback IL-10 upregulation limits excessive responses [44]. It also inhibits M2 markers like CEBPβ and SMAD2, enhancing microglial phagocytosis and inflammatory response [45, 46]. The Neat1 modulates inflammasomes by binding to NLRP3, AIM2, and NLRC4, promoting caspase-1 processing and pyroptosis. In response to stimuli like hypoxia or reactive oxygen species, Neat1 translocates from paraspeckles to cytoplasm in regulating inflammasome activity [47–50].

7.3 Epigenetic Suppression of Autophagy in Aging

In AD, poor neurogenesis and potential neurodegeneration are linked to hypermethylation of the hippocampus ATG5 promoter. Reduced autophagy is linked to this epigenetic alteration in microglia. The intricate relationships between DNA methylation and hydroxymethylation, as well as additional epigenetic changes, coordinate the dynamic regulation of gene expression in AD. These affect key mechanisms such as aberrant protein aggregation, neuroinflammation, and synaptic plasticity [51, 52]. To be specific, transcriptional silence and decreased expression of genes linked to brain function and memory formation result from hypermethylation of particular genomic areas, such as gene promoters and enhancers. Next, transposable element expression can be unlocked by hypomethylation at certain loci, resulting in dysregulated gene networks and genomic instability [53]. In primary neurons exposed to sevoflurane, HDAC inhibitor suberoylanilide hydroxamic acid decreased p62 expression and raised the ratio of LC3-II/I and H3 and H4 histone acetylation. In addition to improving cognitive impairment, HDAC inhibitors also increased histone acetylation, alleviated autophagy deficits, and inhibited the activation of NLRP3 inflammasome in hippocampal regions of elderly mice stimulated with sevoflurane [54].

HDAC6 deacetylates α-tubulin, affecting axonal transport and protein aggregate clearance. It increases tubulin acetylation, enhances motor protein recruitment, and improves transport of neurotrophic factors like brain-derived neurotrophic factor in Huntington's disease [55]. However, HDAC6 inhibition has shown mixed results in animal models [56]. Sirtuin 1 regulates autophagy by deacetylating LC3, promoting its nuclear export and interaction with autophagic machinery during nutrient stress [57]. In Amyotrophic Lateral Sclerosis, miR-34a impairs autophagy by suppressing transcription factor EB. H19 enhances mTOR signaling in diabetic cardiomyopathy by inhibiting autophagy via EZH2-mediated DIRAS family GTPase 3 silencing [58]. H19 is also upregulated in atrial fibrillation, while miR-29a/b-3p is downregulated, which promotes the production of extracellular matrix proteins and the proliferation of fibroblasts through the VEGFA/TGF-β axis. Therapy for atrial fibrillation may be possible by focusing on the H19/miR-29a/b-3p pathway.

7.4 Epigenetic Coupling of Neuroinflammation and Autophagy

The lysosomal pathway breaks down cellular components through a conserved cellular mechanism known as autophagy [59]. This process can be triggered by a variety of cellular cues, including infections, hypoxia, and nutritional deficiencies [60]. There is growing evidence that autophagy is associated with many brain-related diseases [61]. Neurons increase autophagy due to low neuronal activity and loss of neurotrophic factors, which are indirectly brought on by inhibiting mTOR signaling in response to amino acid deficiency [62, 63]. Mammalian cells display three major and distinct forms of autophagy: microautophagy, macroautophagy, and chaperone-mediated autophagy [64]. mTOR, a serine/threonine kinase involved in the regulation of cellular metabolism, is one of the main goals of which is inhibiting autophagy [65]. By stimulating phagophore development, Unc-51-like autophagy activating kinase 1 starts the formation of autophagosomes. Atg9 trafficking and activation of the VPS34 complex facilitate this process [66]. Furthermore, mTOR is a protein kinase that is essential to two different signaling complexes that control translation and synaptic plasticity. It is also involved in long-term memory and brain function, and its dysregulation has been connected to a number of neurological conditions [67]. PGE2 and miR-21 control the polarization of macrophages between pro-inflammatory M1 and anti-inflammatory M2 states. By blocking M2 gene expression by targeting STAT3, miR-21 reveals important molecular mechanisms that may direct novel anti-inflammatory treatments [68]. Neuropathic pain induces long-lasting changes in dorsal root ganglia neurons, potentially mediated by time-dependent upregulation of DNA methyltransferases, suggesting epigenetic regulation as a key factor in maintaining altered protein expression and pain sensitivity. By therapeutically reversing AD-associated neuropathic pain-related microglial M1-like pro-inflammatory polarization to an M2-like state in an AD mouse model, microglia-specific DNMT1 knockdown can induce minute pain and behavioral abnormalities [69, 70]. Both the expression of iNOS and pro-inflammatory cytokine release (IL-1β,

TNF-α, and IL-6) are affected by PI3K/Akt activation. Lipopolysaccharide-induced neuroinflammation and inflammatory markers such as TNF-α and IL-6 are reduced upon blocking the PI3K/Akt pathway [71, 72]. LRP1 stimulation by the apoE-mimic peptide COG1410 induces M2 microglial polarization through the Shc1/PI3K/Akt pathway to decrease white matter damage and enhance neurological outcomes in subarachnoid hemorrhage, and they are potential targets for novel pain treatments, although microglia also mediate neuropathic pain through signaling pathways such as JAK/STAT, MAPKs, PI3K/Akt, and NF-κB [73, 74].

7.5 Advanced Epigenetic Therapeutics from Mechanisms to Multimodal Strategies

The main instruments for genome and epigenome editing at first were zinc-finger and transcription activator-like effector-based systems. More effective and readily programmable targeted DNA (de)methylation was subsequently made possible by the development of CRISPR/dCas9. One of the main concerns with epigenome editing is still off-target effects. Notably, it has been demonstrated that neuronal abnormalities in fragile X syndrome can be restored by Cas9-mediated DNA methylation of FMR1 [75]. DNA methylation editing has potential for treating hereditary diseases and provides an accurate method of controlling gene expression. HDAC6 inhibition normalizes T and B cell development in systemic lupus erythematosus while lowering glomerulonephritis, anti-dsDNA levels, and inflammatory cytokines. Likewise, following traumatic brain injury, microRNAs play a role in secondary damage. Notably, miR-155 is a potential therapeutic target because it causes persistent neuroinflammation, and when it is suppressed, post-traumatic damage is lessened and recovery is enhanced [76].

It is also necessary to consider the delivery of genome editing through the use of CRISPR/Cas9. Typically, viral and nonviral vectors are the two primary types of CRISPR/Cas9 delivery systems [77, 78]. Unfortunately, the use of viral vectors is limited by their intrinsic properties, which include high cost, limited packing capacity, and safety. Nonviral vectors are primarily based on physical methods and synthetic chemical and natural materials, such as liposomes, electroporation, microinjection, nanoparticles, and gold nanoparticles [79]. Therapy and mechanistic gene research may benefit from CRISPR/Cas9's capacity to modify the characteristics of cells and organs [80].

For example, CRISPR/Cas9 in vivo editing is used in human clinical trials to treat genetic illnesses, including Leber congenital amaurosis (NCT03872479) and transthyretin amyloidosis (NCT04601051), which included delivery vectors, adeno-associated viruses, and proprietary lipid nanoparticles [81, 82]. The CRISPR/Cas9 system should be precisely and safely delivered to cells and organs for gene editing. In human clinical trials, extracellular vesicles with therapeutic payloads are presently being assessed for the treatment of illness. For example, mesenchymal stem cell (MSC)-derived extracellular vesicles loaded with miRNA targeting mutant KRAS have been studied for the treatment of pancreatic cancer

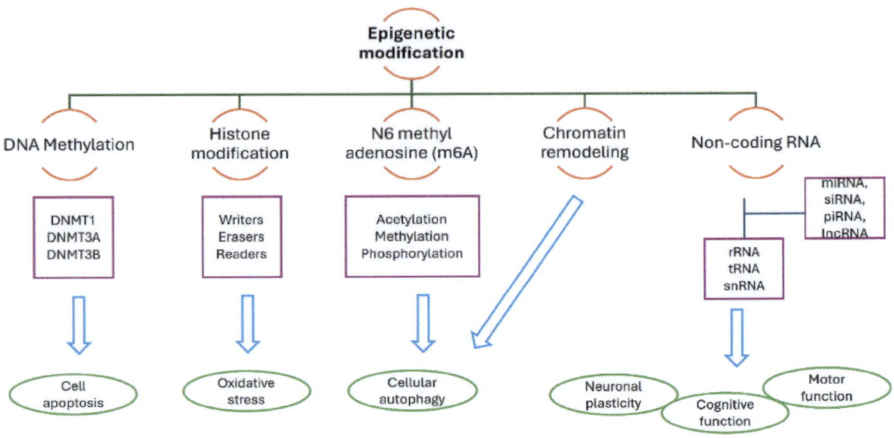

Fig. 7.2 Epigenetic modulation on neuronal function and neurodegeneration. Epigenetic regulators modulate neuronal apoptosis, stress, autophagy, plasticity, cognition, and motor function via DNA, histone, and RNA modifications

(NCT03608631), and MSC-derived extracellular vesicles transfected with miR-124 are being studied as a treatment for ischemic stroke (NCT03384433) [83]. As of right now, no clinical trials are using extracellular vesicles to deliver CRISPR/Cas9. However, extracellular vesicles are becoming more and more acknowledged as promising carriers—or "trucks with GPS"—for the targeted delivery of CRISPR/Cas9, providing benefits over viral and other delivery vectors because of their superior biocompatibility, loading capacity, and intrinsic physiological properties [84]. The epigenetic regulation of neurodegeneration and neuronal functions is depicted in Fig. 7.2.

7.6 Conclusion

Epigenetic modifications have a considerable impact on gene regulation and neurodevelopment. Deregulation of autophagic processes is responsible for the onset of neuroinflammation and neurodegenerative diseases. Microglial NLRP3 hypomethylation and neuronal ATG5 hypermethylation are just two instances of epigenetic modification that illustrate the intricate relationship between autophagy and neuroinflammation. While promising directions for exact gene editing are offered by cutting-edge technologies such as CRISPR/dCas9 and HDAC inhibitors, miRNA therapeutics are hindered by the limitations of delivery and off-targeting. The synergistic use of IL-1β blockers with HDAC6 inhibitors could be a more efficient therapeutic modality to alter neuroinflammatory processes.

References

1. Li Q, Zhu Q. The role of demethylase AlkB homologs in cancer. Front Oncol. 2023;13: 1153463.
2. Smith ZD, Meissner A. DNA methylation: roles in mammalian development. Nat Rev Genet. 2013;14(3):204–20.
3. Lio CWJ, Yue X, Lopez-Moyado IF, Tahiliani M, Aravind L, Rao A. TET methylcytosine oxidases: new insights from a decade of research. J Biosci. 2020;45:21.
4. West AC, Johnstone RW. New and emerging HDAC inhibitors for cancer treatment. J Clin Invest. 2014;124(1):30–9.
5. Marmorstein R, Zhou MM. Writers and readers of histone acetylation: structure, mechanism, and inhibition. Cold Spring Harb Perspect Biol. 2014;6(7):a018762.
6. Capone V, Della Torre L, Carannante D, Babaei M, Altucci L, Benedetti R, Carafa V. HAT1: landscape of biological function and role in cancer. Cells. 2023;12(7):1075.
7. Vahid F, Zand H, Nosrat–Mirshekarlou E, Najafi R, Hekmatdoost A. The role dietary of bioactive compounds on the regulation of histone acetylases and deacetylases: a review. Gene. 2015;562(1):8–15.
8. Wang Z, Zang C, Cui K, Schones DE, Barski A, Peng W, Zhao K. Genome-wide mapping of HATs and HDACs reveals distinct functions in active and inactive genes. Cell. 2009;138(5): 1019–31.
9. Falkenberg KJ, Johnstone RW. Histone deacetylases and their inhibitors in cancer, neurological diseases and immune disorders. Nat Rev Drug Discov. 2014;13(9):673–91.
10. Lei I, Tian S, Gao W, Liu L, Guo Y, Tang P, Chen E, Wang Z. Acetyl-CoA production by specific metabolites promotes cardiac repair after myocardial infarction via histone acetylation. elife. 2021;10:e60311.
11. Zhang Z, Ding S, Wang Z, Zhu X, Zhou Z, Zhang W, Yang X, Ge J. Prmt1 upregulated by Hdc deficiency aggravates acute myocardial infarction via NETosis. Acta Pharm Sin B. 2022;12(4): 1840–55.
12. Yao G, Li J, Wang J, Liu S, Li X, Cao X, Chen H, Xu Y. Improved resting-state functional dynamics in post-stroke depressive patients after Shugan Jieyu capsule treatment. Front Neurosci. 2020;14:297.
13. Devaux Y, Dankiewicz J, Salgado-Somoza A, Stammet P, Collignon O, Gilje P, Gidlöf O, Zhang L, Vausort M, Hassager C, Wise MP, Kuiper M, Friberg H, Cronberg T, Erlinge D, Nielsen N. Association of Circulating MicroRNA-124-3p levels with outcomes after out-of-hospital cardiac arrest. JAMA Cardiol. 2016;1(3):305.
14. Treiber T, Treiber N, Meister G. Regulation of microRNA biogenesis and function. Thromb Haemost. 2012;107(04):605–10.
15. Makeyev EV, Zhang J, Carrasco MA, Maniatis T. The MicroRNA miR-124 promotes neuronal differentiation by triggering brain-specific alternative pre-mRNA splicing. Mol Cell. 2007;27 (3):435–48.
16. Ponomarev ED, Veremeyko T, Barteneva N, Krichevsky AM, Weiner HL. MicroRNA-124 promotes microglia quiescence and suppresses EAE by deactivating macrophages via the C/EBP-α–PU.1 pathway. Nat Med. 2011;17(1):64–70.
17. Zhang L, Dong L, Li Y, Hong Z, Wei W. miR-21 represses FasL in microglia and protects against microglia-mediated neuronal cell death following hypoxia/ischemia. Glia. 2012;60(12): 1888–95.
18. Sheedy FJ. Turning 21: induction of miR-21 as a key switch in the inflammatory response. Front Immunol. 2015;6:19.
19. Sheedy FJ, Palsson-McDermott E, Hennessy EJ, Martin C, O'Leary JJ, Ruan Q, Johnson DS, Chen Y, O'Neill LAJ. Negative regulation of TLR4 via targeting of the proinflammatory tumor suppressor PDCD4 by the microRNA miR-21. Nat Immunol. 2010;11(2):141–7.
20. Djebali S, Davis CA, Merkel A, Dobin A, Lassmann T, Mortazavi A, Tanzer A, Lagarde J, Lin W, Schlesinger F, Xue C, Marinov GK, Khatun J, Williams BA, Zaleski C, Rozowsky J,

Röder M, Kokocinski F, Abdelhamid RF, Alioto T, Antoshechkin I, Baer MT, Bar NS, Batut P, Bell K, Bell I, Chakrabortty S, Chen X, Chrast J, Curado J, Derrien T, Drenkow J, Dumais E, Dumais J, Duttagupta R, Falconnet E, Fastuca M, Fejes-Toth K, Ferreira P, Foissac S, Fullwood MJ, Gao H, Gonzalez D, Gordon A, Gunawardena H, Howald C, Jha S, Johnson R, Kapranov P, King B, Kingswood C, Luo OJ, Park E, Persaud K, Preall JB, Ribeca P, Risk B, Robyr D, Sammeth M, Schaffer L, See LH, Shahab A, Skancke J, Suzuki AM, Takahashi H, Tilgner H, Trout D, Walters N, Wang H, Wrobel J, Yu Y, Ruan X, Hayashizaki Y, Harrow J, Gerstein M, Hubbard T, Reymond A, Antonarakis SE, Hannon G, Giddings MC, Ruan Y, Wold B, Carninci P, Guigó R, Gingeras TR. Landscape of transcription in human cells. Nature. 2012;489(7414):101–8.

21. Wan P, Su W, Zhang Y, Li Z, Deng C, Li J, Jiang N, Huang S, Long E, Zhuo Y. LncRNA H19 initiates microglial pyroptosis and neuronal death in retinal ischemia/reperfusion injury. Cell Death Differ. 2020;27(1):176–91.
22. Xu L, Zhang Z, Xie T, Zhang X, Dai T. Inhibition of BDNF-AS provides neuroprotection for retinal ganglion cells against ischemic injury. PLoS One. 2016;11(12):e0164941.
23. Wang LQ, Zhou HJ. LncRNA MALAT1 promotes high glucose-induced inflammatory response of microglial cells via provoking MyD88/IRAK1/TRAF6 signaling. Sci Rep [Internet]. 2018;8(1):8346. Available from: https://www.nature.com/articles/s41598-018-26421-5
24. Tan X, Sun L, Chen J, Chen ZJ. Detection of microbial infections through innate immune sensing of nucleic acids. Ann Rev Microbiol. 2018;72(1):447–78.
25. Heneka MT, Kummer MP, Stutz A, Delekate A, Schwartz S, Vieira-Saecker A, Griep A, Axt D, Remus A, Tzeng TC, Gelpi E, Halle A, Korte M, Latz E, Golenbock DT. NLRP3 is activated in Alzheimer's disease and contributes to pathology in APP/PS1 mice. Nature. 2013;493(7434): 674–8.
26. Azizi G, Navabi SS, Al-Shukaili A, Seyedzadeh MH, Yazdani R, Mirshafiey A. The role of inflammatory mediators in the pathogenesis of Alzheimer's disease. Sultan Qaboos Univ Med J. 2015;15(3):e305–16.
27. Fang P, Chen C, Zheng F, Jia J, Chen T, Zhu J, Chang J, Zhang Z. NLRP3 inflammasome inhibition by histone acetylation ameliorates sevoflurane-induced cognitive impairment in aged mice by activating the autophagy pathway. Brain Res Bull. 2021;172:79–88.
28. Halle A, Hornung V, Petzold GC, Stewart CR, Monks BG, Reinheckel T, Fitzgerald KA, Latz E, Moore KJ, Golenbock DT. The NALP3 inflammasome is involved in the innate immune response to amyloid-β. Nat Immunol. 2008;9(8):857–65.
29. Daily KP, Badr A, Eltobgy M, Estfanous S, Whitham O, Tan MH, Carafice C, Krause K, McNamara A, Hamilton K, Houle S, Gupta S, Gupta GA, Madhu S, Fitzgerald J, Saadey AA, Laster B, Yan P, Webb A, Zhang X, Pietrzak M, Kokiko-Cochran ON, Ghoneim HE, Amer AO. DNA hypomethylation promotes the expression of CASPASE-4 which exacerbates inflammation and amyloid-β deposition in Alzheimer's disease. Alzheimer's Res Ther. 2024;16(1):29.
30. Sarkar S, Malovic E, Harishchandra DS, Ghaisas S, Panicker N, Charli A, Palanisamy BN, Rokad D, Jin H, Anantharam V, Kanthasamy A, Kanthasamy AG. Mitochondrial impairment in microglia amplifies NLRP3 inflammasome proinflammatory signaling in cell culture and animal models of Parkinson's disease. NPJ Parkinsons Dis. 2017;3(1):30.
31. Junn E, Lee KW, Jeong BS, Chan TW, Im JY, Mouradian MM. Repression of α-synuclein expression and toxicity by microRNA-7. Proc Natl Acad Sci. 2009;106(31):13052–7.
32. Sun Q, Wang S, Chen J, Cai H, Huang W, Zhang Y, Wang L, Xing Y. MicroRNA-190 alleviates neuronal damage and inhibits neuroinflammation via Nlrp3 in MPTP-induced Parkinson's disease mouse model. J Cell Physiol. 2019;234(12):23379–87.
33. Mao Z, Liu C, Ji S, Yang Q, Ye H, Han H, Xue Z. The NLRP3 Inflammasome is involved in the pathogenesis of Parkinson's disease in rats. Neurochem Res. 2017;42(4):1104–15.
34. Broz P, Dixit VM. Inflammasomes: mechanism of assembly, regulation and signalling. Nat Rev Immunol. 2016;16(7):407–20.

35. Titze-de-Almeida R, Titze-de-Almeida SS. miR-7 replacement therapy in Parkinson's disease. Curr Gene Ther. 2018;18(3):143–53.
36. Kohli RM, Zhang Y. TET enzymes, TDG and the dynamics of DNA demethylation. Nature. 2013;502(7472):472–9.
37. He F, Wu H, Zhou L, Lin Q, Cheng Y, Sun YE. Tet2-mediated epigenetic drive for astrocyte differentiation from embryonic neural stem cells. Cell Death Discov. 2020;6(1):30.
38. Durham BS, Grigg R, Wood IC. Inhibition of histone deacetylase 1 or 2 reduces induced cytokine expression in microglia through a protein synthesis independent mechanism. J Neurochem. 2017;143(2):214–24.
39. Zhao Y, Mu H, Huang Y, Li S, Wang Y, Stetler RA, Bennett MVL, Dixon CE, Chen J, Shi Y. Microglia-specific deletion of histone deacetylase 3 promotes inflammation resolution, white matter integrity, and functional recovery in a mouse model of traumatic brain injury. J Neuroinflammation. 2022;19(1):201.
40. Wahane S, Zhou X, Zhou X, Guo L, Friedl MS, Kluge M, Ramakrishnan A, Shen L, Friedel CC, Zhang B, Friedel RH, Zou H. Diversified transcriptional responses of myeloid and glial cells in spinal cord injury shaped by HDAC3 activity. Sci Adv. 2021;7(9):eabd8811.
41. Husmann D, Gozani O. Histone lysine methyltransferases in biology and disease. Nat Struct Mol Biol. 2019;26(10):880–9.
42. Lin K, Gao W, Chen N, Yang S, Wang H, Wang R, Xie F, Meng J, Lam EWF, Li S, Cheng W, Chen P, Wu H, Yan J, Jin D, Jin B. Chronic inflammation pathway NF-κB cooperates with epigenetic reprogramming to drive the malignant progression of glioblastoma. Int J Biol Sci. 2022;18(15):5770–86.
43. Cardoso AL, Guedes JR, Pereira de Almeida L, Pedroso de Lima MC. miR-155 modulates microglia-mediated immune response by down-regulating SOCS-1 and promoting cytokine and nitric oxide production. Immunology. 2012;135(1):73–88.
44. Benveniste EN, Qin H. Type I interferons as anti-inflammatory mediators. Sci STKE. 2007;2007(416):pe70.
45. Ruffell D, Mourkioti F, Gambardella A, Kirstetter P, Lopez RG, Rosenthal N, Nerlov C. A CREB-C/EBPβ cascade induces M2 macrophage-specific gene expression and promotes muscle injury repair. Proc Natl Acad Sci. 2009;106(41):17475–80.
46. Louafi F, Martinez-Nunez RT, Sanchez-Elsner T. MicroRNA-155 targets SMAD2 and modulates the response of macrophages to transforming growth factor-β. J Biol Chem. 2010;285(53):41328–36.
47. Sasaki YTF, Ideue T, Sano M, Mituyama T, Hirose T. MENε/β noncoding RNAs are essential for structural integrity of nuclear paraspeckles. Proc Natl Acad Sci. 2009;106(8):2525–30.
48. Sunwoo H, Dinger ME, Wilusz JE, Amaral PP, Mattick JS, Spector DL. *MEN ε/β* nuclear-retained non-coding RNAs are up-regulated upon muscle differentiation and are essential components of paraspeckles. Genome Res. 2009;19(3):347–59.
49. Zhang F, Wu L, Qian J, Qu B, Xia S, La T, Wu Y, Ma J, Zeng J, Guo Q, Cui Y, Yang W, Huang J, Zhu W, Yao Y, Shen N, Tang Y. Identification of the long noncoding RNA NEAT1 as a novel inflammatory regulator acting through MAPK pathway in human lupus. J Autoimmun. 2016;75:96–104.
50. Martinon F. Signaling by ROS drives inflammasome activation. Eur J Immunol. 2010;40(3): 616–9.
51. Chen Z, Wu M, Lai Q, Zhou W, Wen X, Yin X. Epigenetic regulation of synaptic disorder in Alzheimer's disease. Front Neurosci. 2022;16:888014.
52. Kaur G, Rathod SSS, Ghoneim MM, Alshehri S, Ahmad J, Mishra A, Alhakamy NA. DNA methylation: a promising approach in Management of Alzheimer's disease and other neurodegenerative disorders. Biology (Basel). 2022;11(1):90.
53. Cui D, Xu X. DNA Methyltransferases, DNA methylation, and age-associated cognitive function. Int J Mol Sci. 2018;19(5):1315.
54. Fang Y, Zou L, He W. miR-30a-5p mitigates autophagy by regulating the Beclin-1/ATG16 pathway in renal ischemia/reperfusion injury. Int J Mol Med. 2021;48(1):144.

55. Dompierre JP, Godin JD, Charrin BC, Cordelières FP, King SJ, Humbert S, Saudou F. Histone deacetylase 6 inhibition compensates for the transport deficit in Huntington's disease by increasing tubulin acetylation. J Neurosci. 2007;27(13):3571–83.
56. Bobrowska A, Paganetti P, Matthias P, Bates GP. Hdac6 Knock-out increases tubulin acetylation but does not modify disease progression in the R6/2 mouse model of Huntington's disease. PLoS One. 2011;6(6):e20696.
57. Huang R, Xu Y, Wan W, Shou X, Qian J, You Z, Liu B, Chang C, Zhou T, Lippincott-Schwartz J, Liu W. Deacetylation of nuclear LC3 drives autophagy initiation under starvation. Mol Cell. 2015;57(3):456–66.
58. Zhuo C, Jiang R, Lin X, Shao M. LncRNA H19 inhibits autophagy by epigenetically silencing of DIRAS3 in diabetic cardiomyopathy. Oncotarget. 2017;8(1):1429–37.
59. Aman Y, Schmauck-Medina T, Hansen M, Morimoto RI, Simon AK, Bjedov I, Palikaras K, Simonsen A, Johansen T, Tavernarakis N, Rubinsztein DC, Partridge L, Kroemer G, Labbadia J, Fang EF. Autophagy in healthy aging and disease. Nat Aging. 2021;1(8):634–50.
60. Klionsky DJ, Petroni G, Amaravadi RK, Baehrecke EH, Ballabio A, Boya P, Bravo-San Pedro JM, Cadwell K, Cecconi F, Choi AMK, Choi ME, Chu CT, Codogno P, Colombo MI, Cuervo AM, Deretic V, Dikic I, Elazar Z, Eskelinen E, Fimia GM, Gewirtz DA, Green DR, Hansen M, Jäättelä M, Johansen T, Juhász G, Karantza V, Kraft C, Kroemer G, Ktistakis NT, Kumar S, Lopez-Otin C, Macleod KF, Madeo F, Martinez J, Meléndez A, Mizushima N, Münz C, Penninger JM, Perera RM, Piacentini M, Reggiori F, Rubinsztein DC, Ryan KM, Sadoshima J, Santambrogio L, Scorrano L, Simon H, Simon AK, Simonsen A, Stolz A, Tavernarakis N, Tooze SA, Yoshimori T, Yuan J, Yue Z, Zhong Q, Galluzzi L, Pietrocola F. Autophagy in major human diseases. EMBO J 2021 40(19) e108863.
61. Hwang JY, Yan J, Zukin RS. Autophagy and synaptic plasticity: epigenetic regulation. Curr Opin Neurobiol. 2019;59:207–12.
62. Dong S, Aguirre-Hernandez C, Scrivo A, Eliscovich C, Arias E, Bravo-Cordero JJ, Cuervo AM. Monitoring spatiotemporal changes in chaperone-mediated autophagy in vivo. Nat Commun. 2020;11(1):645.
63. Tang G, Gudsnuk K, Kuo SH, Cotrina ML, Rosoklija G, Sosunov A, Sonders MS, Kanter E, Castagna C, Yamamoto A, Yue Z, Arancio O, Peterson BS, Champagne F, Dwork AJ, Goldman J, Sulzer D. Loss of mTOR-dependent macroautophagy causes autistic-like synaptic pruning deficits. Neuron. 2014;83(5):1131–43.
64. Morita M, Gravel SP, Hulea L, Larsson O, Pollak M, St-Pierre J, Topisirovic I. mTOR coordinates protein synthesis, mitochondrial activity and proliferation. Cell Cycle. 2015;14 (4):473–80.
65. Laplante M, Sabatini DM. mTOR signaling in growth control and disease. Cell. 2012;149(2): 274–93.
66. Zachari M, Ganley IG. The mammalian ULK1 complex and autophagy initiation. Essays Biochem. 2017;61(6):585–96.
67. Hoeffer CA, Klann E. mTOR signaling: at the crossroads of plasticity, memory and disease. Trends Neurosci. 2010;33(2):67–75.
68. Wang Z, Brandt S, Medeiros A, Wang S, Wu H, Dent A, Serezani CH. MicroRNA 21 is a homeostatic regulator of macrophage polarization and prevents prostaglandin E2-mediated M2 generation. PLoS One. 2015;10(2):e0115855.
69. Pollema-Mays SL, Centeno MV, Apkarian AV, Martina M. Expression of DNA methyltransferases in adult dorsal root ganglia is cell-type specific and up regulated in a rodent model of neuropathic pain. Front Cell Neurosci. 2014;8:217.
70. Jin J, Guo J, Cai H, Zhao C, Wang H, Liu Z, Ge ZM. M2-like microglia polarization attenuates neuropathic pain associated with Alzheimer's disease. J Alzheimer's Dis. 2020;76(4):1255–65.
71. Cianciulli A, Calvello R, Porro C, Trotta T, Salvatore R, Panaro MA. PI3k/Akt signalling pathway plays a crucial role in the anti-inflammatory effects of curcumin in LPS-activated microglia. Int Immunopharmacol. 2016;36:282–90.

72. Gao Y, Cui M, Zhong S, Feng C, Nwobodo AK, Chen B, Song Y, Wang Y. Dihydroartemisinin ameliorates LPS-induced neuroinflammation by inhibiting the PI3K/AKT pathway. Metab Brain Dis. 2020;35(4):661–72.
73. Popiolek-Barczyk K, Mika J. Targeting the microglial signaling pathways: new insights in the modulation of neuropathic pain. Curr Med Chem. 2016;23(26):2908–28.
74. Peng J, Pang J, Huang L, Enkhjargal B, Zhang T, Mo J, Wu P, Xu W, Zuo Y, Peng J, Zuo G, Chen L, Tang J, Zhang JH, Jiang Y. LRP1 activation attenuates white matter injury by modulating microglial polarization through Shc1/PI3K/Akt pathway after subarachnoid hemorrhage in rats. Redox Biol. 2019;21:101121.
75. Liu XS, Wu H, Krzisch M, Wu X, Graef J, Muffat J, Hnisz D, Li CH, Yuan B, Xu C, Li Y, Vershkov D, Cacace A, Young RA, Jaenisch R. Rescue of Fragile X Syndrome Neurons by DNA methylation editing of the FMR1 gene. Cell. 2018;172(5):979–992.e6.
76. Henry RJ, Doran SJ, Barrett JP, Meadows VE, Sabirzhanov B, Stoica BA, Loane DJ, Faden AI. Inhibition of miR-155 limits Neuroinflammation and improves functional recovery after experimental traumatic brain injury in mice. Neurotherapeutics. 2019;16(1):216–30.
77. Yip B. Recent advances in CRISPR/Cas9 delivery strategies. Biomolecules. 2020;10(6):839.
78. Yin H, Kauffman KJ, Anderson DG. Delivery technologies for genome editing. Nat Rev Drug Discov. 2017;16(6):387–99.
79. Li L, Hu S, Chen X. Non-viral delivery systems for CRISPR/Cas9-based genome editing: challenges and opportunities. Biomaterials. 2018;171:207–18.
80. Hille F, Richter H, Wong SP, Bratovič M, Ressel S, Charpentier E. The biology of CRISPR-Cas: backward and forward. Cell. 2018;172(6):1239–59.
81. Gillmore JD, Gane E, Taubel J, Kao J, Fontana M, Maitland ML, Seitzer J, O'Connell D, Walsh KR, Wood K, Phillips J, Xu Y, Amaral A, Boyd AP, Cehelsky JE, McKee MD, Schiermeier A, Harari O, Murphy A, Kyratsous CA, Zambrowicz B, Soltys R, Gutstein DE, Leonard J, Sepp-Lorenzino L, Lebwohl D. CRISPR-Cas9 in vivo gene editing for transthyretin amyloidosis. N Engl J Med. 2021;385(6):493–502.
82. Pickar-Oliver A, Gersbach CA. The next generation of CRISPR–Cas technologies and applications. Nat Rev Mol Cell Biol. 2019;20(8):490–507.
83. O'Brien K, Breyne K, Ughetto S, Laurent LC, Breakefield XO. RNA delivery by extracellular vesicles in mammalian cells and its applications. Nat Rev Mol Cell Biol. 2020;21(10):585–606.
84. Liang Y, Iqbal Z, Wang J, Xu L, Xu X, Ouyang K, Zhang H, Lu J, Duan L, Xia J. Cell-derived extracellular vesicles for CRISPR/Cas9 delivery: engineering strategies for cargo packaging and loading. Biomater Sci. 2022;10(15):4095–106.

Part II

Neurodegenerative Diseases and Mechanistic Insights

Autophagy and Neuroinflammatory Pathways in Alzheimer's Disease

8

Rohan Gupta, Jogendra Singh Pawar, Lohitha Gujjari, and Smita Kumari

Abstract

Alzheimer's disease (AD) is a neurodegenerative disorder characterized by proteostatic collapse and chronic neuroinflammation, yet the mechanistic crosstalk between autophagy failure and inflammatory signaling remains underexplored in therapeutic frameworks. This chapter addresses this gap by dissecting how amyloid-β plaques and tau tangles disrupt autophagic flux, following NLRP3 inflammasome activation and cytokine-driven inflammation, driving neuronal vulnerability. By elucidating bidirectional pathways such as caspase-1-mediated cleavage of autophagy proteins and autophagy-dependent regulation of inflammasomes, this chapter underscores the necessity of moving beyond singular therapeutic targets toward multimodal strategies. Here, we emphasized molecular mechanisms by integrating preclinical and clinical

R. Gupta
Department of Biotechnology & Bioengineering, School of Biosciences & Technology, Galgotias University, Greater Noida, Uttar Pradesh, India
e-mail: rohan.gupta@galgotiasuniversity.edu.in

J. S. Pawar
The James Cancer Hospital and Solove Research Institute, Comprehensive Cancer Center, The Ohio State University, Columbus, OH, USA
e-mail: pawar.72@osu.edu

L. Gujjari
Dorothy M. Davis Heart and Lung Research Institute, Division of Pulmonary Critical Care & Sleep Medicine, Department of Internal Medicine, Wexner Medical Center, The Ohio State University, Columbus, OH, USA

S. Kumari (✉)
Division of Medicinal Chemistry and Pharmacognosy, College of Pharmacy, The Ohio State University, Columbus, OH, USA
e-mail: kumari.45@osu.edu

P. Khanal et al. (eds.), *Neuroinflammation and Autophagy in Neurodegeneration*,
https://doi.org/10.1007/978-981-95-3945-1_8

evidence to bridge cellular biology with translational innovation. Focus has been made on emerging therapeutics to illustrate their synergistic potential in disrupting AD's self-perpetuating pathology. This chapter balances mechanistic depth with clinical relevance to offer a roadmap for developing combinatory therapies.

Keywords

Alzheimer's disease · Autophagy · CRISPR · Amyloid-beta · Neuroinflammation

8.1 Introduction to Alzheimer's Disease Pathology

Alzheimer's Disease (AD) is a degenerative neurologic disorder with a progressive loss of cognitive function and memory. Its pathology includes the complex inter-communication among amyloid-beta (Aβ) plaques, tau neurofibrillary tangles (NFTs), synaptic pathology, and neuroinflammatory mechanisms in neuronal loss. The insidious onset and gradual progression of the disease underscore the importance of understanding its multifaceted mechanisms to achieve therapeutic breakthroughs [1, 2]. Aβ peptides are produced by a sequential cleavage of APP by both β- and γ-secretases. Importantly, $A\beta_{42}$ forms amyloid plaques due to its hydrophobic C-terminal residues, favoring β-sheets adoption and aggregation. The process of $A\beta_{42}$ aggregation is a nucleation-dependent polymerization reaction in which monomers assemble into oligomers, and protofibrils form insoluble fibrils that accrete extracellularly as plaques [3]. Disease-causing mutations of APP, PSEN1, and PSEN2 also increase production or deposition of $A\beta_{42}$, resulting in early-onset familial AD. The aggregation process is also controlled by local factors such as metal ions, pH, and APOE isoforms [4, 5]. Furthermore, abnormal hyperphosphorylated tau causes loss of association with microtubules and formation of paired helical filaments and NFTs. The most important kinases for tau hyperphosphorylation are GSK-3β and CDK5. GSK-3β phosphorylates tau on multiple epitopes, and p25-mediated CDK5 activation further perpetuates pathological phosphorylation [6, 7]. Pathological tau, which separates from axonal microtubules and mislocalizes from pre- and postsynaptic compartments, is believed to cause synaptic dysfunction. However, determining the pathological role of tau in synapse degeneration and cognition is critical. Due to the failure of tau-mediated interventions for AD, targeting synaptic structures may be an attractive therapeutic strategy [8]. Synaptic degeneration is the most prominent pathological correlation with cognitive impairments in AD [2]. Soluble oligomeric Aβ aggregates condense to produce senile plaques and demonstrate neurotoxic effects on neuronal synapses. Moreover, Aβ oligomers can impair long-term potentiation. Interestingly, oligomeric species have been shown to induce synaptic loss and cognitive impairment [9]. Also, soluble Aβ oligomers interact with glutamatergic receptors, leading to calcium influx, mitochondrial impairment, and dendritic spine collapse through excitotoxicity [10]. In parallel, hyperphosphorylated tau gets mislocalized to dendritic compartments, negatively impacting axonal transport and release of synaptic vesicles [11]. This proteostasis disruption causes abnormalities in presynaptic vesicular pools and postsynaptic receptor density in long-term potentiation failures required for memory encoding. Development of synaptic damage precedes neuronal death, causing network desynchronization in hippocampal and cortical circuits. The resulting dysfunction of synaptic integrity shows a direct correlation with cognitive impairment measures and thus identifies proteotoxicity as a main cause of neural circuit dysfunction in AD [12].

Under the Dual-Hit Hypothesis in AD, neurodegeneration is synergistically caused by the intersection of autophagy failure, leading to proteostatic collapse and chronic neuroinflammation. In early AD, autophagy, a critical cellular process for degrading misfolded proteins and damaged organelles, is impaired, which results in the accumulation of toxic Aβ and aggregates of hyperphosphorylated tau [13, 14]. The interplay between autophagy dysregulation and chronic neuroinflammation is crucial in AD pathology. Autophagy is an essential cellular process for misfolded proteins and damaged organelles, and its failure results in the accumulation of toxic aggregates like Aβ and hyperphosphorylated tau [15]. This proteostasis disruption not only impairs neuronal homeostasis but also acts as a persistent stimulus for microglia activation and subsequent neuroinflammation [16]. The autophagic processes impair protein and damaged cellular components cleaning efficacy, which increase immune responses along with the release of pro-inflammatory cytokines [17]. Chronic neuroinflammation worsens autophagic flux dysfunction through inflammasome activation by creating a vicious cycle supporting both proteinopathy and neuroinflammation. Therefore, the interlink between autophagy dysregulation and neuroinflammation is an important driver of neurodegeneration in AD, and therapeutic interventions restoring autophagic function may have the twin benefit of reducing protein aggregation and inflammatory damage [18, 19]. Genetic evidence also provides strong evidence for interaction between autophagy and neuroinflammation in AD. For example, the APOE4 allele impedes Aβ clearance, skews microglial responses, and disrupts lipid homeostasis [20]. In addition, TREM2 is a critical microglial receptor that is critical for phagocytic function and immune modulation, impairing microglial efficacy for Aβ and cellular debris removal and simultaneously modifying inflammatory signaling [21–23].

Currently, Aβ- and tau-centric therapeutic interventions have shown limited efficacy as they fail to address multifaceted AD pathogenesis. However, approaches that restore autophagic flux, dampen chronic inflammation, and personalize interventions based on genetic profiles may offer more effective disease modification in AD [24]. Current therapeutic approaches against Aβ and tau in AD include immunotherapies, β- and γ-secretase inhibitors, and agents that regulate tau phosphorylation. These therapies aim to reduce Aβ plaques and tau tangles delaying disease progression. However, although large number of clinical trials conducted on Aβ, resulting therapeutic benefits have been relatively limited, failing to demonstrate significant clinical benefit or to change the course of the disease [25, 26]. In early AD stages, monoclonal antibodies like aducanumab and lecanemab showed some efficacy in slowing cognitive decline, but reduction in Aβ plaque burden does not consistently result in significant cognitive improvement [27]. Similarly, therapeutic approaches targeting tau have not yet shown significant clinical efficacy, although tau pathology has a more direct involvement in neuronal degeneration and cognitive impairment than Aβ. The simultaneous development of Aβ and tau pathologies along with neuroinflammation, synaptic dysfunction, and other molecular processes illustrates the AD pathogenesis. Thus, by targeting the different interlinked processes that underlie neurodegeneration in AD, this multimodal approach is a more integrated therapeutic strategy [28–30].

8.2 Autophagy Dysfunction in AD: Mechanisms and Consequences

8.2.1 Autophagic Flux Disruption

Aβ oligomers impair lysosomal proteolytic activity by inhibiting cathepsin D, confirming that intracellular $A\beta_{42}$ accumulation prevents lysosomal acidification and oxidizes cathepsin D, reducing its efficacy to degrade soluble Aβ species [31]. This creates a self-perpetuating cycle because impaired cathepsin D activity promotes Aβ aggregation, impairing lysosomal function [32], which manifests as a buildup of poorly acidified autolysosomes and autophagic vacuoles, giving rise to the "PANTHOS" structures formation, perikaryal rosettes densely filled with A-β-positive autophagic vesicles that are essential for plaque formation and cell death [33]. Concurrently, tau pathology contributes to autophagosome stalling through protein aggregation [34]. Hyperphosphorylated and acetylated tau proteins undergo insoluble aggregation that resists chaperone-mediated autophagy and inhibits cargo binding, resulting in inefficient aggregate encapsulation into autophagosomes [35]. Furthermore, tau aggregates sequester elements of autophagic machinery, which inhibits autophagosome maturation and their fusion with lysosomes, and the autophagosomal blockade affects the cell's efficacy to remove misfolded proteins and defective organelles [36]. In addition, dysregulation of nutrient-sensing pathways in AD neurons by mTORC1 hyperactivation suppresses autophagy initiation, reducing new autophagosome formation and impairing the cell ability to eliminate accumulating protein aggregates [37]. These events create a feed-forward process in which the defective autophagic–lysosomal machinery induces progressive synaptic failure, oxidative stress, and eventual neuronal death.

8.2.2 Stage-Specific Autophagy Defects

Stage-specific autophagy deficits have increasingly been implicated as significant players in AD pathogenesis, with breakdowns in the initiation, elongation, and degradation phases of the autophagy process. At the initiation phase, the efficacy of the ULK1 complex is lost in AD, resulting in impaired autophagosome formation in response to exposure to cellular stressors [38]. This inhibition is exacerbated by

AMPK signaling defects that further decrease ULK1 activation and autophagy initiation. During elongation, LC3 is lipidated to form LC3-II, which is essential for autophagosome membrane expansion and cargo engulfment. LC3-II is later degraded following fusion with lysosomes [39]. Decreased LC3-II levels with impaired autophagosome maturation are associated with compromised efficacy to engulf and clear cytotoxic aggregates [40]. However, the most critical defect can be found at the degradation stage, where lysosomal pH is abnormally alkaline, which inhibits the lysosomal hydrolases activity. Lysosomal alkalization results in undigested autophagic vacuoles and electron-dense autolysosomes in dystrophic neurites [41]. Further, AMPK inhibits mTORC1, thereby alleviating the mTORC1-mediated inhibitory phosphorylation of ULK1 and ATG13 [42]. Furthermore, endoplasmic reticulum stress inhibits autophagosome formation in AD. When endoplasmic reticulum stress is prolonged, the unfolded protein response is initiated, and IRE1α cleaves XBP1 mRNA to form XBP1s. The transcription factor regulates protein folding, protein degradation, and autophagy-related genes. The transient presence of the IRE1α/XBP1 pathway might induce autophagy initiation by upregulating autophagy-related genes. Long-term or unresolved endoplasmic reticulum stress inhibits the latter phase of autophagy through autophagosome formation and merging with lysosomes [43].

8.2.3 Downstream Pathological Outcomes

The pathological features downstream in AD are typified by mitochondrial activity disruption and synaptic vesicle accumulation due to impaired autophagic-lysosomal clearance mechanisms. Mitochondrial activity disruption is stimulated by inadequate mitophagy, resulting in damaged mitochondria buildup overproduction of reactive oxygen species (ROS) that oxidize protein [44], lipid, and nucleic acid, exacerbating cellular stress and bioenergetic deficits. The increase in ROS triggers harmful cascades, such as GSK3β-mediated tau hyperphosphorylation, promoting neurofibrillary tangle formation and further damaging mitochondrial dynamics [45, 46]. Additionally, ROS-mediated oxidative damage to ATP synthase subunits negatively impacts oxidative phosphorylation by reducing ATP synthesis and maintaining a bioenergetic failure state [47]. The inability to efficiently clear damaged mitochondria through mitophagy exacerbates neuronal vulnerability, triggering a feedforward cycle of oxidative stress and metabolic failure [48]. Additionally, synaptic dysfunction is marked by synaptic vesicle accumulation through impaired synaptic autophagy (synaptophagy). Aβ oligomers disrupt synaptic vesicle dynamics by interfering with proteins like synaptophysin-1 and VAMP2, which are involved in synaptic vesicle docking, fusion, and recycling. This leads to pathological neurotransmitter release, glutamate excitotoxicity, and eventual synaptic vesicle reserve depletion contributes to synaptic loss [49]. Autophagic vacuoles that comprise synaptic vesicles occlude dystrophic neurites, reflecting lysosomal proteolytic impairment and defective autophagosome maturation. Interestingly, autophagosome accumulation correlates with early synaptic pathology and precedes neuronal injury

[48]. This failure of synaptic clearance is exacerbated by endoplasmic reticulum stress and lysosomal alkalization, disrupting vesicular transport and blocking autophagosome-lysosome fusion [11].

8.3 Triggers and Enhancers of Neuroinflammatory Pathways in AD

8.3.1 Inflammasome Activation

NLRP3 inflammasome is an intracellular multiprotein complex that triggers inflammation through a tightly regulated process in response to Aβ [50]. This process involves priming and activation with subsequent post-transcriptional and post-translational modifications [51]. NLRP3 priming is triggered by signals from Aβ fibrils, which are detected by innate immune cells in the brain through notable pattern recognition receptors, TLR4 and its co-receptor CD14 [52]. This recognition activates nuclear factor kappa B (NF-κB) signaling, upregulating NLRP3, and pro-inflammatory cytokines [53]. This transcriptional upregulation constitutes the "priming" phase of inflammasome activation, which prepares the cell for response to subsequent danger signals [54]. The priming step alone does not activate the inflammasome, but it is crucial to establish a pro-inflammatory state within microglia [55]. The association of Aβ with TLR4 and its co-receptor CD14 facilitates ROS production and mitochondrial stress. It makes cells more sensitive to secondary signals such as lysosomal rupture or ionic imbalance and enabling full NLRP3 inflammasome activation [56]. The priming state is maintained by long-term exposure to Aβ, and this contributes to a feed-forward loop of inflammation and neurodegeneration in AD [57]. Thus, TLR4/CD14-mediated NLRP3 inflammasome priming represents a critical link between Aβ pathology and innate immune activation. Targeting this pathway could interrupt neuroinflammation, reduce Aβ deposition and associated neurotoxicity by inhibiting neuronal pyroptosis, offering a promising therapeutic strategy for AD [58].

Chronic neuroinflammation is a key pathological hallmark of AD, and a central player in this process is the NLRP3 inflammasome [59]. After priming and activation of NLRP3 triggered by Aβ aggregates, its components assemble into a multiprotein complex that includes the adaptor protein ASC [60]. ASC oligomerizes (Table 8.1) to form large intracellular aggregates known as ASC specks, which act as scaffolds for pro-caspase 1 recruitment and activation [61]. This assembly is essential to initiate autocatalytic cleavage of pro-caspase-1 into its enzymatic active form, caspase-1 [62]. This cleaves the pro-inflammatory cytokine precursor pro-IL-1β and pro-IL-18 into their biologically active forms IL-1β and IL-18, which are then released into the extracellular space [63]. These cytokines are potent inflammatory mediators that subsequently activate glial cells, thereby sustaining and amplifying the inflammatory environment in the AD brain [64]. In addition to cytokine maturation, active caspase-1 induces pyroptosis and membrane rupture, leading to neuronal loss [65]. In AD, ASC specks can persist extracellularly and seed further

Table 8.1 Molecular triggers and potential therapeutic approaches in AD

Factors	Mechanism/effects	Role in AD	Downstream effects	Therapeutic implications
NLRP3 inflammasome priming (*Aβ fibril-induced TLR4/CD14 signaling*)	Upregulating NLRP3 and pro-IL-1β, with ROS and lysosomal damage, triggers inflammasome assembly	Initiates microglial inflammation	IL-1β/IL-18 releases, pyroptosis	Inhibit NLRP3 priming and activation, block TLR4/CD14 signal or NF-κB activation
ASC speak formation (*ROS, lysosomal rupture*)	ASC oligomerizes upon inflammasome activation, recruits, and activates caspase-1	Increases inflammasome signaling	Caspase-1 activation, pyroptosis	Block ASC oligomerization and caspase-1 inhibition
AIM2 inflammasome (*Cytosolic DNA released from damaged neurons*)	AIM2 activates the inflammasome and caspase-1 activation	DNA-driven neuroinflammation	Cytokine release and pyroptosis	Target AIM2 to address DNA-triggered inflammation
IL-1β-astrocyte activation (*Microglial IL-1β secretion*)	Microglial IL-1β triggers reactive astrocytes releasing C1q/C3, promoting synapse loss and cognitive decline	Synapse pruning and cognitive dysfunction	Complement-mediated synapse pruning	Inhibit IL-1β signal for synaptic integrity
Tumor necrosis factor-α (TNF-α)/NF-κB signaling (*TNF-α release from reactive microglia and astrocytes*)	TNF-α released by reactive glia activates NF-κB signaling, driving sustained pro-inflammatory gene expression	Sustained glial inflammation in Aβ/Tau-mediated neurodegeneration	NF-κB promotes inflammation and Aβ production	Target TNF-α and NF-κB to interrupt chronic glial inflammation and neurodegeneration
cGAS-STING pathway (*Mitochondrial DNA leakage in defective autophagy*)	Cytosolic DNA, due to autophagy dysfunction, activates STING, inducing type I interferons	Cytosolic DNA sensing and type I IFN signaling in low-grade chronic neuroinflammation	STING activates TBK1/IRF3, inducing type I IFN and inflammatory gene expression	Modulate cGAS-STING signal and prevent chronic sterile inflammation
S100B/RAGE axis (*S100B secretion by reactive astrocytes*)	Activate NF-κB/MAPK to induce cytokine release, oxidative stress	DAMP-driven inflammation and synaptic dysfunction through astrocyte signaling	RAGE activation amplifies NF-κB signaling and ROS-driven inflammation	Target S100B-RAGE interaction to attenuate glial activation and oxidative damage

inflammasome activation, reinforcing a neuro-inflammatory self-perpetuating cycle [66]. This continuous loop involving NLRP3 worsens AD pathology, highlighting these molecular events as promising targets for therapeutic intervention [67].

Moreover, there are other alternative inflammasome pathways in addition to well well-characterized NLRP3 inflammasome. For example, AIM2 detects cytosolic DNA from damaged cells, contributing to inflammasome activation in response to endogenous danger signals [68]. AIM2 recognizes that DNA originated from damaged neurons under oxidative and metabolic stress, which are commonly observed in AD [69]. This serves as a damage-associated molecular pattern (DAMP) by activating AIM2 inflammasome in mitochondria [70]. During neuronal damage, mitochondrial and/or nuclear DNA leak into the cytosol, triggering AIM2 activation. AIM2 oligomerizes upon sensing cytosolic DNA and recruits the adaptor protein ASC [71]. This stimulates inflammasome complex assembly and caspase-1 activation. Like NLRP3, activated AIM2 inflammasome converts pro-IL-1β and pro-IL-18 to mature, pro-inflammatory forms and induces pyroptosis [72]. Hence, sustained AIM2 activation can increase neuroinflammation, exacerbate neuronal damage, and contribute to AD progression [73]. Thus, understanding the AIM2 role underscores immune responses in AD and suggests that targeting multiple inflammasome pathways may be important in attenuating AD [74].

8.3.2 Cytokine-Driven Inflammation

In AD, astrocyte activation by pro-inflammatory cytokines plays a pivotal role in driving neurodegeneration through maladaptive immune responses [75]. For example, IL-1β binds to astrocyte receptors and induces intracellular signaling cascades leading to their transformation into a reactive, neurotoxic phenotype [76]. These reactive astrocytes secrete complement components, which are the mediators of the classical complement pathway [77]. These proteins induce maladaptive complement tagging of synapses [78]. Complement-marked synapses become targets for microglial phagocytosis through complement receptors, resulting in excessive complement-mediated synapse loss. This mechanism triggers early synaptic dysfunction and cognitive deterioration [79]. Notably, C3 levels are consistently elevated in both cerebrospinal fluid and brain tissue of AD patients, correlating with disease progression [80]. The IL-1β-mediated astrocyte response represents a harmful feedback loop in which inflammation leads to synapse destruction [81]. Targeting IL-1β signaling, complement activation, or astrocyte reactivity may help to preserve synaptic integrity and slow neurodegeneration in AD [82]. Additionally, TNF-α also plays a crucial role in chronic neuroinflammation in AD [83]. It is produced from microglia and astrocytes activation, exerting its effects through the activated NF-κB signaling pathway [84], playing a pivotal role in regulating inflammatory gene transcription responses [85]. Moreover, TNF-α translocates signaling molecules into the nucleus [86]. After translocating into the nucleus, NF-κB promotes sustained transcription of a broad range of pro-inflammatory genes, cytokines, adhesion molecules, chemokines, and enzymes such as iNOS and COX-2 [87]. In AD, chronic

TNF-α/NF-κB activation in glial cells drives persistent inflammation, exacerbating neuronal dysfunction and accelerating disease progression [88]. In the AD brain, microglia and astrocytes adopt a persistently reactive state, leading to sustained inflammatory mediators [89]. This chronic stimulation contributes to neuronal damage and disrupts repair mechanisms and synaptic plasticity [90]. Moreover, NF-κB signaling augments the expression of APP and BACE1, thus promoting Aβ production and tau hyperphosphorylation [91]. Sustained TNF-α/NF-κB activation disseminates inflammation, leading to neuronal damage, glial reactivity, and cognitive impairment [92].

8.3.3 Non-Canonical Inflammatory Pathways

In AD, the non-canonical cGAS-STING signaling pathway contributes to neuroinflammation when autophagy is impaired [93]. Autophagy is the primary mechanism by which neurons maintain homeostasis by clearing misfolded proteins and damaged organelles [94]. In AD, aging, Aβ toxicity, and tau pathology trigger autophagic dysfunction leading to the accumulation of damaged mitochondria [95]. These dysfunctional mitochondria enhance membrane permeability and release mitochondrial DNA into the cytosol [96]. This process is perceived as a DAMP and is then recognized by cGAS to produce cGAMP, which binds and activates STING on the endoplasmic reticulum membrane [97]. Next, STING activation induces TBK1 and IRF3 signaling, leading to the transcription of type I interferons and pro-inflammatory genes [98]. This sustained inflammatory response promotes further neuronal stress and contributes to neurodegenerative progression [99]. Unlike classical inflammasome-mediated pathways, the cGAS-STING axis bridges innate immunity and damaged DNA responses [100]. In AD, persistent cGAS-STING activation from mitochondrial dysfunction and autophagy impairment may initiate low-grade neuroinflammation, exacerbating neuronal injury [101, 102]. Additionally, non-canonical pathways like the S100B/RAGE axis also play a key role in sustaining chronic neuroinflammation and amplifying glial activation [103]. S100B is involved in neurotrophic functions and homeostasis [104], and in the aged brain, elevated extracellular S100B levels serve as DAMP and bind to RAGE, which is highly expressed on microglia, neurons, and endothelial cells [105]. This further activates NF-κB and MAPK pathways, promoting pro-inflammatory cytokines production and establishing a feed-forward loop of inflammation by activating microglia and astrocytes [106]. In addition, S100B/RAGE signaling enhances oxidative stress by promoting ROS production and mitochondrial dysfunction [107]. RAGE activation is also associated with increased APP processing and tau phosphorylation, which are the key features of AD pathology [108]. Notably, the S100B/RAGE axis links astrocytes and microglia, with astrocyte-derived S100B priming microglial activation [109]. This bidirectional glial interaction exacerbates neurodegeneration and cognitive impairment [110]. Thus, the role of this pathway shows the significance of non-canonical DAMP-mediated pathways in driving persistent neuroinflammation and presents a potential therapeutic target [111].

8.4 Autophagy–Inflammation Feedback Loops

The interaction between inflammasomes and autophagy constitutes a complex regulatory network essential for sustaining cellular homeostasis (Fig. 8.1). For example, caspase-1 targets Beclin-1 and ATG5, which are essential components in the initiation and elongation phases of autophagosome formation [112]. Beclin-1 is a component of the class III PI3K complex, which is essential in the nucleation of autophagic vesicles. In contrast, ATG5 is involved in forming a conjugation complex that is critical to expanding the autophagosome membrane. The proteolytic cleavage of these proteins impairs autophagosome formation and enhances inflammasome signaling by facilitating damaged organelles and misfolded proteins [113]. This illustrates crosstalk between autophagy and inflammation, where NLRP3 activation inhibits a crucial cellular clearance pathway, worsening neurodegenerative pathology. Moreover, autophagy acts as a negative regulator of inflammasome activity, thereby limiting excessive inflammatory responses. Moreover, p62/SQSTM1 serves as a crucial molecular mediator in this process. To detail, P62 has been demonstrated to bind to ASC specks, thereby marking them for autophagic degradation [114]. This restricts the duration and spread of inflammasome signaling by eliminating the structural core necessary for caspase-1 activation. Furthermore, enhancing autophagy or preventing its suppression represents a promising therapeutic strategy for mitigating neuroinflammation [115]. The interaction between the NLRP3 inflammasome and autophagy highlights a reciprocal regulatory mechanism exhibiting important consequences for AD

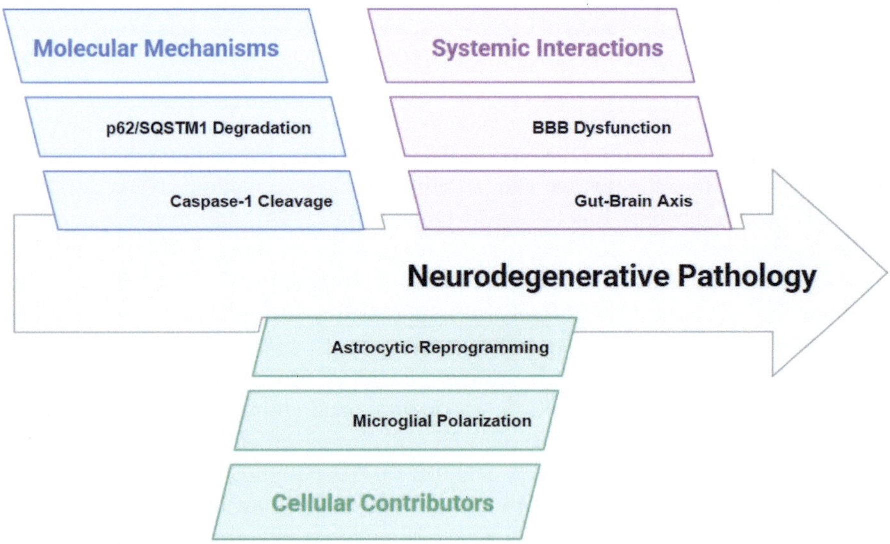

Fig. 8.1 Depicting autophagy–neuroinflammation imbalance in neurodegeneration. Neurodegenerative pathology is driven by molecular mechanisms, systemic interactions, and cellular contributors such as astrocytic reprogramming and microglial polarization

progression. Moreover, inflammasome activation can impair autophagy through caspase-1-mediated cleavage of Beclin-1 and ATG5. Conversely, its inhibition may facilitate inflammatory resolution [16].

Microglia exhibit dynamic polarization states that affect neurodegenerative processes. The cells can be phagocytic, neuroprotective phenotype, or an inflammatory, neurotoxic phenotype based on the signaling context. The TREM family is a crucial modulator of microglial phenotype. TREM2 signaling enhances the phagocytic and homeostatic state of microglia, aiding Aβ and cellular debris clearance, thus preserving neural tissue integrity [116]. Next, TREM1 activation enhances pro-inflammatory responses, leading microglia to adopt a pathogenic state marked by elevated cytokine production and oxidative stress [117]. The TREM2 and TREM1 equilibrium activity is essential in influencing the inflammatory state of central nervous system (CNS). Further, astrocytes undergo functional reprogramming in neuroinflammatory states. Under physiological conditions, astrocytes exhibit a neuroprotective phenotype, which is defined by their homeostatic functions, such as trophic factor secretion and synapse maintenance. In response to inflammatory stimuli, cytokines IL-1α, TNF-α, and C1q released by activated microglia, astrocytes undergo a transition to a reactive phenotype [118].

Neuroinflammation extends beyond the CNS, with interactions from the peripheral immune system significantly contributing due to blood–brain barrier (BBB) dysfunction. In AD, the compromised BBB integrity facilitates peripheral immune cells such as monocytes and T-cells. Cells that are typically excluded migrate in response to chemokines such as CCL2 and CXCL10, which are secreted by activated microglia and astrocytes [119]. Upon CNS infiltration, monocytes differentiate into macrophage-like cells, which intensify inflammation by releasing pro-inflammatory cytokines. Concurrently, $CD8^+$ and $CD4^+$ T-cells participate in neuronal damage through cytotoxic actions and cytokine production [120]. This infiltration exacerbates neuroinflammatory processes and may hasten cognitive decline. The gut–brain axis is another significant factor in neuroinflammation, especially via microbial metabolites like lipopolysaccharide. Lipopolysaccharide is a main constituent of gram-negative bacterial cell walls, may enter systemic circulation due to leaky gut, and subsequently activate microglia through TLR4 signaling [121]. This increases microglial responsiveness to secondary stimuli, including Aβ, and hence, gut-derived lipopolysaccharide establishes dysbiosis and neurodegeneration. Peripheral immune cell infiltration and microbiota-derived factors emphasize the systemic aspect of neuroinflammation, underscoring the necessity of preserving both BBB integrity and gut microbial equilibrium to avert CNS pathology [122].

8.5 Advanced Multimodal Therapeutic Strategies

Autophagy-focused interventions have a potential therapeutic strategy for complex polygenic neurodegenerative diseases like AD. mTOR-independent inducers like metformin enhance AMPK activation, which suppresses mTOR and p70 S6 kinase, hence activating autophagy as indicated by increased LC3-II and reduced p62 levels

[123]. Similarly, EP300 acetyltransferase competitive inhibitor spermidine enhances autophagic flux by inhibiting autophagy-related proteins like ATG5 and LC3 acetylation [124]. Concurrently, curcumin derivatives interact with TFEB, inducing its nuclear translocation and lysosomal biogenesis independently of mTOR inhibition. Next, Trehalose induces low-grade stress that disrupts mTORC1 signaling and activates TFEB, stimulating autophagy–lysosome biogenesis [125, 126]. Selective autophagy inducers focus degradation on specific cargo, as observed with HDAC6 inhibitors such as ACY-738. These compounds augment aggrephagy by stimulating microtubule-dependent trafficking mechanisms, guiding ubiquitinated protein aggregates towards aggresomes for autophagic degradation [127].

MCC950 is a small-molecule inhibitor that is highly specific and effective at preventing NLRP3 activation by blocking ASC oligomerization. This is achieved through MCC950 binding to NACHT domain, which prevents ATP hydrolysis and subsequent conformational changes required for inflammasome assembly [128, 129]. Moreover, OLT1177 hinders the NLRP3 and NEK7 interaction, which is essential for inflammasome activation. However, this drug directly disrupts NLRP3 oligomerization, suppresses IL-1β and IL-18 release, without impacting potassium efflux and pro-IL-1β expression levels. The therapeutic modification of microglial function by anti-TREM2 antibodies presents a novel approach to increase microglial phagocytic efficacy. Moreover, it has been demonstrated that agonistic anti-TREM2 antibodies can restore microglial activation and phagocytosis, facilitating amyloid plaques and apoptotic cells [130]. Additionally, the astrocytes reactivation with LXR agonists, such as GW3965, restores neurotoxic astrocytes and promotes a neuroprotective astrocyte phenotype [131].

The combined synergistic application of rapamycin and senolytic compounds like dasatinib and quercetin holds potential to target senescent glial cells in neurodegenerative disorders. Rapamycin inhibits the SASP, reduces chronic inflammation, and prevents cellular senescence progression. Senolytics selectively induce apoptosis in senescent cells by inhibiting anti-apoptotic mechanisms, eliminating non-functional cells from brain [132]. For example, dasatinib and quercetin can effectively reduce senescent glia burden, as determined by a reduction of SA-βGal+ and pS6+ cell populations in the brain. Rapamycin and senolytics are supposed to be combined synergistically as they act on different but complementary sides of cellular aging. To highlight, rapamycin regulates SASP and autophagy, while senolytics remove the primary causes of pro-inflammatory signals [133]. Additionally, APOE4 gene editing can offer a novel approach for reducing inflammation and enhancing cell cleanup issues [134]. APOE4 is the genetic risk for late-onset AD, and its presence complicates Aβ clearance, enhances tau pathology, and alters brain cell function. CRISPR/Cas9 editing can correct gene expression, enhance cell cleanup, and reduce inflammation, thereby alleviating several AD-related issues [135, 136]. Moreover, nanotechnology approaches represent a cutting-edge strategy to inhibit pro-inflammatory targets in the CNS, including IL-1β and P2X7R. Lipid nanoparticles prevent siRNA from degradation, improve bioavailability, and allow tailored delivery to brain cells. Recent advances include lipid nanoparticle development encapsulating siRNA against P2X7R, which have demonstrated effective gene silencing and reduced neuroinflammation in preclinical models [137, 138].

8.6 Translational Challenges and Emerging Frontiers

Molecular imaging technology advancements have brought about the formulation of promising positron emission tomography (PET) tracers that allow in vivo assessment of disease processes. For example, PET ligands showing specificity for autophagic flux through LC3B-associated targets can be developed, allowing autophagy kinetics imaging to understand disease onset, progression, and therapeutic response [139]. Next-generation TSPO ligand such as [18F]LW223, with enhanced metabolic stability and specificity for activated microglia, showed accurate neuroinflammation assessment [140]. Additionally, cerebrospinal fluid quantification of inflammatory mediators such as IL-18, lysosomal enzymes such as cathepsin D, and conventional AD biomarkers like p-tau/Aβ_{42} ratios improves patient stratification and longitudinal monitoring of disease course. Such multimodal biomarkers hold great promise for early diagnosis and therapeutic response monitoring [141, 142].

Despite their extensive utilization, transgenic rodent models of APP/PS1 mice poorly recapitulate the complete range of human AD pathophysiology. For example, these models largely focus on amyloid pathology but do not exhibit strong tau pathology and do not replicate the temporal course of human disease, which restricts their predictive validity for clinical results [143]. Additionally, rodent neuroimmune and glial responses are considerably different from humans, making translational interpretation complex. To address these caveats, iPSC-derived 3D cultures and brain organoids have emerged as better platforms to model patient-specific disease phenotypes and glial-neuronal interactions. These organoids replicate major features of AD, which include amyloid aggregation, tau phosphorylation, and neuroinflammation in a human genetic background, yielding a more physiologically pertinent system for mechanistic investigations and drug screening [144, 145].

New therapeutic strategies are increasingly incorporating biological rhythms, epigenetics, and artificial intelligence. Circadian control of lysosomal and autophagic processes suggests synchronizing autophagy boosters with endogenous peaks can maximize efficacy and minimize side effects [146]. The use of epigenetic manipulation with histone deacetylase inhibitors has demonstrated promise in reestablishing dysregulated autophagy and inflammatory gene networks, offering a reversible strategy to reprogram aberrant gene expression [147]. Moreover, artificial intelligence-assisted drug discovery platforms enable the prediction and design of multitarget compounds for simultaneously modulating autophagy and neuroinflammation pathways [148].

8.7 Conclusion

AD stands as a complex neurodegenerative disorder with evidence implicating autophagy and chronic neuroinflammation dysregulation as essential, interdependent drivers of its progression. This chapter highlighted how disruptions in cellular clearance mechanisms, predominantly the autophagy–lysosomal pathway, converge with immune responses to form a self-sustaining cycle of proteotoxic stress, mitochondrial dysfunction, and neuronal loss. Key molecular players such as the NLRP3 inflammasome, caspase-1, and pro-inflammatory cytokines interact closely with impaired autophagic flux, further undermining neuronal homeostasis. Defective mitophagy and synaptic autophagy contribute to bioenergetic failure and impaired neurotransmission. Additionally, microglial and astrocytic reprogramming exacerbate synaptic degradation and inflammation. Furthermore, systemic contributors, including BBB leakage and gut–brain axis disturbances, further compound disease pathology. Due to the limitations of monotherapies targeting amyloid and tau, the field is shifting toward integrative approaches that simultaneously restore autophagy, modulate immune responses, and account for genetic risk factors. Understanding the reciprocal regulation between autophagy and inflammation offers a combined framework for future therapeutic innovation. Precision medicine approaches that leverage molecular insights and emerging technologies hold the potential to meaningfully alter AD pathogenesis.

References

1. Long JM, Holtzman DM. Alzheimer disease: an update on pathobiology and treatment strategies. Cell. 2019;179:312–39. https://doi.org/10.1016/j.cell.2019.09.001.
2. Metaxas A, Finsen B. Editorial: Mechanisms, markers and therapeutics of synaptic pathology in Alzheimer's disease. Front Neurosci. 2022;16:1022447. https://doi.org/10.3389/fnins.2022.1022447.
3. Zhang C, Browne A, Divito JR, et al. Amyloid-β production via cleavage of amyloid-β protein precursor is modulated by cell density. J Alzheimers Dis. 2010;22:683–984. https://doi.org/10.3233/JAD-2010-100816.
4. Lanoiselée H-M, Nicolas G, Wallon D, et al. APP, PSEN1, and PSEN2 mutations in early-onset Alzheimer disease: a genetic screening study of familial and sporadic cases. PLoS Med. 2017;14:e1002270. https://doi.org/10.1371/journal.pmed.1002270.
5. Tcw J, Goate AM. Genetics of β-amyloid precursor protein in Alzheimer's disease. Cold Spring Harb Perspect Med. 2017;7:a024539. https://doi.org/10.1101/cshperspect.a024539.
6. Iqbal K, Liu F, Gong C-X. Tau and neurodegenerative disease: the story so far. Nat Rev Neurol. 2016;12:15–27. https://doi.org/10.1038/nrneurol.2015.225.
7. Rawat P, Sehar U, Bisht J, et al. Phosphorylated tau in Alzheimer's disease and other tauopathies. Int J Mol Sci. 2022;23:12841. https://doi.org/10.3390/ijms232112841.
8. Wu M, Zhang M, Yin X, et al. The role of pathological tau in synaptic dysfunction in Alzheimer's diseases. Transl Neurodegener. 2021;10:45. https://doi.org/10.1186/s40035-021-00270-1.
9. Koffie RM, Hyman BT, Spires-Jones TL. Alzheimer's disease: synapses gone cold. Mol Neurodegener. 2011;6:63. https://doi.org/10.1186/1750-1326-6-63.

10. Danysz W, Parsons CG. Alzheimer's disease, β-amyloid, glutamate, NMDA receptors and memantine—searching for the connections. Br J Pharmacol. 2012;167:324–52. https://doi.org/10.1111/j.1476-5381.2012.02057.x.
11. Meftah S, Gan J. Alzheimer's disease as a synaptopathy: evidence for dysfunction of synapses during disease progression. Front Synaptic Neurosci. 2023;15:1129036. https://doi.org/10.3389/fnsyn.2023.1129036.
12. Huo A, Wang J, Li Q, et al. Molecular mechanisms underlying microglial sensing and phagocytosis in synaptic pruning. Neural Regen Res. 2024;19:1284–90. https://doi.org/10.4103/1673-5374.385854.
13. Hellén M, Weert I, Müller SA, et al. Inflammation-induced lysosomal dysfunction in human iPSC-derived microglia is exacerbated by APOE 4/4 genotype. J Neuroinflammation. 2025;22:147. https://doi.org/10.1186/s12974-025-03470-y.
14. Zhu X, Raina AK, Perry G, Smith MA. Alzheimer's disease: the two-hit hypothesis. Lancet Neurol. 2004;3:219–26. https://doi.org/10.1016/S1474-4422(04)00707-0.
15. Cheng X, Wei Y, Qian Z, Han L. Autophagy balances neuroinflammation in Alzheimer's disease. Cell Mol Neurobiol. 2023;43:1537–49. https://doi.org/10.1007/s10571-022-01269-6.
16. Lu R, Zhang L, Yang X. Interaction between autophagy and the NLRP3 inflammasome in Alzheimer's and Parkinson's disease. Front Aging Neurosci. 2022;14:1018848. https://doi.org/10.3389/fnagi.2022.1018848.
17. Chen Y, Chen J, Xing Z, et al. Autophagy in neuroinflammation: a focus on epigenetic regulation. Aging Dis. 2024;15:739–54. https://doi.org/10.14336/AD.2023.0718-1.
18. Yang Q, Yang C, Lv H, et al. Autophagy regulation attenuates neuroinflammation and cognitive decline in an Alzheimer's disease mouse model with chronic cerebral hypoperfusion. Inflammation. 2024;48:541–56. https://doi.org/10.1007/s10753-024-02043-0.
19. Xu Y, Propson NE, Du S, et al. Autophagy deficiency modulates microglial lipid homeostasis and aggravates tau pathology and spreading. Proc Natl Acad Sci USA. 2021;118:e2023418118. https://doi.org/10.1073/pnas.2023418118.
20. Shi Y, Yamada K, Liddelow SA, et al. ApoE4 markedly exacerbates tau-mediated neurodegeneration in a mouse model of tauopathy. Nature. 2017;549:523–7. https://doi.org/10.1038/nature24016.
21. Li Y, Xu H, Wang H, et al. TREM2: potential therapeutic targeting of microglia for Alzheimer's disease. Biomed Pharmacother. 2023;165:115218. https://doi.org/10.1016/j.biopha.2023.115218.
22. Hickman SE, El Khoury J. TREM2 and the neuroimmunology of Alzheimer's disease. Biochem Pharmacol. 2014;88:495–8. https://doi.org/10.1016/j.bcp.2013.11.021.
23. Ulland TK, Colonna M. TREM2—a key player in microglial biology and Alzheimer disease. Nat Rev Neurol. 2018;14:667–75. https://doi.org/10.1038/s41582-018-0072-1.
24. Cummings J, Zhou Y, Lee G, et al. Alzheimer's disease drug development pipeline: 2023. Alzheimers Dement. 2023;9:e12385. https://doi.org/10.1002/trc2.12385.
25. Zhang Y, Chen H, Li R, et al. Amyloid β-based therapy for Alzheimer's disease: challenges, successes and future. Signal Transduct Target Ther. 2023;8:248. https://doi.org/10.1038/s41392-023-01484-7.
26. Tondo G, De Marchi F, Bonardi F, et al. Novel therapeutic strategies in Alzheimer's disease: pitfalls and challenges of anti-amyloid therapies and beyond. J Clin Med. 2024;13:3098. https://doi.org/10.3390/jcm13113098.
27. Aljuhani M, Ashraf A, Edison P. Evaluating clinical meaningfulness of anti-β-amyloid therapies amidst amyloid-related imaging abnormalities concern in Alzheimer's disease. Brain Commun. 2024;6:fcae435. https://doi.org/10.1093/braincomms/fcae435.
28. Pluta R, Ulamek-Koziol M. Tau protein-targeted therapies in Alzheimer's disease: current state and future perspectives. In: Alzheimer's disease: drug discovery. Exon Publications; 2020. p. 69–82.

29. Angioni D, Middleton L, Bateman R, et al. Challenges and opportunities for novel combination therapies in Alzheimer's disease: a report from the EU/US CTAD Task Force. J Prev Alzheimers Dis. 2025;12:100163. https://doi.org/10.1016/j.tjpad.2025.100163.
30. Lansdall CJ. An effective treatment for Alzheimer's disease must consider both amyloid and tau. Biosci Horiz. 2014;7:hzu002. https://doi.org/10.1093/biohorizons/hzu002.
31. Mançano ASF, Pina JG, Froes BR, Sciani JM. Autophagy-lysosomal pathway impairment and cathepsin dysregulation in Alzheimer's disease. Front Mol Biosci. 2024;11:1490275. https://doi.org/10.3389/fmolb.2024.1490275.
32. Uddin MS, Stachowiak A, Mamun AA, et al. Autophagy and Alzheimer's disease: from molecular mechanisms to therapeutic implications. Front Aging Neurosci. 2018;10:4. https://doi.org/10.3389/fnagi.2018.00004.
33. Lee J-H, Yang D-S, Goulbourne CN, et al. Faulty autolysosome acidification in Alzheimer's disease mouse models induces autophagic build-up of Aβ in neurons, yielding senile plaques. Nat Neurosci. 2022;25:688–701. https://doi.org/10.1038/s41593-022-01084-8.
34. Tangavelou K, Bhaskar K. The mechanistic link between tau-driven proteotoxic stress and cellular senescence in Alzheimer's disease. Int J Mol Sci. 2024;25:12335. https://doi.org/10.3390/ijms252212335.
35. Caballero B, Bourdenx M, Luengo E, et al. Acetylated tau inhibits chaperone-mediated autophagy and promotes tau pathology propagation in mice. Nat Commun. 2021;12:2238. https://doi.org/10.1038/s41467-021-22501-9.
36. Ando K, Houben S, Homa M, et al. Alzheimer's disease: tau pathology and dysfunction of endocytosis. Front Mol Neurosci. 2021;13:583755. https://doi.org/10.3389/fnmol.2020.583755.
37. Tramutola A, Lanzillotta C, Di Domenico F. Targeting mTOR to reduce Alzheimer-related cognitive decline: from current hits to future therapies. Expert Rev Neurother. 2017;17:33–45. https://doi.org/10.1080/14737175.2017.1244482.
38. Tran M, Reddy PH. Defective autophagy and mitophagy in aging and Alzheimer's disease. Front Neurosci. 2021;14:612757. https://doi.org/10.3389/fnins.2020.612757.
39. Liu J, Li L. Targeting autophagy for the treatment of Alzheimer's disease: challenges and opportunities. Front Mol Neurosci. 2019;12:203. https://doi.org/10.3389/fnmol.2019.00203.
40. Guan X, Iyaswamy A, Sreenivasmurthy SG, et al. Mechanistic insights into selective autophagy subtypes in Alzheimer's disease. Int J Mol Sci. 2022;23:3609. https://doi.org/10.3390/ijms23073609.
41. Nixon RA, Yang D-S. Autophagy failure in Alzheimer's disease—locating the primary defect. Neurobiol Dis. 2011;43:38–45. https://doi.org/10.1016/j.nbd.2011.01.021.
42. Zachari M, Ganley IG. The mammalian ULK1 complex and autophagy initiation. Essays Biochem. 2017;61:585–96. https://doi.org/10.1042/EBC20170021.
43. Ren H, Zhai W, Lu X, Wang G. The cross-links of endoplasmic reticulum stress, autophagy, and neurodegeneration in Parkinson's disease. Front Aging Neurosci. 2021;13:691881. https://doi.org/10.3389/fnagi.2021.691881.
44. Birla H, Minocha T, Kumar G, et al. Role of oxidative stress and metal toxicity in the progression of Alzheimer's disease. Curr Neuropharmacol. 2020;18:552–62. https://doi.org/10.2174/1570159X18666200122122512.
45. Chakravorty A, Jetto CT, Manjithaya R. Dysfunctional mitochondria and mitophagy as drivers of Alzheimer's disease pathogenesis. Front Aging Neurosci. 2019;11:311. https://doi.org/10.3389/fnagi.2019.00311.
46. Misrani A, Tabassum S, Yang L. Mitochondrial dysfunction and oxidative stress in Alzheimer's disease. Front Aging Neurosci. 2021;13:617588. https://doi.org/10.3389/fnagi.2021.617588.
47. Mary A, Eysert F, Checler F, Chami M. Mitophagy in Alzheimer's disease: molecular defects and therapeutic approaches. Mol Psychiatry. 2023;28:202–16. https://doi.org/10.1038/s41380-022-01631-6.

48. Zhang W, Xu C, Sun J, et al. Impairment of the autophagy–lysosomal pathway in Alzheimer's diseases: pathogenic mechanisms and therapeutic potential. Acta Pharm Sin B. 2022;12:1019–40. https://doi.org/10.1016/j.apsb.2022.01.008.
49. Marsh J, Alifragis P. Synaptic dysfunction in Alzheimer's disease: the effects of amyloid beta on synaptic vesicle dynamics as a novel target for therapeutic intervention. Neural Regen Res. 2018;13:616–23. https://doi.org/10.4103/1673-5374.230276.
50. Gupta S, Cassel SL, Sutterwala FS, Dagvadorj J. Regulation of the NLRP3 inflammasome by autophagy and mitophagy. Immunol Rev. 2025;329:e13410. https://doi.org/10.1111/imr.13410.
51. McGroarty J, Salinas S, Evans H, et al. Inflammasome-mediated neuroinflammation: a key driver in Alzheimer's disease pathogenesis. Biomolecules. 2025;15:676. https://doi.org/10.3390/biom15050676.
52. Ayyubova G, Madhu LN. Microglial NLRP3 inflammasomes in Alzheimer's disease pathogenesis: from interaction with autophagy/mitophagy to therapeutics. Mol Neurobiol. 2025;62:7124–43. https://doi.org/10.1007/s12035-025-04758-z.
53. Wang Z, Weaver DF. Microglia and microglial-based receptors in the pathogenesis and treatment of Alzheimer's disease. Int Immunopharmacol. 2022;110:109070. https://doi.org/10.1016/j.intimp.2022.109070.
54. Broz P, Monack DM. Molecular mechanisms of inflammasome activation during microbial infections. Immunol Rev. 2011;243:174–90. https://doi.org/10.1111/j.1600-065X.2011.01041.x.
55. Savage CD, Lopez-Castejon G, Denes A, Brough D. NLRP3-inflammasome activating DAMPs stimulate an inflammatory response in glia in the absence of priming which contributes to brain inflammation after injury. Front Immunol. 2012;3:288. https://doi.org/10.3389/fimmu.2012.00288.
56. von Bernhardi R, Eugenín-von Bernhardi L, Eugenín J. Microglial cell dysregulation in brain aging and neurodegeneration. Front Aging Neurosci. 2015;7:124. https://doi.org/10.3389/fnagi.2015.00124.
57. Hanslik KL, Ulland TK. The role of microglia and the Nlrp3 inflammasome in Alzheimer's disease. Front Neurol. 2020;11:570711. https://doi.org/10.3389/fneur.2020.570711.
58. Li Z, Gong C. NLRP3 inflammasome in Alzheimer's disease: molecular mechanisms and emerging therapies. Front Immunol. 2025;16:1583886. https://doi.org/10.3389/fimmu.2025.1583886.
59. Pennisi M, Crupi R, Di Paola R, et al. Inflammasomes, hormesis, and antioxidants in neuroinflammation: role of NRLP3 in Alzheimer disease. J Neurosci Res. 2017;95:1360–72. https://doi.org/10.1002/jnr.23986.
60. Agrawal I, Jha S. Comprehensive review of ASC structure and function in immune homeostasis and disease. Mol Biol Rep. 2020;47:3077–96. https://doi.org/10.1007/s11033-020-05345-2.
61. Hulse J, Bhaskar K. Crosstalk between the NLRP3 inflammasome/ASC speck and amyloid protein aggregates drives disease progression in Alzheimer's and Parkinson's disease. Front Mol Neurosci. 2022;15:805169. https://doi.org/10.3389/fnmol.2022.805169.
62. Hoss F, Rodriguez-Alcazar JF, Latz E. Assembly and regulation of ASC specks. Cell Mol Life Sci. 2017;74:1211–29. https://doi.org/10.1007/s00018-016-2396-6.
63. Barker BR, Taxman DJ, Ting JP-Y. Cross-regulation between the IL-1β/IL-18 processing inflammasome and other inflammatory cytokines. Curr Opin Immunol. 2011;23:591–7. https://doi.org/10.1016/j.coi.2011.07.005.
64. Rogers J. The inflammatory response in Alzheimer's disease. J Periodontol. 2008;79:1535–43. https://doi.org/10.1902/jop.2008.080171.
65. Kesavardhana S, Malireddi RKS, Kanneganti T-D. Caspases in cell death, inflammation, and pyroptosis. Annu Rev Immunol. 2020;38:567–95. https://doi.org/10.1146/annurev-immunol-073119-095439.

66. Chee SEJ, Solito E. The impact of ageing on the CNS immune response in Alzheimer's disease. Front Immunol. 2021;12:738511. https://doi.org/10.3389/fimmu.2021.738511.
67. Liang T, Zhang Y, Wu S, et al. The role of NLRP3 inflammasome in Alzheimer's disease and potential therapeutic targets. Front Pharmacol. 2022;13:845185. https://doi.org/10.3389/fphar.2022.845185.
68. Yu D, Zheng S, Sui L, et al. The role of AIM2 in inflammation and tumors. Front Immunol. 2024;15:1466440. https://doi.org/10.3389/fimmu.2024.1466440.
69. Li T, Lu L, Pember E, et al. New insights into neuroinflammation involved in pathogenic mechanism of Alzheimer's disease and its potential for therapeutic intervention. Cells. 2022;11:1925. https://doi.org/10.3390/cells11121925.
70. Xu C, Jing W, Liu C, et al. Cytoplasmic DNA and AIM2 inflammasome in RA: where they come from and where they go? Front Immunol. 2024;15:1343325. https://doi.org/10.3389/fimmu.2024.1343325.
71. Zhang G, Wei H, Zhao A, et al. Mitochondrial DNA leakage: underlying mechanisms and therapeutic implications in neurological disorders. J Neuroinflammation. 2025;22:34. https://doi.org/10.1186/s12974-025-03363-0.
72. Piancone F, La Rosa F, Marventano I, et al. The role of the inflammasome in neurodegenerative diseases. Molecules. 2021;26:953. https://doi.org/10.3390/molecules26040953.
73. Yang K, Wang X, Pan H, et al. The roles of AIM2 in neurodegenerative diseases: insights and therapeutic implications. Front Immunol. 2024;15:1441385. https://doi.org/10.3389/fimmu.2024.1441385.
74. Nguyen TT, Wei S, Nguyen TH, et al. Mitochondria-associated programmed cell death as a therapeutic target for age-related disease. Exp Mol Med. 2023;55:1595–619. https://doi.org/10.1038/s12276-023-01046-5.
75. Junyi L, Yueyang W, Bin L, et al. Gut microbiota mediates neuroinflammation in Alzheimer's disease: unraveling key factors and mechanistic insights. Mol Neurobiol. 2025;62:3746–63. https://doi.org/10.1007/s12035-024-04513-w.
76. Biswas K. Microglia mediated neuroinflammation in neurodegenerative diseases: a review on the cell signaling pathways involved in microglial activation. J Neuroimmunol. 2023;383:578180. https://doi.org/10.1016/j.jneuroim.2023.578180.
77. Pekna M, Pekny M. The complement system: a powerful modulator and effector of astrocyte function in the healthy and diseased central nervous system. Cells. 2021;10:1812. https://doi.org/10.3390/cells10071812.
78. Wen L, Bi D, Shen Y. Complement-mediated synapse loss in Alzheimer's disease: mechanisms and involvement of risk factors. Trends Neurosci. 2024;47:135–49. https://doi.org/10.1016/j.tins.2023.11.010.
79. Wedervang-Resell K, Szabo A. The immunopsychiatry of early-onset psychosis. In: Adolescent psychosis. Elsevier; 2023. p. 153–75.
80. Daborg J, Andreasson U, Pekna M, et al. Cerebrospinal fluid levels of complement proteins C3, C4 and CR1 in Alzheimer's disease. J Neural Transm (Vienna). 2012;119:789–97. https://doi.org/10.1007/s00702-012-0797-8.
81. Beder N, Belkhelfa M, Leklou H. Involvement of inflammasomes in the pathogenesis of Alzheimer's disease. J Alzheimer's Dis. 2024;102:11–29. https://doi.org/10.1177/13872877241283677.
82. Stanca S, Rossetti M, Bongioanni P. Astrocytes as neuroimmunocytes in Alzheimer's disease: a biochemical tool in the neuron-glia crosstalk along the pathogenetic pathways. Int J Mol Sci. 2023;24:13880. https://doi.org/10.3390/ijms241813880.
83. Figiel I. Pro-inflammatory cytokine TNF-alpha as a neuroprotective agent in the brain. Acta Neurobiol Exp (Wars). 2008;68:526–34. https://doi.org/10.55782/ane-2008-1720.
84. Singh A, Ansari VA, Mahmood T, et al. Neurodegeneration: microglia: Nf-kappab signaling pathways. Drug Res. 2022;72:496–9. https://doi.org/10.1055/a-1915-4861.
85. Medzhitov R, Horng T. Transcriptional control of the inflammatory response. Nat Rev Immunol. 2009;9:692–703. https://doi.org/10.1038/nri2634.

86. Tseng W-Y, Huang Y-S, Lin H-H, et al. TNFR signalling and its clinical implications. Cytokine. 2018;101:19–25. https://doi.org/10.1016/j.cyto.2016.08.027.
87. Jha NK, Jha SK, Kar R, et al. Nuclear factor-kappa β as a therapeutic target for Alzheimer's disease. J Neurochem. 2019;150:113–37. https://doi.org/10.1111/jnc.14687.
88. Amelimojarad M, Amelimojarad M, Cui X. The emerging role of brain neuroinflammatory responses in Alzheimer's disease. Front Aging Neurosci. 2024;16:1391517. https://doi.org/10.3389/fnagi.2024.1391517.
89. Kaur D, Sharma V, Deshmukh R. Activation of microglia and astrocytes: a roadway to neuroinflammation and Alzheimer's disease. Inflammopharmacology. 2019;27:663–77. https://doi.org/10.1007/s10787-019-00580-x.
90. Massaad CA, Klann E. Reactive oxygen species in the regulation of synaptic plasticity and memory. Antioxid Redox Signal. 2011;14:2013–54. https://doi.org/10.1089/ars.2010.3208.
91. Chen C-H, Zhou W, Liu S, et al. Increased NF-κB signalling up-regulates BACE1 expression and its therapeutic potential in Alzheimer's disease. Int J Neuropsychopharmacol. 2012;15:77–90. https://doi.org/10.1017/S1461145711000149.
92. Rather MA, Khan A, Alshahrani S, et al. Inflammation and Alzheimer's disease: mechanisms and therapeutic implications by natural products. Mediators Inflamm. 2021;2021:9982954. https://doi.org/10.1155/2021/9982954.
93. Govindarajulu M, Ramesh S, Beasley M, et al. Role of cGAS-STING signaling in Alzheimer's disease. Int J Mol Sci. 2023;24:8151. https://doi.org/10.3390/ijms24098151.
94. Nikoletopoulou V, Papandreou M-E, Tavernarakis N. Autophagy in the physiology and pathology of the central nervous system. Cell Death Differ. 2015;22:398–407. https://doi.org/10.1038/cdd.2014.204.
95. Reddy PH, Oliver DM. Amyloid beta and phosphorylated tau-induced defective autophagy and mitophagy in Alzheimer's disease. Cells. 2019;8:488. https://doi.org/10.3390/cells8050488.
96. De Gaetano A, Solodka K, Zanini G, et al. Molecular mechanisms of mtDNA-mediated inflammation. Cells. 2021;10:2898. https://doi.org/10.3390/cells10112898.
97. Hopfner K-P, Hornung V. Molecular mechanisms and cellular functions of cGAS-STING signalling. Nat Rev Mol Cell Biol. 2020;21:501–21. https://doi.org/10.1038/s41580-020-0244-x.
98. Fan X, Diao W, Wang H, et al. Interferon regulatory factors as a potential therapeutic target for neuroinflammation: a focus on Alzheimer's disease. Int J Mol Sci. 2025;26:2906. https://doi.org/10.3390/ijms26072906.
99. Griffin WS, Sheng JG, Royston MC, et al. Glial-neuronal interactions in Alzheimer's disease: the potential role of a "cytokine cycle" in disease progression. Brain Pathol. 1998;8:65–72. https://doi.org/10.1111/j.1750-3639.1998.tb00136.x.
100. Liu J, Zhou J, Luan Y, et al. cGAS-STING, inflammasomes and pyroptosis: an overview of crosstalk mechanism of activation and regulation. Cell Commun Signal. 2024;22:22. https://doi.org/10.1186/s12964-023-01466-w.
101. Andronie-Cioara FL, Ardelean AI, Nistor-Cseppento CD, et al. Molecular mechanisms of neuroinflammation in aging and Alzheimer's disease progression. Int J Mol Sci. 2023;24:1869. https://doi.org/10.3390/ijms24031869.
102. Decout A, Katz JD, Venkatraman S, Ablasser A. The cGAS-STING pathway as a therapeutic target in inflammatory diseases. Nat Rev Immunol. 2021;21:548–69. https://doi.org/10.1038/s41577-021-00524-z.
103. Michetti F, Donno C, D'Ambrosi N. The S100B-RAGE pathway is dysregulated in the ALS-linked neuroinflammatory process. Springerplus. 2015;4:P27. https://doi.org/10.1186/2193-1801-4-S1-P27.
104. Michetti F, Clementi ME, Di Liddo R, et al. The S100B protein: a multifaceted pathogenic factor more than a biomarker. Int J Mol Sci. 2023;24:9605. https://doi.org/10.3390/ijms24119605.

105. Li W, Chen Q, Peng C, et al. Roles of the receptor for advanced glycation end products and its ligands in the pathogenesis of Alzheimer's disease. Int J Mol Sci. 2025;26:403. https://doi.org/10.3390/ijms26010403.
106. Swaroop S, Sengupta N, Suryawanshi AR, et al. HSP60 plays a regulatory role in IL-1-β-induced microglial inflammation via TLR4-p38 MAPK axis. J Neuroinflammation. 2016;13:27. https://doi.org/10.1186/s12974-016-0486-x.
107. Langeh U, Singh S. Targeting S100B protein as a surrogate biomarker and its role in various neurological disorders. Curr Neuropharmacol. 2021;19:265–77. https://doi.org/10.2174/1570159X18666200729100427.
108. Cai Z, Liu N, Wang C, et al. Role of RAGE in Alzheimer's disease. Cell Mol Neurobiol. 2016;36:483–95. https://doi.org/10.1007/s10571-015-0233-3.
109. Goswami D, Anuradha U, Angati A, et al. Pharmacological and pathological relevance of S100 proteins in neurological disorders. CNS Neurol Disord Drug Targets. 2023;22:1403–16. https://doi.org/10.2174/1871527322666221128160653.
110. Salmina AB. Neuron-glia interactions as therapeutic targets in neurodegeneration. J Alzheimer's Dis. 2009;16:485–502. https://doi.org/10.3233/JAD-2009-0988.
111. Chen R, Zou J, Zhong X, et al. The DAMP theory: concepts, evidence, and implications. 2024. https://doi.org/10.22541/au.173372824.45261156/v1.
112. Wang Y, Ning X, Gao P, et al. Inflammasome activation triggers caspase-1-mediated cleavage of cGAS to regulate responses to DNA virus infection. Immunity. 2017;46:393–404. https://doi.org/10.1016/j.immuni.2017.02.011.
113. Shi J, Zhao Y, Wang Y, et al. Inflammatory caspases are innate immune receptors for intracellular LPS. Nature. 2014;514:187–92. https://doi.org/10.1038/nature13683.
114. Barrow ER, Valionyte E, Baxter CR, et al. Discovery of SQSTM1/p62-dependent P-bodies that regulate the NLRP3 inflammasome. Cell Rep. 2024;43:113935. https://doi.org/10.1016/j.celrep.2024.113935.
115. Zhao S, Li X, Wang J, Wang H. The role of the effects of autophagy on NLRP3 inflammasome in inflammatory nervous system diseases. Front Cell Dev Biol. 2021;9:657478. https://doi.org/10.3389/fcell.2021.657478.
116. Ulland TK, Song WM, Huang SC-C, et al. TREM2 maintains microglial metabolic fitness in Alzheimer's disease. Cell. 2017;170:649–663.e13. https://doi.org/10.1016/j.cell.2017.07.023.
117. Deczkowska A, Weiner A, Amit I. The physiology, pathology, and potential therapeutic applications of the TREM2 signaling pathway. Cell. 2020;181:1207–17. https://doi.org/10.1016/j.cell.2020.05.003.
118. Liddelow SA, Guttenplan KA, Clarke LE, et al. Neurotoxic reactive astrocytes are induced by activated microglia. Nature. 2017;541:481–7. https://doi.org/10.1038/nature21029.
119. Mildner A, Schlevogt B, Kierdorf K, et al. Distinct and non-redundant roles of microglia and myeloid subsets in mouse models of Alzheimer's disease. J Neurosci. 2011;31:11159–71. https://doi.org/10.1523/JNEUROSCI.6209-10.2011.
120. Gate D, Saligrama N, Leventhal O, et al. Clonally expanded CD8 T cells patrol the cerebrospinal fluid in Alzheimer's disease. Nature. 2020;577:399–404. https://doi.org/10.1038/s41586-019-1895-7.
121. Zhao Y, Cong L, Jaber V, Lukiw WJ. Microbiome-derived lipopolysaccharide enriched in the perinuclear region of Alzheimer's disease brain. Front Immunol. 2017;8:1064. https://doi.org/10.3389/fimmu.2017.01064.
122. Rob M, Yousef M, Lakshmanan AP, et al. Microbial signatures and therapeutic strategies in neurodegenerative diseases. Biomed Pharmacother. 2025;184:117905. https://doi.org/10.1016/j.biopha.2025.117905.
123. Gao C, Fang L, Zhang H, et al. Metformin induces autophagy via the AMPK-mTOR signaling pathway in human hepatocellular carcinoma cells. Cancer Manag Res. 2020;12:5803–11. https://doi.org/10.2147/CMAR.S257966.

124. Pietrocola F, Lachkar S, Enot DP, et al. Spermidine induces autophagy by inhibiting the acetyltransferase EP300. Cell Death Differ. 2015;22:509–16. https://doi.org/10.1038/cdd.2014.215.
125. Song J-X, Sun Y-R, Peluso I, et al. A novel curcumin analog binds to and activates TFEB in vitro and in vivo independent of MTOR inhibition. Autophagy. 2016;12:1372–89. https://doi.org/10.1080/15548627.2016.1179404.
126. Basyuda H, Arikawa K, Morishita Y, et al. Graft versus host disease in a human T cell leukemia virus-I carrier. Nihon Kyobu Geka Gakkai Zasshi. 1989;37:579–83.
127. Lamark T, Johansen T. Aggrephagy: selective disposal of protein aggregates by macroautophagy. Int J Cell Biol. 2012;2012:736905. https://doi.org/10.1155/2012/736905.
128. Coll RC, Robertson AAB, Chae JJ, et al. A small-molecule inhibitor of the NLRP3 inflammasome for the treatment of inflammatory diseases. Nat Med. 2015;21:248–55. https://doi.org/10.1038/nm.3806.
129. Zahid A, Li B, Kombe AJK, et al. Pharmacological inhibitors of the NLRP3 inflammasome. Front Immunol. 2019;10:2538. https://doi.org/10.3389/fimmu.2019.02538.
130. Zhao P, Xu Y, Fan X, et al. Discovery and engineering of an anti-TREM2 antibody to promote amyloid plaque clearance by microglia in 5XFAD mice. MAbs. 2022;14:2107971. https://doi.org/10.1080/19420862.2022.2107971.
131. Song X, Wu W, Warner M, Gustafsson J-Å. Liver X receptor regulation of glial cell functions in the CNS. Biomedicine. 2022;10:2165. https://doi.org/10.3390/biomedicines10092165.
132. Ribierre T, Bacq A, Donneger F, et al. Targeting pathological cells with senolytic drugs reduces seizures in neurodevelopmental mTOR-related epilepsy. Nat Neurosci. 2024;27:1125–36. https://doi.org/10.1038/s41593-024-01634-2.
133. Chaib S, Tchkonia T, Kirkland JL. Cellular senescence and senolytics: the path to the clinic. Nat Med. 2022;28:1556–68. https://doi.org/10.1038/s41591-022-01923-y.
134. Khan MS, Qureshi N, Khan R, et al. CRISPR/Cas9-based therapeutics as a promising strategy for management of Alzheimer's disease: progress and prospects. Front Cell Neurosci. 2025;19:1578138. https://doi.org/10.3389/fncel.2025.1578138.
135. Raulin A-C, Doss SV, Trottier ZA, et al. ApoE in Alzheimer's disease: pathophysiology and therapeutic strategies. Mol Neurodegener. 2022;17:72. https://doi.org/10.1186/s13024-022-00574-4.
136. Raffaele I, Cipriano GL, Anchesi I, et al. CRISPR/Cas9 and iPSC-based therapeutic approaches in Alzheimer's disease. Antioxidants. 2025;14:781. https://doi.org/10.3390/antiox14070781.
137. Rungta RL, Choi HB, Lin PJ, et al. Lipid nanoparticle delivery of siRNA to silence neuronal gene expression in the brain. Mol Ther Nucleic Acids. 2013;2:e136. https://doi.org/10.1038/mtna.2013.65.
138. Kiaie SH, Zangi AR, Sheibani M, et al. Novel synthesized ionizable lipid for LNP-mediated P2X7siRNA to inhibit migration and induce apoptosis of breast cancer cells. Purinergic Signal. 2024;20:533–46. https://doi.org/10.1007/s11302-024-09989-8.
139. Wilson H, Politis M, Rabiner EA, Middleton LT. Novel PET biomarkers to disentangle molecular pathways across age-related neurodegenerative diseases. Cells. 2020;9:2581. https://doi.org/10.3390/cells9122581.
140. Liao K, Chen J, Ma J, et al. Preclinical characterization of [18F]D2-LW223: an improved metabolically stable PET tracer for imaging the translocator protein 18 kDa (TSPO) in neuroinflammatory rodent models and non-human primates. Acta Pharmacol Sin. 2025;46:393–403. https://doi.org/10.1038/s41401-024-01375-9.
141. Anoop A, Singh PK, Jacob RS, Maji SK. CSF biomarkers for Alzheimer's disease diagnosis. Int J Alzheimers Dis. 2010;2010:606802. https://doi.org/10.4061/2010/606802.
142. Sarty IM, Picard C, Labonte A, et al. Longitudinal association between CSF markers of Alzheimer disease and inflammation in preclinical Alzheimer disease. Alzheimers Dement. 2024;20:e087640. https://doi.org/10.1002/alz.087640.

143. Collins HM, Greenfield S. Rodent models of Alzheimer's disease: past misconceptions and future prospects. Int J Mol Sci. 2024;25:6222. https://doi.org/10.3390/ijms25116222.
144. Castiglione H, Vigneron P-A, Baquerre C, et al. Human brain organoids-on-chip: advances, challenges, and perspectives for preclinical applications. Pharmaceutics. 2022;14:2301. https://doi.org/10.3390/pharmaceutics14112301.
145. Choi SH, Kim YH, Hebisch M, et al. A three-dimensional human neural cell culture model of Alzheimer's disease. Nature. 2014;515:274–8. https://doi.org/10.1038/nature13800.
146. Ma D, Li S, Molusky MM, Lin JD. Circadian autophagy rhythm: a link between clock and metabolism? Trends Endocrinol Metab. 2012;23:319–25. https://doi.org/10.1016/j.tem.2012.03.004.
147. Coppedè F. The potential of epigenetic therapies in neurodegenerative diseases. Front Genet. 2014;5:220. https://doi.org/10.3389/fgene.2014.00220.
148. Zhavoronkov A. Artificial intelligence for drug discovery, biomarker development, and generation of novel chemistry. Mol Pharm. 2018;15:4311–3. https://doi.org/10.1021/acs.molpharmaceut.8b00930.

Inflammation and Autophagic Dysfunction in Parkinson's Disease

9

Hannah Lalengzuali Fanai, Jagdish Chand, and Akil Akhtar

Abstract

Parkinson's disease (PD) is a complex neurodegenerative disorder driven by the loss of dopamine-producing neurons in the substantia nigra and a convergence of cellular dysfunctions, including α-synuclein accumulation, mitochondrial and lysosomal failure, and chronic inflammation. This chapter explains PD as a systems-level disorder. In PD, disrupted protein clearance through chaperone-mediated autophagy (CMA), impaired mitochondrial quality control, and overactive immune responses form a self-reinforcing cycle of damage. Modified α-synuclein blocks CMA by overloading LAMP2A, leading to protein build-up and lysosomal stress, while mitochondrial defects, caused by abnormal fission–fusion dynamics and impaired mitophagy, lead to excess reactive oxygen species (ROS). Oxidative by-products from dopamine metabolism and lipid breakdown place additional strain on mitochondria, impairing the function of key components like Parkin and complex I. As cellular damage progresses, mitochondrial fragments and reactive oxygen species (ROS) act as distress signals, triggering toll-like receptor pathways and initiating inflammatory responses through glial cells. This inflammation, in turn, disrupts synaptic communication and

H. L. Fanai
Department of Pharmaceutical Chemistry, JSS College of Pharmacy, JSS Academy of Higher Education & Research, Ooty, The Nilgiris, Tamil Nadu, India

J. Chand (✉)
Department of Pharmaceutical Chemistry, JSS College of Pharmacy, JSS Academy of Higher Education & Research, Ooty, The Nilgiris, Tamil Nadu, India

Silicon Script Sciences Private Limited, Ghorahi, Dang, Nepal

A. Akhtar
Emory Vaccine Center, Department of Microbiology and Immunology, Emory University, Atlanta, GA, USA

P. Khanal et al. (eds.), *Neuroinflammation and Autophagy in Neurodegeneration*,
https://doi.org/10.1007/978-981-95-3945-1_9

contributes to neuronal vulnerability. In response, emerging treatments are being developed to address these underlying cellular disturbances, such as compounds that enhance autophagy, drugs that restore lysosomal function, targeted immune therapies, and advanced delivery systems using CRISPR technology. This chapter underscores the importance of tailoring these interventions to the diverse genetic backgrounds of PD patients and recommends future research into spatial gene expression and organelle coordination as pathways toward more effective, individualized therapies.

9.1 Introduction

Parkinson's disease (PD) is a progressive neurodegenerative disorder characterized by the loss of dopaminergic neurons in the substantia nigra pars compacta (SNc) and the consequent depletion of striatal dopamine levels [1]. Clinically, PD presents with cardinal motor symptoms, including bradykinesia, rigidity, resting tremor, and postural instability, along with a range of non-motor manifestations such as sleep disturbances, mood disorders, cognitive impairment, and autonomic dysfunction [2]. In addition to the deficiency of dopamine, PD is also conceptualized nowadays as a complex syndrome that has a multifaceted, genetically and environmentally dependent interaction with cellular responses of stress. The misfolded protein of α-synuclein has been seen as the central assumption of PD pathology, and that aggregation plays a central role in its pathogenesis, leading to the formation of intracellular Lewy bodies and Lewy neurites, which are the aggregates of the misfolded protein [3]. The hypothesis of PD is proteinopathy, and the aggregates are a crucial part, meaning that the abnormal processing, aggregation, and clearance of α-synuclein spread into the whole brain via neurotoxic processes. Protein aggregation is only one of the factors that alter cellular homeostasis in PD, as the dysfunction of mitochondria and lysosomal impairment are also critical factors in PD pathogenesis. Affected neurons have an impaired complex I activity of mitochondria, increased reactive oxygen species (ROS) generation, and impaired bioenergetics [4]. These defects facilitate oxidative insults on proteins, lipids, and nucleic acids, worsening cell susceptibility. Simultaneously, lysosomal and autophagy processes that are critical during the degradation of long-lived proteins and damaged organelles are disrupted. Important aspects of the autophagy–lysosome pathway, such as autophagosome formation, acidification of the lysosomes, and degradation of enzymes, have very serious deficiencies in PD. The result of this is the accumulation of dysfunctional mitochondria and soluble aggregates of proteins, and the development of a toxic intracellular pathway results in accelerated neuronal degeneration. A third axis of PD pathophysiology, which is complementary and exacerbates the harmful effects of proteinopathy and organelle failure, is neuroinflammation. Activated microglia are the central nervous system resident immune cells, which have consistently been shown to be increased in PD postmortem tissue as well as animal models [5]. The release of pro-inflammatory cytokines such as tumor necrosis factor-alpha (TNF-α), interleukin-1β, and interferon-γ by these cells further derails mitochondrial functioning, along with thwarting autophagic traffic. Additionally, misfolded protein α-synuclein may assume a damage-associated molecular pattern (DAMP) state, which activates toll-like receptors (TLRs) on glial cells and causes an innate immune response to be maintained [6]. This creates a negative feedback loop where protein aggregates trigger the inflammation process, which worsens cellular stress and damages degradative mechanisms. This study will use this crosstalk to provide an insight into immune dysregulation, bidirectional crosstalk of autophagic dysfunction, and mitochondrial-ROS cascades amidst the scenario of dopaminergic neurodegeneration. Much focus is placed on the symbiotic nature of these systems that mitochondrial system damage can lead not just to ROS generation but to an auto-regulated process, mitophagy, involving the autophagic degradation of impaired mitochondria [7]. The inability of mitophagy leads to defective organelle survival, which further increases oxidative stress and inflammatory signals. In parallel, the continued neuroinflammation as a result of GAD impairs the autophagy signaling pathways, for example, mTOR and AMPK, causing lysosomal membrane permeabilization and discharge of enzymes. The ensuing cellular environment assumes an increasingly adverse one that eventually leads to apoptotic and necroptotic neuronal death [8]. This chapter integrates recent knowledge of neuropathological observations, genetic modeling, and molecular analyses to give a complete interpretation of the inner connections between autophagic decline, mitochondrial disease, and immune activation, which is a cause of neurodegeneration in PD. Life at the molecular nodes of these systems, where they converge, can be deciphered, and possibly promising therapeutic targets can be found that can somehow restore homeostatic balance and relax neuroinflammation. Thereby, this chapter is making a smaller meal in the overall support of reframing PD not just as a disorder of dopaminergic transmission but rather as a paradigmatic neurodegenerative disease through mutual distortion in the processes of proteostasis, organellar structure, and innate immunity.

9.2 Molecular Mechanisms of Pathogenesis

9.2.1 α-Synuclein Aggregation and Autophagy–Lysosomal Dysfunction

Clustering of α-synuclein is considered a key event in the neurodegenerative process of PD and has been used as a point of proteostatic failure and lysosomal damage [9]. Different post-translational modifications of α-synuclein, such as serine 129 phosphorylation and ubiquitination, cooperate in the establishment of oligomers and the aggregation of α-synuclein into fibrils [10]. Phosphorylated α-synuclein has greater resistance to degradation combined with an increased tendency toward the formation of β-sheet-rich oligomers that are neurotoxic and disrupt proteostasis. The proteins and their accumulation are regulated through the lysosome-associated membrane protein type 2A (LAMP2A)-mediated pathway that inhibits the accumulation of damaged or misfolded proteins by the selective movement of cytosolic proteins into lysosomes via chaperone-mediated autophagy [11]. The pathological accumulation of α-synuclein at LAMP2A receptor saturates the receptor, and this outcompetes the transport of other CMA ligands in PD, hence, efficient transport of the toxic protein aggregates to the lysosome. Direct lysosomal dysregulation caused by α-synuclein clumps also disrupts the lysosomal pathway that is crucial in neuronal homeostasis.

α-synuclein in fibrillar form has been found to cause the inhibition of vacuolar-type H + -ATPase (V-ATPase). The challenge with deregulation of lysosomal pH is that it not only affects the activities of the enzymes but also the recognition and breakdown of cargo [12]. In sporadic PD, this is applicable because glucocerebrosidase, which is a lysosomal hydrolase involved in glycolipid degradation, suffers oxidative modification, which, besides affecting its actual potential catalytic activity, additionally impairs its enzymatic activity [13]. The impairment of GBA1 activity due to the dysfunctional GBA1 affects the clearance of α-synuclein, which contributes to a feedback loop of a failed lysosome and the accumulation of α-synuclein. This vicious cycle has been seen to contribute significantly to neuronal vulnerability and degeneration of PD.

Besides environmental and idiopathic factors, the autophagic and lysosomal malfunction in PD is worsened by the presence of genetic mutations. Defective transport of polyamines occurs due to a mutation in ATP13A2, a P5-type ATPase in lysosomes, which leads to impaired integrity of lysosomal membranes and thus makes the affected cells more vulnerable to proteotoxic stress [14]. The shortage of ATP13A2 has been linked to autophagic vacuoles and inefficient lysosome degradation. A parallel comes through VPS35, which is a central protein in the retromer complex involved in ferrying sorting receptors back to the Golgi through endosomal trafficking-mutations in VPS35 cause mislocalization of lysosomal enzymes [15]. This trafficking pathway inhibition interferes with lysosomal proteolytic activity and increases α-synuclein pathology. All these molecular imbalances characterize an ultimate mechanistic landscape as one where the pathological dynamics of α-synuclein propagate and lysosomal defense with inherited propensities converge to compromise the proteostasis of the neurons.

9.2.2 Mitochondrial Dysfunction and ROS Propagation

The central hallmark of PD is mitochondrial dysfunction, which is one of the main precursors of cellular stress and loss. The SNc neurons are vulnerable to changes in mitochondrial movements because their high metabolic requirements necessitate and depend upon oxidative phosphorylation [16]. It has robust impairment with regard to the balance in mitochondrial fission and fusion in PD. There is hyperactivation of dynamin-related protein 1 (DRP1) that results in insufficient fragmentation of the mitochondria. Simultaneously, the fusion mediators like optic atrophy 1 (OPA1) and mitofusin 2 (MFN2) are readily downregulated, affecting mitochondrial elongation

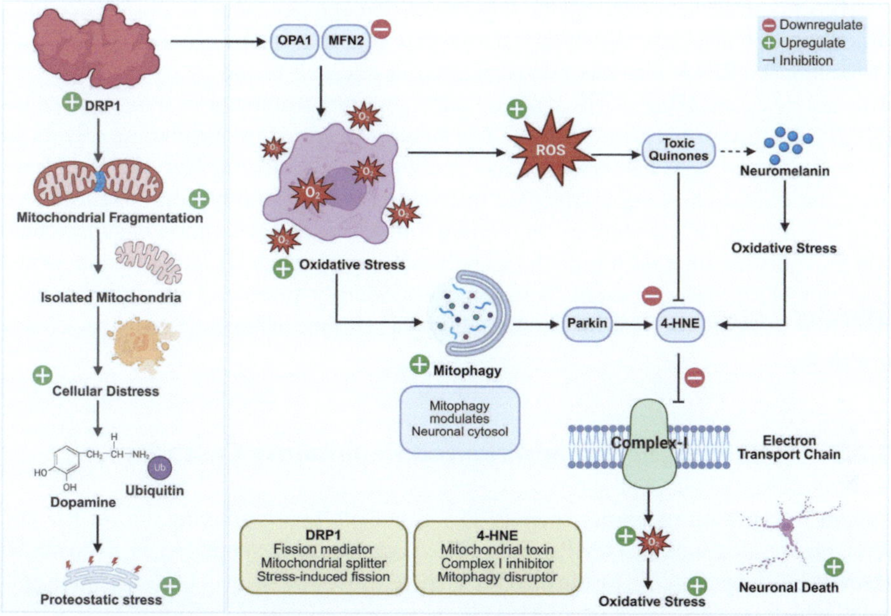

Fig. 9.1 Mitochondrial dysfunction and proteostatic collapse in PD. Oxidation of dopamine produces reactive quinones and neuromelanin, which disrupt the ubiquitin-proteasome system and increase proteostatic stress. Excessive DRP1 has been shown to cause mitochondrial fragmentation, and lowering of OPA1/MFN2 inhibits fusion, giving rise to resistant to mitophagy ROS-generating mitochondria. Oxidative stress accumulation leads to the production of 4-hydroxynonenal (4-HNE), inactivating the mitochondrial complex I activity in addition to suppressing Parkin E3 ligase. This process increases ROS and degenerates dopaminergic neurons in PD. Created using BioRender (https://www.biorender.com/)

and network integrity. The consequence of this imbalance is the creation of small, isolated mitochondria, which are poor producers of energy and which are susceptible to oxidative damage. Besides, these mitochondria are impervious to conventional mitophagy, thus piling up in the neuronal cytosol to cause cellular distress [17]. The redox state of dopaminergic neurons is further aggravated by auto-oxidation of dopamine, the formation of cytotoxic quinones, and neuromelanin. These quinones may conjugate with proteins (including components of the ubiquitin-proteasome system) and thus interfere with proteasomal proteolysis and increase proteostatic stress. Moreover, another by-product of lipid peroxidation, 4-hydroxynonenal, alters subunits of mitochondrial complex I, worsening the functioning of the electron transport chain [18]. The E3 ubiquitin ligase property of Parkin, which is an essential regulator of mitophagy, is also inhibited by 4-hydroxynonenal, and thus, this factor maintains the process of accumulating dysfunctional mitochondria. The combination of these oxidative injuries impairs the integrity of mitochondria and strengthens the feed-forward loop of energy deficiency and ROS expansion (Fig. 9.1).

Mitochondrial-generated ROS may cause damage not only to the intracellular structures but also are signaling factors that regulate innate immune responses. Mitochondrial DNA that has been oxidized is released to the cytosol, which gets detected by cyclic GMP-AMP synthase, activating the stimulator of interferon genes (STING) pathway in astrocytes [19]. The signal cascade exerts promoting effects on the formation of type I interferons and upregulates pro-inflammatory genes. Moreover, the oxidation of superannuated micromembrane-bound protein NADPH oxidase 2 in microglia and neurons target tissues results in the generation of superoxide, which works together with mitochondrial ROS to perpetuate NLRP3 inflammasome activation [20]. This crosstalk between the oxidative stress and the inflammatory signaling protects the neurons and enables the chronic neuroinflammatory effects in PD.

9.2.3 TLR2/4 Signaling and Neuroinflammatory Cascades

Pattern recognition receptors such as TLRs mediate neuroinflammation in PD and recognize endogenous DAMPs [21, 22]. Aggregated, α-synuclein is a powerful DAMP binding directly to the leucine-rich repeat domain of TLR2 on microglia cells, which induces cascades of MyD88-dependent signals. Conformational activation is related to the translocation of nuclear factor-kB and pro-inflammatory cytokines. Similarly, high mobility group box 1, a nuclear toll, whose liberation is caused by either necrotic or dying neurons, attaches itself to TLR4 and expands the reaction to inflammation, with the help of MyD88 and TRIF adaptations [23]. Such processes trigger an exuberant inflammatory microenvironment that encourages extended neuronal degradation and strengthens glia responsiveness. Recruitment of downstream signaling mechanisms attributed to the activation of TLRs includes several converging signaling pathways that enhance immune dysregulation. Interferon regulatory factor 3 activation depends on TRIF and promotes the production of interferon-β and the induction of the complement proteins, including C3, which facilitates synaptic pruning and neuronal stress [24]. Moreover, aberrant expression of microRNA-155 in microglia silences suppressor of cytokine signaling 1, maintaining the JAK/STAT pathway status and prolonging the process of producing inflammatory cytokines [25]. This epigenetic reprogramming of microglia phenotype further enhances chronic inflammation and triggers the development of neurodegenerative disorders.

In addition to microglia, astrocytes, as well as peripheral immune cells, are major contributors to neuroinflammation in PD. Astrocytes produce chemokines, including CCL2, that enable the migration of peripheral monocytes through a damaged blood–brain barrier (BBB) and further escalate the local immune cell infiltration. Neurons in themselves are not inert observers. Relatively, they secrete their CX3CL1 (fractalkine) binding to CX3CR1 receptor on the microglia, affecting its synaptic inspector, facilitating synaptic stripping. The effect of CX3CL1 signaling pathway dysregulation includes the interruption of communication between neurons with glia, as well as the disruption of synaptic integrity [26]. Overall, these immune cascades draw attention to the multifactorial and systemic features of PD inflammation, where the cellular, molecular, and immune mechanisms are interacting towards the continuation of neurodegeneration.

9.3 Dopaminergic Neuron Vulnerability

9.3.1 Metabolic and Structural Susceptibility

Oxidative stress affecting dopaminergic SNc neurons is especially vulnerable because of a highly specific metabolic profile and morphological characteristics. Misregulation of iron homeostasis is one of the decisive factors. In normal cells, iron control is strict, so that it does not cause the synthesis of toxic reactive species [27]. However, the autophagy, ferritinophagy, which is a specialization of autophagy, has a pathological upregulation in PD. This generates an intracellular ferroptosis iron overload, which can promote the chemistry of Fenton, where ferrous iron (Fe_2^+) promotes the formation of hydroxyl radicals that damage lipid, protein, and DNA. Simultaneously, there has also been reported upregulation of divalent metal transporter 1 in SNc neurons, increasing the entry of redox-active iron in mitochondria even further [28]. This regional formation of iron increases the production of mitochondrial ROS, disrupts the stability of the membrane potential, and distorts the capacity of ATP synthesis, leaving dopaminergic neurons in metabolic dysfunction.

The degeneration of this neuron is also worsened by structural vulnerabilities of these neurons. SNc neurons have dense axonal arborization, and their ability to support adequate exchange of organelles and proteins through axonal transport, ensures dopaminergic neurotransmission, neuronal resilience, and defense against the degeneration-related stress, thus maintaining motor and cognitive abilities. In PD, the retrograde transport process that may be facilitated mainly by the dynein–dynactin complex is impaired [29]. This breakdown results in increased autophagosomes parenchyma and impaired mitochondria in the distal axonal swellings, further restricting their breakdown over time and setting the stage for localized energy shortages. In addition, the tau protein that stabilizes the microtubules is hyperphosphorylated on pathological epitopes (e.g., AT8), thereby losing microtubule binding affinities and leading to structural instability of the microtubules. The same microtubule destabilization is used to further damage the axonal transport and the initiation and progression of mitophagy [30]. The combination of oxidative damage caused by iron and that of disruption of intracellular trafficking makes it possible to consider a facet of biogenetic vulnerability that predisposes dopaminergic neurons towards undergoing degeneration.

9.3.2 Autophagy–Mitochondria Crosstalk in Substantia Nigra Pars Compacta SNc

Mitochondria act in support of an endoplasmic reticulum (ER) complex interaction that is critically important in calcium signaling, lipid metabolism, and organelle turnover. This interaction also occurs in dopaminergic neurons, in which mitochondria-associated ER membranes (MAMs) act as a structural and functional bridge that maintains a Ca_2^+ flux required to initiate mitophagy [31]. PD has been linked with loss-of-function mutations and downregulation of ER-resident protein disulfide isomerase PDIA3/ERp57 that compromises MAM integrity in PD. This structural disintegration hampers the formation of the IP3R-GRP75-VDAC1 complex that transfers Ca_2^+ between the ER and the mitochondria. This mitochondrial Ca_2^+ (Fig. 9.2) load also impairs the depolarization-stimulated migration of PINK1 and Parkin, central mediators of mitophagy, by impeding the successful removal of

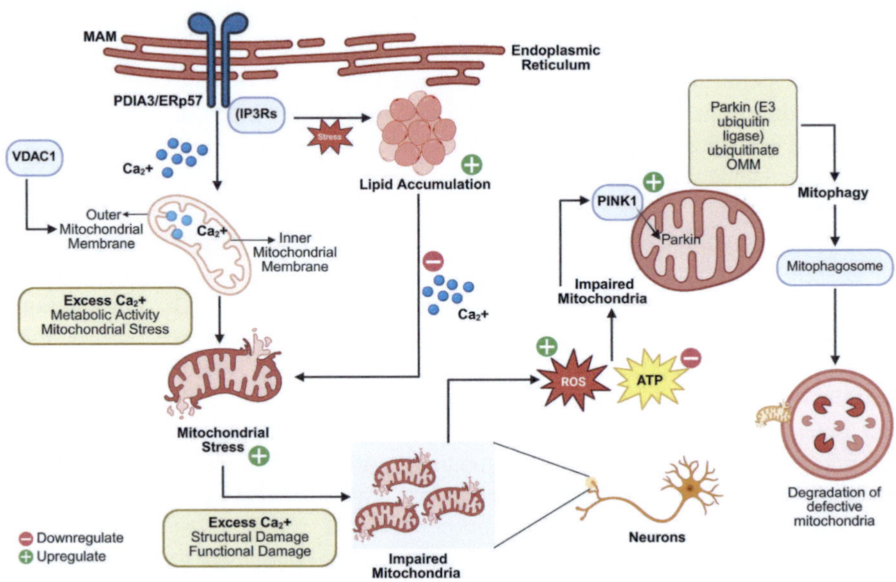

Fig. 9.2 Mitochondrial dysfunction, dependent calcium signaling, and mitophagy in PD. A crosstalk between the ER and the mitochondria in dopaminergic neurons through the MAMs. The ER protein PDIA3 contributes under physiological conditions to the stabilization of the IP3R-GRP75-VDAC1 complex, existing to enable regulated transfer of Ca_2^+ into mitochondria, with the support of bioenergetics. In PD, MAMs' perturbation damages this complex and causes impaired Ca_2^+ entry, excess lipid, and mitochondrial stress. Defective mitochondria containing excessive Ca_2^+ and producing ROS activate PINK1 stabilization and Parkin enrolment, introducing mitophagy to remove them. Nonetheless, a permanent mitochondrial impairment along with its poor clearance leads to the accumulation of mitophagosomes, proteostatic stress, and neuroinflammation. This chain enhances neurodegeneration of dopaminergic neurons, making the failure of ER-mitochondria signaling pathological in PD. Created using BioRender (https://www.biorender.com/)

defective mitochondria and favoring their impoverishment in the soma and processes of the neurons [32].

A secondary insult adding to the mitochondrial-autophagy dysregulation is that of lipid peroxidation. Peroxisomes are also of utmost importance in the detoxification of lipid hydroperoxides and the production of plasmalogens, but they are often damaged in PD [33]. Plasmalogen synthesis is reduced by the downregulation of an important peroxisomal enzyme, acyl-CoA oxidase 1, and polyunsaturated fatty acid-rich membranes are more susceptible to oxidative damage. One of these targets is cardiolipin, which is a phospholipid present in the mitochondria on the inner membrane [34]. When subjected to the rigor of distress, oxidized cardiolipin transports outward to the outer mitochondrial membrane and becomes a signal of LC3-II recruitment at the onset of mitophagy. In PD, massive cardiolipin oxidation disrupts this signal transduction and results in defective mitophagosome assembly and poor mitochondrial clearance [35]. The outcome is a vicious cycle where compromised lipid metabolism, disrupted mitochondrial function, and autophagic deficiency corroborate each other and expedite the loss of SNc brain cells.

9.3.3 Inflammatory Synaptopathy and Network Collapse

PD neuronal soma is not the only structure where neuroinflammatory processes occur, since they also involve synaptic compartments, which correspond to the progressive loss of dopaminergic networks. The classical complement cascade is used by the microglia, as the major immune effectors in the central nervous system, in synaptic surveillance and pruning. Phosphorylated forms of α-synuclein aggregates in PD form complex with the synaptic membranes, which act as the attachment site for C1q deposition [36]. The result of such an event of opsonization triggers a cascade of complement that culminates in the covalent tagging of dopaminergic synapses with C3b. These tagged terminals become contacted by microglia expressing complement receptor 3, which triggers phagocytosis, and is detected in the striatum of PD models and human postmortem tissue [37]. These inappropriate pruning cuts off functional synapses and disturbs neurotransmission, and hastens the progression of motor symptoms.

At the same time, the astrocytes also contribute to the excitotoxic stress via impairment of glutamate homeostasis. TNF-α triggers the internalization of astrocytic glutamate transporter GLT-1, which is the major way of clearing glutamate in synaptic regions. The net effect is enhancement of extracellular glutamate, causing prolonged engagement of neuronal NMDA receptors. This hyperactivation allows extracellular Ca_2^+ to flow into cells and induces mitochondrial permeability shift, which destroys the mitochondrial membrane potential [38]. The energetic crisis and the release of ROS thereafter trigger the apoptotic routes and also increase local inflammation. By doing so, neuroinflammation regulates the desynchronization of synapses along with the establishment of excitotoxic pathways, interfering with neural pathways [39]. Collectively, these processes constitute a pathway of inflammatory synaptopathy whereby immune signaling, a drive of excitatory imbalance, and failure of mitochondria converge to destroy the architecture of a dopaminergic network in PD.

9.4 Advanced Multimodal Therapeutic Strategies

9.4.1 Autophagy–Lysosome-Targeted Interventions

Disease-modifying therapies such as autophagy-lysosome pathway restoration can be a potential and novel therapeutic approach in PD when the defects in proteostasis leading to α-synuclein aggregation are central to PD. Selective autophagy that transports proteins into lysosomes using chaperone activity is CMA, and is seriously impaired in PD because of competitive blockage of LAMP2A receptor by pathological misfolded α-synuclein [40]. Pharmacological enhancement of CMA is promising, with the AR7 analogues offering a therapeutic intervention to stabilize LAMP2A multimerization. Additionally, exposure to heat shock transcription factor 1-activating small molecules, for example, geranylgeranylacetone, promotes the expression of heat shock protein, which plays a vital role in the identification and transfer of CMA substrates [41]. These interventions are meant to re-establish selectivity and the throughput in CMA, which is beneficial in reducing α-synuclein toxicity during the emerging phases of lysosomal impairment.

Lysosomal reconstitution of degradative capacity is a point of the overall autophagic reaction. Recombinant glucocerebrosidase is predominantly difficult to give in instances of sporadic and genetic PD syndrome linked with GBA1 mutations because of ER-associated degradation pathways [42]. Delivery systems using nanoparticles to encapsulate acid-stable derivatives of recombinant GBA1 minimize degradation and correct lysosomal enzyme activity. A noncompetitive V-ATPase inhibitor (saliphenylhalamide) is useful as a model of lysosomal malfunction but cannot be prescribed as a restorative agent to autophagic flow in PD [43]. Rather, treatment approaches are to be personalized to increase V-ATPase-mediated or lysosomal biogenesis. Therapeutic management of autophagy-lysosome complex is a potential course of action in advanced/multimodal therapy of PD. These representatives also supplement luminal acidification to maximize the enzymatic reaction of lysosomal hydrolase. Therefore, the selective treatment of both enzyme replacement and pH balance can be considered one of the key levers of lysosomal restoration in PD.

There is further refinement of selective autophagy routes, exclusively aggrephagy and mitophagy. They can use designer ubiquitin variants that are high-avidity binders of p62/SQSTM1 to redirect the cargo adaptor to 62-SQSTM1, and this redirects them into autophagosomes [44]. Moreover, photodynamic strategies are evident to exert spatiotemporal regulation of mitophagy by using mitochondria-directed photosensitizers coupled with rapamycin-based drugs. They shift to oxidative depolarization of the agent, mTOR inhibition in the presence of light, and thus recruit LC3-II and assembly of the autophagosome [45]. These precision modes help in elucidating therapeutic potential in regulating the selectivity of autophagy, where an off-target scavenging of protein is minimal, with degradation of pathogenic substrates.

9.4.2 Precision Anti-inflammatory Strategies

Verifying the microglial reprogramming of pro-inflammatory to phagocytic condition is one of the primary aspects of neuroinflammatory regulation. Microglial phagocytic competence is an important regulator, where triggering receptor expressed on myeloid cells 2 (TREM2) plays a vital role. This is a strategy being explored using agonistic antibodies (AL002a) that enhance TREM2 signaling, clearance of extracellular α-synuclein and debris, and do not activate inflammatory cytokines [46]. Also, galectin-3, a β-galactoside-binding lectin involved in NLRP3 inflammasome priming, is increased in PD. The anti-inflammatory effects of such inhibitors as TD139 are true in obstructing glycan-amenity areas of galectin-3 and, thus, preventing inflammasome activation as well as microglia maintenance [47].

Biologics that interfere with cytokine activity are more specific to conserve the beneficial effects of signaling, whilst reversing disease-causing signaling cascades. Dominant-negative TNF-α receptors like XPro1595 noncompetitively inactivate soluble TNF trimer that mediates neurotoxicosis, but leave the TNFR2-driven neuroprotective effects [48]. Equally, IL-1β signaling has been found to worsen neuroinflammation and vulnerability of dopaminergic cells and can be neutralized through the use of IL-1β traps like rilonacept. When designed as transferrin receptor-binding domains, these fusion proteins facilitate BBB penetration and target cytokine sequestration [49]. Such a high degree of accuracy indicates a possibly high degree of therapeutic window by altering small nodes of immune system to cause a small-scale systemic interruption.

In addition to interventions on glial modulation, a wider innate immunity signaling can be investigated. The pathway involving cyclic GMP–AMP synthase-stimulator of interferon genes is observed in PD by influencing type I interferon response implemented by astrocytes [50]. The STING inhibitors, such as the H-151, may relieve chronic release of the IFN-β and glial toxicity. Moreover, peripheral immune cells infiltration into the substantia nigra mediated by chemokines can be inhibited through CCR2 antagonists like INCB3344, and this will diminish the monocyte-derived inflammation [51]. Such strategies are a testament to the treatment prospects of obstructing peripheral–central immune communication and restoring the neuroimmune microenvironment in PD.

9.4.3 Multimodal Synergy and Delivery Innovations

Due to the complicated interaction between proteostasis malfunction, mitochondrial impairment, and immune dysregulation, combinatorial treatments targeting several pathomechanisms are emerging. Transcription factor EB is a transcriptional master regulator of lysosomal biogenesis, and its activation, together with the treatment of ursodeoxycholic acid, shows synergism of lysosomal membrane repair and reinstitution of autophagic flux [52]. Likewise, when leucine-rich repeat kinase 2 (LRRK2) is inhibited synergistically with DNL201, and when senescent microglia are removed with dasatinib, it can reduce neuroinflammation, improve mitochondrial and lysosomal functioning, and enhance neuronal survival [53]. These twin tactics utilize mutually cooperative processes to frighten away therapeutic over-saturation and resistance.

However, precise administration is still a pillar in the contemporary PD medication. Simultaneous delivery of N-acetylcysteine (a ROS scavenger) and FK506 (a calcineurin inhibitor that suppresses excessive mitophagy) in liposomes containing a Trojan-horse would lead to a decrease in oxidative stress, a reinstatement of mitophagy flux, microglial activation, and mitochondrial functioning and neuronal survival [54]. The systems utilize receptor-mediated transcytosis through the BBB and allow subcellular targeting. At the same time, brain-derived exosomes loaded with CRISPR-Cas9 ribonucleoproteins provide a low-immunogenicity platform of genome editing [55]. The possibility of reversing the disease *in vivo* using these programmable vectors by targeting the SNCA or LRRK2 locus, therefore, opens up genetic disease-centered hope.

At the systems level, it is necessary to facilitate a speedy discovery of synergistic drug combinations using artificial intelligence-mediated drug repurposing. As an example, Ambroxol (glucocerebrosidase chaperone) in combination with exenatide (GLP-1 receptor agonist) can locally attack the lysosomal and metabolic pathway and has the potential to halt disease progression [56]. At the same time, computational individualized models called digital twins, which involve the use of genomics, proteomics, and clinical information to predict a therapeutic response and adapt patient stratification, are also being developed. The innovations not only lead to an increased level of precise mechanistic control but also open the route to personalized, adaptive treatment paradigms applicable in PD.

9.5 Conclusion and Translational Perspectives

PD is increasingly being considered as a systems-level neurodegenerative disorder, in which aggregation of α-synuclein provides a triggering event in a pathological cascade involving lysosomal dysfunction, mitochondrial dysfunction, and innate immune activation. The gradual buildup of misfolded α-synuclein interferes with the CMA and overwhelms the autophagy lysosomal pathway, causing accumulation of undegraded substrates and destabilization of the lysosomal membrane. It is a lysosomal stress that interferes with mitophagy, which eventually leads to the retention of non-functional mitochondria that generate ROS. ROS-mediated harm and mitochondrial DNA discharge act as DAMPs, activating constant neuroinflammation through TLR and inflammasome signaling. These result in dopaminergic neuronal passage, especially inside the SNc. However, clinical interpretation remains challenging due to pronounced heterogeneity across PD subtypes, as GBA-associated PD exemplifies early lysosomal deficiency, whereas LRRK2-

associated PD is often characterized by atypical vesicular trafficking. Besides, endeavors to pharmacologically invigorate autophagy must deal with the paradox of the intemperate debris era, which can increase neuroinflammatory reactions in the event that clearance capacity is overpowered. Moving forward, high-resolution strategies such as spatial transcriptomics hold promise for portraying neuron-glia crosstalk, offering insights into the multifactorial drivers. Moreover, focusing on the mitochondria–lysosome contact destinations may offer a novel hub for reestablishing organellar homeostasis and moderating neurodegenerative disease.

References

1. Ramesh S, Arachchige ASPM. Depletion of dopamine in Parkinson's disease and relevant therapeutic options: a review of the literature. AIMSN. 2023;10:200–31.
2. Vellata C, Belli S, Balsamo F, Giordano A, Colombo R, Maggioni G. Effectiveness of Telerehabilitation on motor impairments, non-motor symptoms and compliance in patients with Parkinson's disease: a systematic review. Front Neurol. 2021;12:627999.
3. Soto C. α-Synuclein seed amplification technology for Parkinson's disease and related synucleinopathies. Trends Biotechnol. 2024;42:829–41.
4. Flores-Ponce X, Velasco I. Dopaminergic neuron metabolism: relevance for understanding Parkinson's disease. Metabolomics. 2024;20:116.
5. Peng S, Chen Y, Wang R, Zhang J. Z-ligustilide provides a neuroprotective effect by regulating the phenotypic polarization of microglia via activating Nrf2-TrxR axis in the Parkinson's disease mouse model. Redox Biol. 2024;76:103324.
6. Furusawa-Nishii E, Solongo B, Rai K, Yoshikawa S, Chiba A, Okuzumi A, et al. α-Synuclein orchestrates Th17 responses as antigen and adjuvant in Parkinson's disease. J Neuroinflammation. 2025;22:38.
7. Liang H, Ma Z, Zhong W, Liu J, Sugimoto K, Chen H. Regulation of mitophagy and mitochondrial function: natural compounds as potential therapeutic strategies for Parkinson's disease. Phytother Res. 2024;38:1838–62.
8. Hussain MS, Moglad E, Afzal M, Sharma S, Gupta G, Sivaprasad GV, et al. Autophagy-associated non-coding RNAs: unraveling their impact on Parkinson's disease pathogenesis. CNS Neurosci Ther. 2024;30:e14763.
9. Morrone Parfitt G, Coccia E, Goldman C, Whitney K, Reyes R, Sarrafha L, et al. Disruption of lysosomal proteolysis in astrocytes facilitates midbrain organoid proteostasis failure in an early-onset Parkinson's disease model. Nat Commun. 2024;15:447.
10. Das D, Mattaparthi VSK. An in-Silico approach to study the effect of phosphorylation (Ps129) on the conformational dynamics of membrane bound A-Synuclein. CCB. 2024;18:144–56.
11. Valdor R, Martinez-Vicente M. The role of chaperone-mediated autophagy in tissue homeostasis and disease pathogenesis. Biomedicine. 2024;12:257.
12. Zhou N, Chen J, Hu M, Wen N, Cai W, Li P, et al. SLC7A11 is an unconventional H+ transporter in lysosomes. Cell. 2025;188:3441–3458.e25.
13. Dai L, Liu M, Ke W, Chen L, Fang X, Zhang Z. Lysosomal dysfunction in α-synuclein pathology: molecular mechanisms and therapeutic strategies. Cell Mol Life Sci. 2024;81:382.
14. Nixon RA, Rubinsztein DC. Mechanisms of autophagy–lysosome dysfunction in neurodegenerative diseases. Nat Rev Mol Cell Biol. 2024;25:926–46.
15. Filippini F, Galli T. Unveiling defects of secretion mechanisms in Parkinson's disease. J Biol Chem. 2024;300:107603.
16. Buck SA, Sanders LH. LRRK2-mediated mitochondrial dysfunction in Parkinson's disease. Biochem J. 2025;482:721–39.

17. Zanfardino P, Amati A, Perrone M, Petruzzella V. The balance of MFN2 and OPA1 in mitochondrial dynamics, cellular homeostasis, and disease. Biomolecules. 2025;15:433.
18. Bartman S, Coppotelli G, Ross JM. Mitochondrial dysfunction: a key player in brain aging and diseases. CIMB. 2024;46:1987–2026.
19. Zhang G, Wei H, Zhao A, Yan X, Zhang X, Gan J, et al. Mitochondrial DNA leakage: underlying mechanisms and therapeutic implications in neurological disorders. J Neuroinflammation. 2025;22:34.
20. Yang D-C, Cheng Y, Lin F. NADPH and Alzheimer's disease and Parkinson's disease. In: Qin Z-H, editor. Biology of nicotinamide coenzymes [Internet]. Singapore: Springer; 2025. p. 641–57. [cited 2025 Jul 5]. Available from: https://link.springer.com/10.1007/978-981-97-9877-3_39.
21. Castro-Gomez S, Heneka MT. Innate immune activation in neurodegenerative diseases. Immunity. 2024;57:790–814.
22. Sharma V, Sharma P, Singh TG. Mechanistic insights on TLR-4 mediated inflammatory pathway in neurodegenerative diseases. Pharmacol Rep. 2024;76:679–92.
23. Acioglu C, Elkabes S. Innate immune sensors and regulators at the blood brain barrier: focus on toll-like receptors and inflammasomes as mediators of neuro-immune crosstalk and inflammation. J Neuroinflammation. 2025;22:39.
24. García-Domínguez M. Neuroinflammation: mechanisms, dual roles, and therapeutic strategies in neurological disorders. CIMB. 2025;47:417.
25. Zhou M, Pang X. Polyphenols and miRNA interplay: a novel approach to combat apoptosis and inflammation in Alzheimer's disease. Front Aging Neurosci. 2025;17:1571563.
26. Lee C-Y, Yang C-H. The role of Fractalkine in diabetic retinopathy: pathophysiology and clinical implications. IJMS. 2025;26:378.
27. Wang J, Zhao J, Zhao K, Wu S, Chen X, Hu W. The role of calcium and iron homeostasis in Parkinson's disease. Brain Sci. 2024;14:88.
28. Fu C, Cao N, Zeng S, Zhu W, Fu X, Liu W, et al. Role of mitochondria in the regulation of ferroptosis and disease. Front Med. 2023;10:1301822.
29. Parveen S, Showkat F, Badesra N, Dar MS, Maqbool T, Dar MJ. Axonal degeneration, impaired axonal transport, and synaptic dysfunction in motor neuron disorder. In: Khan A, Rather MA, Ashraf GM, editors. Mechanism and genetic susceptibility of neurological disorders [internet]. Singapore: Springer; 2024. p. 199–229. [cited 2025 Jul 7]. Available from: https://link.springer.com/10.1007/978-981-99-9404-5_8.
30. Kolarova M, García-Sierra F, Bartos A, Ricny J, Ripova D. Structure and pathology of tau protein in Alzheimer disease. Int J Alzheimers Dis. 2012;2012:1–13.
31. Yu H, Sun C, Gong Q, Feng D. Mitochondria-associated endoplasmic reticulum membranes in breast cancer. Front Cell Dev Biol [Internet]. 2021;9. [cited 2025 Jul 7]. Available from: https://www.frontiersin.org/articles/10.3389/fcell.2021.629669/full
32. Barazzuol L, Giamogante F, Brini M, Calì T. PINK1/Parkin mediated Mitophagy, Ca2+ Signalling, and ER–mitochondria contacts in Parkinson's disease. IJMS. 2020;21:1772.
33. Antelo-Cea DA, Martínez-Rojas L, Cabrerizo-Ibáñez I, Roudi Rashtabady A, Hernández-Alvarez MI. Regulation of mitochondrial and Peroxisomal metabolism in female obesity and type 2 diabetes. IJMS. 2024;25:11237.
34. Ventura R, Martínez-Ruiz I, Hernández-Alvarez MI. Phospholipid membrane transport and associated diseases. Biomedicine. 2022;10:1201.
35. Li X-X, Tsoi B, Li Y-F, Kurihara H, He R-R. Cardiolipin and its different properties in Mitophagy and apoptosis. J Histochem Cytochem. 2015;63:301–11.
36. Mackie P, Koshy J, Bhogade M, Hammoor T, Hachmeister W, Lloyd G, et al. Complement C1q-dependent engulfment of alpha-synuclein induces ENS-resident macrophage exhaustion and accelerates Parkinson's-like gut pathology [internet]. Cold Spring Harbor Laboratory; 2023. [cited 2025 Jul 7]. Available from: http://biorxiv.org/lookup/doi/10.1101/2023.10.24.563832

37. Boero E, Gorham RD, Francis EA, Brand J, Teng LH, Doorduijn DJ, et al. Purified complement C3b triggers phagocytosis and activation of human neutrophils via complement receptor 1. Sci Rep [Internet]. 2023;13. [cited 2025 Jul 7]. Available from: https://www.nature.com/articles/s41598-022-27279-4
38. Olmos G, Lladó J. Tumor necrosis factor alpha: a link between Neuroinflammation and Excitotoxicity. Mediat Inflamm. 2014;2014:1–12.
39. Lin M, Liu N, Qin Z, Wang Y. Mitochondrial-derived damage-associated molecular patterns amplify neuroinflammation in neurodegenerative diseases. Acta Pharmacol Sin. 2022;43:2439–47.
40. Huang J, Wang J. Selective protein degradation through chaperone-mediated autophagy: implications for cellular homeostasis and disease (Review). Mol Med Rep [Internet]. 2024;31. [cited 2025 Jul 7]. Available from: http://www.spandidos-publications.com/10.3892/mmr.2024.13378
41. Kaushik S, Cuervo AM. The coming of age of chaperone-mediated autophagy. Nat Rev Mol Cell Biol. 2018;19:365–81.
42. Menozzi E, Toffoli M, Schapira AHV. Targeting the GBA1 pathway to slow Parkinson disease: insights into clinical aspects, pathogenic mechanisms and new therapeutic avenues. Pharmacol Ther. 2023;246:108419.
43. Song Q, Meng B, Xu H, Mao Z. The emerging roles of vacuolar-type ATPase-dependent Lysosomal acidification in neurodegenerative diseases. Transl Neurodegener [Internet]. 2020;9. [cited 2025 Jul 7]. Available from: https://translationalneurodegeneration.biomedcentral.com/articles/10.1186/s40035-020-00196-0
44. Yang X, Cao X, Zhu Q. p62/SQSTM1 in cancer: phenomena, mechanisms, and regulation in DNA damage repair. Cancer Metastasis Rev [Internet] 2025 44. [cited 2025 Jul 7]. Available from: https://link.springer.com/10.1007/s10555-025-10250-w
45. Ma L, Han T, Zhan Y. Mechanism and role of mitophagy in the development of severe infection. Cell Death Discov [Internet]. 2024;10. [cited 2025 Jul 7]. Available from: https://www.nature.com/articles/s41420-024-01844-4
46. Melchiorri D, Merlo S, Micallef B, Borg J-J, Dráfi F. Alzheimer's disease and neuroinflammation: will new drugs in clinical trials pave the way to a multi-target therapy? Front Pharmacol [Internet]. 2023;14:1196413.
47. Bouffette S, Botez I, De Ceuninck F. Targeting galectin-3 in inflammatory and fibrotic diseases. Trends Pharmacol Sci. 2023;44:519–31.
48. Fischer R, Kontermann RE, Pfizenmaier K. Selective targeting of TNF receptors as a novel therapeutic approach. Front Cell Dev Biol [Internet]. 2020;8. [cited 2025 Jul 7]. Available from: https://www.frontiersin.org/article/10.3389/fcell.2020.00401/full
49. Gryka-Marton M, Grabowska AD, Szukiewicz D. Breaking the barrier: the role of Proinflammatory cytokines in BBB dysfunction. IJMS. 2025;26:3532.
50. Sharma V, Verma R, Sharma P, Singh TG. Therapeutic modulation of cGAS-STING pathway in neurodegeneration. Brain Res. 2025;1864:149784.
51. Tanaka M, Yasuda H, Nakamura S, Shimazawa M. H-151, a selective STING inhibitor, has potential as a treatment for Neovascular age-related macular degeneration. Invest Ophthalmol Vis Sci. 2024;65:16.
52. Xu J, Zhang X-Q, Zhang Z. Transcription factor EB agonists from natural products for treating human diseases with impaired autophagy-lysosome pathway. Chin Med [Internet]. 2020;15. [cited 2025 Jul 7]. Available from: https://cmjournal.biomedcentral.com/articles/10.1186/s13020-020-00402-1
53. Thakur G, Kumar V, Lee KW, Won C. Structural insights and development of LRRK2 inhibitors for Parkinson's disease in the last decade. Genes. 2022;13:1426.

54. He H, Liu X, Lv L, Liang H, Leng B, Zhao D, et al. Calcineurin suppresses AMPK-dependent cytoprotective autophagy in cardiomyocytes under oxidative stress. Cell Death Dis. 2014;5: e997.
55. Stahl EC, Sabo JK, Kang MH, Allen R, Applegate E, Kim SE, et al. Genome editing in the mouse brain with minimally immunogenic Cas9 RNPs. Mol Ther. 2023;31:2422–38.
56. Bendikov-Bar I, Maor G, Filocamo M, Horowitz M. Ambroxol as a pharmacological chaperone for mutant glucocerebrosidase. Blood Cell Mol Dis. 2013;50:141–5.

Glial Activation and Autophagic Imbalance in Amyotrophic Lateral Sclerosis

10

Shamanand P. Mallapur, Sathgowda Patil, Vishal S. Patil, and Poonam Yadav

Abstract

Amyotrophic lateral sclerosis is a progressive neurodegenerative disorder characterized by selective motor neuron degeneration driven by intersecting pathological processes, including impaired proteostasis, defective autophagy, oxidative stress, excitotoxicity, and chronic neuroinflammation. A key axis involves the pathological interplay between autophagy-deficient neurons and reactive glial cells, involving astrocyte-derived microRNAs such as miR-155 that suppress neuronal ATG5 expression, and microglial cytokines inhibit the ULK1/FOXO1 complex, collectively impairing autophagosome initiation. Concurrently, elevated neuronal reactive oxygen species oxidize ATG4B, disrupting autophagosome lysosome fusion and exacerbating intracellular toxic substrate accumulation. Damaged neurons release mitochondrial DNA, cathepsins, and CX3CL1, further activating glial cells and perpetuating a self-reinforcing cycle of neurotoxicity. These intercellular interactions contribute to progressive motor neuron vulnerability and highlight the central role of

S. P. Mallapur · S. Patil
Department of Pharmacology, KLE College of Pharmacy, Belagavi, KLE Academy of Higher Education and Research, Deemed-to-be-University, Belagavi, Karnataka, India
e-mail: shamanandmallapur@klepharm.edu; sathgowdapatil@klepharm.edu

V. S. Patil (✉)
Department of Pharmacology, KLE College of Pharmacy, Belagavi, KLE Academy of Higher Education and Research, Deemed-to-be-University, Belagavi, Karnataka, India

Department of Pharmacognosy, KLE College of Pharmacy, Belagavi, KLE Academy of Higher Education and Research, Deemed-to-be-University, Belagavi, Karnataka, India
e-mail: vishalpatil@klepharm.edu

P. Yadav
Department of Pharmacology and Chemical Biology, School of Medicine, Emory University School of Medicine, Emory University, Atlanta, GA, USA

P. Khanal et al. (eds.), *Neuroinflammation and Autophagy in Neurodegeneration*,
https://doi.org/10.1007/978-981-95-3945-1_10

autophagy neuroinflammation crosstalk in amyotrophic lateral sclerosis pathogenesis. Although challenges such as limited central nervous system drug penetration and insufficient biomarkers remain, therapeutic strategies targeting this axis using gene therapy, nanocarriers, engineered exosomes, and optogenetic tools offer potential to restore cellular homeostasis and mitigate disease progression.

Keywords

Autophagy dysfunction · Glial activation · Motor neuron degeneration · Neuroinflammation · Protein aggregation

10.1 Introduction

Amyotrophic lateral sclerosis (ALS) is a fatal, progressive neurodegenerative disease marked by the selective loss of upper and lower motor neurons, leading to muscle weakness, paralysis, and respiratory failure. Although the majority of cases are sporadic, approximately 5–10% are familial and are associated with pathogenic mutations in genes such as SOD1, TARDBP, FUS, and the hexanucleotide repeat expansion in C9ORF72, which is the most common genetic cause identified to date. These genetic discoveries have significantly advanced our understanding of ALS pathogenesis, revealing that disruptions in RNA processing, protein homeostasis, axonal transport, and nucleocytoplasmic trafficking are dominant to disease progression. Importantly, ALS is increasingly recognized as a multisystem disorder involving not only motor neurons but also glial cells and non-neuronal pathways that contribute to neurodegeneration. Despite this progress, effective therapies remain elusive, emphasizing the need for continued translational research into the genetic and molecular mechanisms [1, 2]. ALS typically presents in mid-to-late adulthood, with an incidence of 1.5–2.7 per 100,000 people annually and a median survival of 3–5 years post-diagnosis. The disease shows a slight male predominance and is more common in individuals of European ancestry, suggesting possible genetic and environmental modifiers of disease risk. Additionally, clinical heterogeneity in onset site, progression rate, and survival outcomes highlights the influence of demographic and population-level factors in ALS manifestation and prognosis [3, 4]. Histopathologically, ALS is characterized by selective motor neuron loss and the accumulation of cytoplasmic aggregates, including phosphorylated and abnormally modified TDP-43 inclusions, observed in nearly 97% of cases [5, 6].

Recent advances in neurodegenerative research highlight that the pathogenesis of ALS is not confined to motor neurons alone but involves a broader network of cellular dysfunction, particularly implicating glial cells and disrupted autophagy. Aberrant accumulation of misfolded mutant SOD1 and dipeptide repeat proteins (DPRs) arising from C9ORF72 expansions has been shown to impair key cellular processes. These include mitochondrial dysfunction and defective nucleocytoplasmic transport, both of which combine to trigger oxidative damage and interfere with normal endoplasmic reticulum function. Such findings point to a multifaceted disease mechanism that extends well beyond neuronal degeneration, emphasizing the importance of glial–neuron interactions and cellular clearance in ALS progression. Astrocytes and microglia further exacerbate neurotoxicity by releasing pro-inflammatory cytokines, reactive oxygen species (ROS), and excitotoxins such as glutamate, which promote motor neuron injury [5–10]. This harmful interaction between astrocytes and microglia sustains chronic neuroinflammation, whereas defective autophagy impedes the protein aggregates and damaged organelles, leading to a proteostatic crisis and neuronal death. Autophagic dysfunction has been shown to precede overt neurodegeneration, suggesting it plays a causative role rather than being a secondary consequence. In ALS models such as $SOD1^{G93A}$ mice, early enhancement of autophagy appears neuroprotective, whereas impaired autophagic flux in later stages exacerbates motor neuron loss [7, 11, 12]. The addition of glial activation, mutant protein toxicity, and autophagic collapse positions ALS as a vital model for exploring the neuroinflammation autophagy axis in neurodegenerative disease.

10.2 Glial Activation in ALS Pathogenesis

Astrocytes play a pivotal role in glutamate homeostasis regulation, and their dysfunction significantly contributes to excitotoxicity in ALS. Here, one of the major contributing factors is excitatory amino acid transporter 2 (EAAT2) downregulation, which is the principal glutamate transporter in astrocytes. This loss of function is partially attributed to epigenetic alterations, notably DNA methylation at the promoter region of EAAT2 gene. In addition, microRNA-mediated regulation, especially by miR-181a, has emerged as a critical mechanism, as it downregulates EAAT2 expression. The resulting decrease in EAAT2 activity compromises extracellular glutamate clearance at synaptic sites, contributing to excitotoxic stress and neuronal vulnerability [13, 14]. In addition to transporter dysfunction, astrocytic mitochondrial impairment leads to calcium buffering deficits, resulting in intracellular calcium overflow. This disrupts calcium signaling and contributes to N-methyl-D-aspartate receptor hyperactivation in motor neurons, augmenting neuronal excitotoxicity. Prolonged mitochondrial calcium overload further promotes ROS generation, exacerbating oxidative damage and mitochondrial membrane destabilization. These mitochondrial defects compromise energy production and accelerate motor neuron vulnerability, underscoring the dominant role of calcium dysregulation in ALS [15]. Moreover, astrocytes facilitate non-cell-autonomous neurodegeneration through the release of exosomes containing misfolded SOD1 proteins, which can be internalized by neighboring neurons, thereby circulating misfolded protein pathology and further exacerbating neuronal dysfunction and death. These exosomes not only carry toxic proteins like mutant SOD1 and TDP-43 but also transport inflammatory mediators and microRNAs that modulate neuronal viability. Accumulating evidence suggests such astrocyte-derived exosomal signaling plays a critical role in disease spread and progression, highlighting potential targets for therapeutic intervention aimed at halting intercellular propagation of neurotoxicity [16–19].

10.3 NLRP3 Inflammasome Signaling in Microglial Cells

Microglia are the resident immune cells of central nervous system (CNS) and play a critical role in ALS pathogenesis through NLRP3 inflammasome activation. Damage-associated molecular patterns (DAMPs) such as extracellular ATP, mutant SOD1 aggregates, and oxidized mitochondrial DNA serve as potent triggers for NLRP3 priming and assembly, leading to caspase-1 activation and pro-inflammatory cytokine release. Sustained activation of NLRP3 inflammasome correlates with disease severity and motor neuron loss, signifying a vital role in disease progression. Furthermore, mutant SOD1 released from damaged neurons can directly stimulate microglial activation, perpetuating a toxic cycle of inflammation and neurodegeneration through increased ROS production and cytokine release [20, 21]. In parallel, C9ORF72 loss-of-function mutations commonly implicated in familial ALS compromise microglial phagocytic function by disrupting lysosomal acidification due to v-ATPase dysfunction, thereby spoiling cellular debris and aggregates [22]. Microglial polarization further modulates disease progression, where, in early disease stages, microglia often exhibit an M2-like, anti-inflammatory phenotype with neuroprotective roles, but over time, a shift toward an M1-like, pro-inflammatory state exacerbates neurotoxicity via sustained cytokines, ROS, and nitric oxide release. This contributes to chronic inflammation and tissue damage, undermining neuronal survival and repair mechanisms. Therapeutic strategies that promote or sustain the M2 phenotype have shown promise in preclinical models by reducing inflammation and slowing disease progression [23–25]. The temporally evolving and dysregulated nature of these immune response spots microglia as central effectors in the intensification of neuroinflammation and promotion of motor neuron degeneration in ALS.

10.4 Autophagic Dysregulation in ALS

10.4.1 Molecular Blockade of Autophagosome-Lysosome Fusion

In ALS, one of the pivotal disruptions in autophagic pathway involves defective fusion of autophagosomes with lysosomes. Here, a key contributing mechanism is the sequestration of VPS41, a vital component of the homotypic fusion and vacuole protein sorting complex, by DPRs produced from *C9ORF72* expansions. This impairs the tethering and fusion steps required for autophagosome maturation. Additionally, mutant SOD1 aggregates disrupt the function of Rab7 GTPase, which is a critical regulator of endolysosomal trafficking. Through abnormal interaction with Rab escort proteins, these aggregates impede vesicle transport and further compromise lysosomal degradation efficiency [26–28]. Organized, these disruptions highlight how converging molecular insults can derail autophagic flux in ALS, exacerbating cellular toxicity. Additionally, selective autophagy processes like mitophagy are disrupted through the inactivation of TANK-binding kinase 1 (TBK1) in C9ORF72-mutant models. TBK1 inactivation impairs dysfunctional mitochondria clearance, promoting mitochondrial ROS accumulation. Next, mutations in TBK1 reduce its kinase activity, leading to defective recruitment of autophagy adaptors necessary for mitophagy initiation. This impairment exacerbates mitochondrial damage and contributes to neuronal vulnerability observed in ALS pathology [29–32]. Together, these disruptions collectively impair cellular proteostasis and accelerate neurodegeneration in ALS.

10.4.2 Systemic Consequences of Autophagic Collapse

The systemic effect of autophagic failure in ALS reaches far beyond impaired intracellular degradation, exerting widespread effects on inter-organelle communication and compromising neuronal integrity. A predominantly perilous significance is the disruption of mitochondria-associated membranes, where the endoplasmic reticulum and mitochondria normally maintain close contact to coordinate calcium signaling and metabolic exchange. The accumulation of undegraded proteins and damaged organelles at these sites interferes with proper tethering, leading to a breakdown in endoplasmic reticulum–mitochondria crosstalk. This decoupling promotes excessive calcium release via inositol 1,4,5-trisphosphate receptors on the endoplasmic reticulum, resulting in both cellular stress and calcium overload, which are two synergistic events that accelerate neuronal degeneration [33, 34]. Concurrently, excessive buildup of misfolded proteins leads to aggresome formation, saturating ubiquitin-proteasome system and collapsing endoplasmic reticulum-associated degradation pathways. This overwhelms the protein quality control mechanisms in cell, triggering the unfolded protein response aimed at restoring homeostasis. However, chronic unfolded protein response can induce apoptotic signaling pathways, leading to neuronal loss in neurodegenerative diseases [35]. At the neuromuscular junction, local autophagy failure activates sterile alpha and TIR motif-containing 1 (SARM1), which is a dominant executor of axonal degeneration, initiating Wallerian-like dieback of motor neuron axons and accelerating functional decline. SARM1 activation triggers a cascade of metabolic and structural changes, including rapid depletion of NAD^+, leading to energetic failure in the axon. Targeting SARM1-mediated pathways offers a promising therapeutic strategy to slow or prevent axonal degeneration in ALS and other neurodegenerative diseases [36, 37]. This cascade of molecular disruptions illustrates how impaired autophagy propagates ALS pathology from subcellular dysfunction to whole-neuron degeneration.

10.5 Glial and Autophagic Interactions in a Self-Perpetuating Cycle

10.5.1 Glia-Driven Autophagy Suppression

Glial cells not only contribute to neuroinflammation but also actively suppress neuronal autophagy. Astrocytes release exosomes enriched with microRNAs such as miR-155-5p, which are utilized by motor neurons and directly downregulate ATG5, thereby disrupting the autophagic flux at an early stage. Moreover, miR-155-5p is a potent modulator of inflammatory pathways, enhancing pro-inflammatory cytokines production such as TNF-α and IL-6. Its elevated expression has been linked to sustained microglial activation, which further exacerbates neuroinflammatory damage in neurodegenerative diseases [38–40]. Simultaneously, activated microglia secrete pro-inflammatory cytokines like TNF-α, which suppresses ULK1 through NF-κB signaling, effectively blocking autophagy initiation [41]. Next, interleukin-1β downregulates FOXO1, which is a transcription factor essential for autophagy gene expression, further supporting this suppression. However, reduced FOXO1 activity leads to decreased transcription of key autophagy-related genes, impairing the cell ability to maintain proteostasis. This cytokine-driven autophagy inhibition contributes to damaged proteins and organelles, aggravating neuronal dysfunction and death [42, 43]. In addition, astrocyte-derived mitochondrial ROS oxidize autophagy-related protein ATG4B, impairing LC3 processing and lipidation, which are essential for autophagosome formation. This oxidative alteration leads to a short-term inhibition of ATG4B enzymatic activity, permitting the enhanced accumulation of lipid-bound LC3 necessary for autophagosome membrane expansion. However, excessive or sustained ROS levels can dysregulate this balance, disrupting autophagy flux and contributing to cellular stress [44, 45]. These glia-derived inhibitory signals converge to paralyze neuronal autophagy, intensifying proteostatic stress and exacerbating neurodegeneration in ALS.

10.5.2 Autophagy-Dependent Glial Priming

Autophagy dysfunction in neurons not only compromises intracellular proteostasis but also serves as a potent trigger for glial priming. One key mechanism involves lysosomal membrane permeabilization, leading to cathepsins B and D release from degenerating neurons. These proteases activate NLRP3 inflammasome in microglia and stimulate astrocytic protease-activated receptors (PAR-1/2), initiating inflammatory signaling cascades [46, 47]. Furthermore, impaired mitophagy results in extracellular release of mitochondrial DAMPs such as oxidized mtDNA and cardiolipin. These molecules engage toll-like receptors (TLR4/9) on microglia, inducing a pro-inflammatory transcriptional response [48, 49]. Autophagic stress also enhances neuronal fractalkine (CX3CL1) shedding, which binds to CX3CR1 receptors on astrocytes, promoting interleukin-6 production and further intensifying

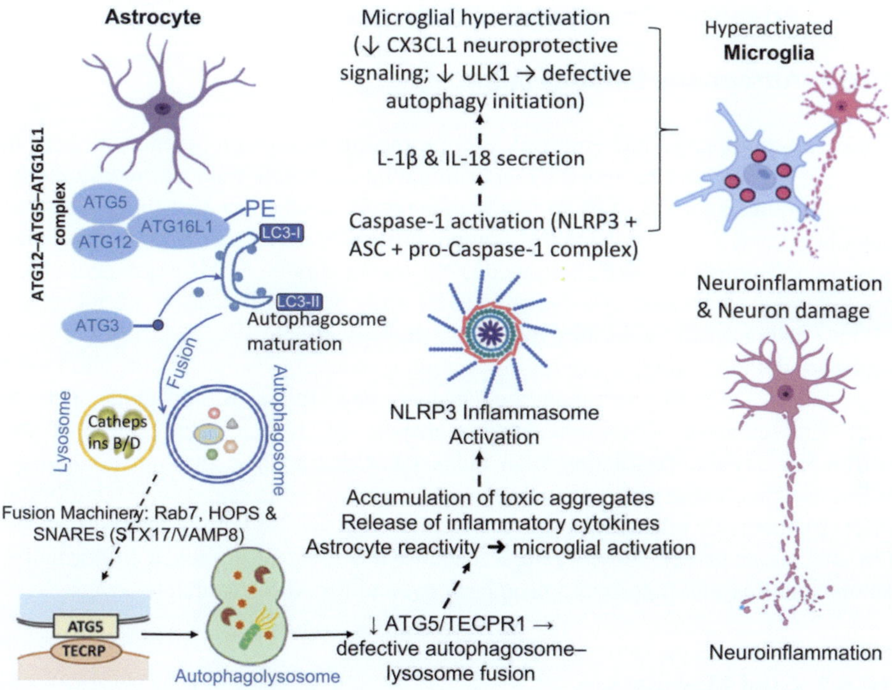

Fig. 10.1 Mechanism of glial cell-mediated autophagy impairment in motor neurons. Glial cells, such as astrocytes and microglia, regulate neuronal homeostasis and autophagy by providing nutrients, antioxidants, and modulating lysosomal clearance mechanisms through direct and indirect interactions with neurons. When glial autophagy is impaired due to genetic mutations, metabolic stress, or inflammatory signaling, this leads to the accumulation of dysfunctional proteins and organelles in both glial cells and neighboring neurons. As a result, neurons experience increased cellular stress and protein aggregation, ultimately leading to neurodegeneration and the progression of ALS disease. Created using BioRender (https://www.biorender.com/)

glial-mediated neuroinflammation. Interestingly, CX3CL1-CX3CR1 signaling plays a crucial role in modulating microglial activation states to maintain a balance between neuroprotective and neurotoxic responses. Dysregulation of this pathway leads to sustained microglial activation and contributes to chronic inflammation and neuronal damage in neurodegenerative conditions [50, 51]. Thus, autophagy-deficient neurons actively prime glial cells, creating a feedback loop of sustained neuroinflammatory damage (Fig. 10.1).

10.6 Advanced Therapeutic Strategies

10.6.1 Autophagy Enhancement

Enhancing autophagy has emerged as a promising therapeutic avenue in ALS in restoring intracellular proteostasis and mitigating aggregate-induced neurotoxicity. Transcription factor EB (TFEB) is a master regulator of lysosomal biogenesis and autophagy and is upregulated by compounds such as trehalose and resveratrol analogs, promoting C9ORF72-derived DPRs and misfolded SOD1 aggregates clearance through enhanced lysosomal activity [52, 53]. Stabilization of the Rab7 GTPase using small molecules like RMC-4630 is effective in restoring vesicular transport deficits, thereby improving autophagosome-lysosome fusion and cargo degradation [54]. In corresponding, histone deacetylase 6 inhibitors such as ACY-738 enhance dynein-mediated transport of protein aggregates to the perinuclear region, facilitating their subsequent degradation through autophagy. When used in combination with autophagy inducers, histone deacetylase 6 inhibition offers a synergistic effect, improving aggregate clearance and motor neuron survival [55, 56]. Collectively, these strategies represent a targeted approach to normalize autophagic flux and represent a promising avenue for ALS treatment.

10.6.2 Glial Modulation

Targeting glial dysfunction represents a critical frontier in ALS therapy, with some strategies aimed at rebalancing the neuroinflammatory microenvironment. Colony-stimulating factor 1 receptor inhibitors like masitinib and PLX3397 selectively deplete pathogenic, pro-inflammatory microglial subsets, though preserving and/or promoting homeostatic microglia survival, thereby attenuating chronic neuroinflammation without suppressing microglial function. Studies with masitinib slowed disease progression and improved motor function, though challenges remain in optimizing dosing regimens to minimize off-target effects while sustaining beneficial microglial populations [57, 58]. Astrocyte NRF2 pathway has shown promise, where enhanced NRF2 signaling restores EAAT2 glutamate transporter expression and re-establishes redox homeostasis by reducing excitotoxic damage to motor neurons [59]. Pharmacological inhibition of NLRP3 inflammasome with MCC950 suppresses IL-1β secretion and caspase-1 activation, reducing neuroinflammation and potentially delaying motor neuron degeneration. In models of neurodegenerative disease, MCC950 promotes non-phlogistic clearance of pathological protein aggregates, enhancing microglial phagocytic function without triggering excessive inflammation [60]. This targeted approach holds promise for modulating innate immune responses and improving neurological outcomes in neurodegenerative diseases. These glial-modulating interventions, either alone or in combination with autophagy-enhancing therapies, offer a promising strategy to disrupt the pathogenic neuron-glia feedback loop in ALS.

10.6.3 Multimodal Synergy

Emerging multimodal therapeutics strategies in ALS aim to integrate synergistic mechanisms into unified treatment platforms. Nanocarrier-based systems, such as lipid nanoparticles, have been engineered to co-deliver autophagy enhancers like trehalose alongside microglial modulators such as masitinib, achieving targeted release in the spinal cord and simultaneously addressing proteostatic failure and neuroinflammation [61, 62]. In parallel, dual gene therapy approaches using AAV9 vectors offer combinatorial benefits; silencing mutant SOD1 through overexpressing TFEB in both motor neurons and astrocytes has been shown to mitigate protein aggregation and boost lysosomal function. TFEB acts as a master regulator of autophagy and lysosomal biogenesis and enhances cellular clearance mechanisms that are typically impaired in ALS. This synergistic strategy addresses multiple pathological facets simultaneously, offering a promising avenue for disease modification [63]. Furthermore, the development of photoswitchable autophagy-inducing compounds, including light-activated rapamycin analogs, enables precise temporal and spatial control over autophagic flux in vivo, minimizing off-target effects and allowing tailored therapeutic modulation during different stages of disease progression. In preclinical studies, nanoparticle-based delivery of rapamycin enhanced tissue targeting, improved bioavailability, and significantly restored autophagic function. This targeted activation of autophagy has been associated with improved muscle strength and cardiac performance, emphasizing the translational potential of controlled autophagy modulation in degenerative diseases [64, 65].

10.7 Mechanistic Case Studies

10.7.1 SOD1-G93A Transgenic Mice

The $SOD1^{G93A}$ transgenic mouse model has contributed to defining spatial and temporal aspects of ALS pathology, offering insights into mechanistically distinct phases of disease progression. Spatial proteomic analyses have revealed a progressive decline in autophagosome–neuron interactions, with a notable disparity between lumbar and cervical spinal cord regions. This regional vulnerability mirrors the typical onset of hindlimb weakness in ALS, suggesting that impaired proteostasis precedes symptom manifestation in a site-specific manner. Accumulation of mutant SOD1 and ubiquitinated protein aggregates in the ventral horns of lumbar spinal cord coincides with reduced proteasomal and autophagic activity, and this correlation emphasizes the pathogenic role of protein clearance failure in early disease stages [66–68]. Additionally, temporal profiling indicates that microglial NLRP3 inflammasome activation occurs before overt autophagic collapse, highlighting a critical early window during which anti-inflammatory interventions may delay downstream neuronal degeneration [69]. In C9ORF72-linked ALS, patient-derived cerebrospinal fluid samples demonstrate elevated levels of poly-glycine-arginine (poly-GR) DPRs biomarkers that inversely correlate with autophagic flux markers, thus providing a stratification tool to identify patients with pronounced autophagy dysfunction. These DPRs are neurotoxic and contribute to impaired protein clearance and nucleolar stress. Notably, their presence in cerebrospinal fluid offers a minimally invasive biomarker platform for disease monitoring and potentially guiding personalized therapeutic interventions [70]. These mechanistic case studies reinforce the interplay between inflammation, proteinopathy, and autophagy and underscore the value of integrating molecular diagnostics into preclinical and clinical ALS research.

10.7.2 C9ORF72 Hexanucleotide Repeat Models

C9ORF72 hexanucleotide repeat expansion models have revealed distinct, autophagy-independent toxic mechanisms contributing to ALS pathogenesis. One key feature is nucleolar stress, where DPRs sequester nucleophosmin, a critical nucleolar protein involved in ribosomal RNA processing and ribosome assembly, thereby impairing protein synthesis capacity in motor neurons [71]. In parallel, these arginine-rich DPRs interact directly with nuclear pore complex proteins such as NUP62, disrupting nucleocytoplasmic transport. This leads to defective mRNA export and mislocalization of nuclear proteins like TDP-43, which is a hallmark of ALS pathology. The *GGGGCC* repeat expansion in C9ORF72 sequesters key transport factors, including RanGAP1 and importins, further impairing nucleocytoplasmic trafficking. These disruptions contribute to widespread cellular dysfunction, including impaired RNA metabolism and nuclear envelope integrity, eventually exacerbating neurodegeneration [72]. Furthermore, DPRs exert translational repression by binding ribosomal subunits and translation initiation factors, thereby inhibiting protein synthesis independent of autophagy pathways. Cryo-EM studies showed poly-PR and poly-GR to occupy ribosomal polypeptide exit tunnel, physically stalling translation elongation. This ribosomal blockade led to reduced production of essential proteins and contributed to cellular stress, further increasing neurodegenerative cascades in C9ORF72-linked ALS [73]. These findings highlight that in C9ORF72-associated ALS, toxic gain-of-function effects of DPRs can independently drive cellular dysfunction, complementing autophagy and proteostasis-related disease mechanisms.

10.8 Challenges and Innovations

ALS research continues to face significant challenges despite major advances in uncovering disease mechanisms. A critical barrier is the limited CNS penetration of promising therapeutics, especially peptide-based autophagy enhancers like Tat-Beclin1, which exhibit poor blood–brain barrier permeability in non-human primate models, limiting their clinical utility [74, 75]. Moreover, non-invasive biomarkers for monitoring autophagic flux or glial activation in real-time in ALS patients hamper both diagnosis and therapeutic evaluation. Although autophagy can be assessed through static markers like LC3 and p62 in tissue samples, these do not accurately reflect dynamic flux, especially in vivo. The development of reliable, real-time indicators of autophagic activity remains a major unmet need for both clinical monitoring and treatment efficacy assessment [76–78]. Rodent models often fail to replicate key features of human ALS pathology, such as widespread TDP-43 mislocalization and the complex spatial heterogeneity of glial responses perceived in human spinal cord tissues. Additionally, many rodent models rely on overexpression of mutant genes like SOD1, which do not represent the majority of sporadic ALS cases. This limits their translational applicability, emphasizing the need for more representative models that incorporate human-specific genetic and pathological features [79]. These boundaries underline the need for innovative approaches to bridge the translational gap between bench and bedside.

To address these obstacles, next-generation technologies are emerging with transformative potential. For example, CRISPR-Cas9 genome editing, delivered through advanced AAV vectors like AAV-PHP.B with spinal cord tropism, enables precise in vivo gene silencing, such as SOD1 knockout specifically in astrocytes or microglia, thereby minimizing off-target neuronal toxicity [80, 81]. Simultaneously, engineered exosomes derived from astrocytes can be loaded with therapeutic cargo (e.g., LAMP2A mRNA) to restore lysosomal function in motor neurons and enhance proteostasis. These exosomes offer high biocompatibility and can cross the blood–brain barrier, making them ideal vehicles for targeted CNS delivery. Moreover, surface modification of exosomes with specific ligands allows precise targeting of diseased neurons or glia, reducing off-target effects and improving therapeutic outcomes [82–84]. Additionally, the development of three-dimensional motor neuron-glia assembloids derived from patient-induced pluripotent stem cells provides a physiologically relevant platform to model disease and conduct high-throughput screening of combination therapies targeting both autophagy and glial modulation. These organoid models successfully recapitulate key features of motor neuron connectivity and glial interaction observed in vivo, including synaptic architecture and myelination. Moreover, they enable longitudinal observation of disease phenotypes under controlled conditions, facilitating mechanistic studies and precision therapeutic testing [85, 86]. These innovations promise to refine our understanding of ALS and expedite the discovery of personalized, disease-modifying treatments.

10.9 Conclusion

ALS is a multifactorial neurodegenerative disorder characterized by progressive motor neuron loss driven by converging mechanisms such as mutant protein toxicity, glial dysfunction, and autophagic failure. Dominant to ALS pathology are misfolded mutant SOD1 and DPRs from C9ORF72 expansions, which disrupt nucleocytoplasmic transport and mitochondrial homeostasis, triggering sustained neuroinflammation. Reactive astrocytes and microglia amplify neuronal stress by releasing pro-inflammatory cytokines, ROS, and promoting excitotoxicity. Simultaneously, impaired autophagy compromises toxic protein aggregates and damaged organelles, contributing to proteostasis collapse and axonal degeneration. The dynamic crosstalk between glia and autophagy pathways, such as glial-derived suppression of neuronal autophagy and autophagy-dependent glial priming, highlights the complex cellular interplay underlying ALS progression. Despite translational barriers like limited CNS drug permeability and suboptimal disease models, emerging strategies, including autophagy enhancers, glial modulators, gene editing technologies, and engineered exosomes, present promising therapeutic avenues. Collectively, these insights reinforce ALS as a prototype to study neuroinflammation–autophagy axis failure and developing targeted, multimodal interventions for neurodegenerative diseases.

References

1. Renton AE, Chiò A, Traynor BJ. State of play in amyotrophic lateral sclerosis genetics. Nat Neurosci. 2014;17(1):17–23. https://doi.org/10.1038/nn.3584.
2. Taylor JP, Brown RH, Cleveland DW. Decoding ALS: from genes to mechanism. Nature. 2016;539(7628):197–206. https://doi.org/10.1038/nature20413.
3. Marin B, Logroscino G, Boumédiene F, Labrunie A, Couratier P, Babron MC, et al. Clinical and demographic factors and outcome of amyotrophic lateral sclerosis in relation to population ancestral origin. Eur J Epidemiol. 2016;31(3):229–45. https://doi.org/10.1007/s10654-015-0090-x.
4. Talbott EO, Malek AM, Lacomis D. The epidemiology of amyotrophic lateral sclerosis. Handb Clin Neurol. 2016;138:225–38. https://doi.org/10.1016/B978-0-12-802973-2.00013-6.
5. Neumann M, Sampathu DM, Kwong LK, Truax AC, Micsenyi MC, Chou TT, et al. Ubiquitinated TDP-43 in frontotemporal lobar degeneration and amyotrophic lateral sclerosis. Science. 2006;314(5796):130–3. https://doi.org/10.1126/science.1134108.
6. Arai T, Hasegawa M, Akiyama H, Ikeda K, Nonaka T, Mori H, et al. TDP-43 is a component of ubiquitin-positive tau-negative inclusions in frontotemporal lobar degeneration and amyotrophic lateral sclerosis. Biochem Biophys Res Commun. 2006;351(3):602–11. https://doi.org/10.1016/j.bbrc.2006.10.093.
7. Nixon RA. The role of autophagy in neurodegenerative disease. Nat Med. 2013;19(8):983–97. https://doi.org/10.1038/nm.3232.
8. Philips T, Rothstein JD. Glial cells in amyotrophic lateral sclerosis. Exp Neurol. 2014;262 (Pt B):111–20. https://doi.org/10.1016/j.expneurol.2014.05.015.
9. Vande Velde C, Miller TM, Cashman NR, Cleveland DW. Selective association of misfolded ALS-linked mutant SOD1 with mitochondria. Science. 2008;319(5863):1335–8. https://doi.org/10.1126/science.1151254.
10. Freibaum BD, Lu Y, Lopez-Gonzalez R, Kim NC, Almeida S, Lee KH, et al. GGGGCC repeat expansion in C9orf72 compromises nucleocytoplasmic transport. Nature. 2015;525(7567): 129–33. https://doi.org/10.1038/nature14973.
11. Rudnick ND, Griffey CJ, Guarnieri P, Gerbino V, Wang X, Piersaint JA, et al. Distinct roles for motor neuron autophagy early and late in the SOD1G93A mouse model of ALS. Proc Natl Acad Sci USA. 2017;114(39):E8294–303. https://doi.org/10.1073/pnas.170429411.
12. Sasaki S. Autophagy in spinal cord motor neurons of sporadic amyotrophic lateral sclerosis. J Neuropathol Exp Neurol. 2011;70(5):349–59. https://doi.org/10.1097/NEN.0b013e3182171c66.
13. Alam MA, Datta PK. Epigenetic regulation of excitatory amino acid transporter 2 in neurological disorders. Front Pharmacol. 2019;10:483008. https://doi.org/10.3389/fphar.2019.01510.

14. Bai Y, Su X, Piao L, Jin Z, Jin R. Involvement of astrocytes and microRNA dysregulation in neurodegenerative diseases: from pathogenesis to therapeutic potential. Front Mol Neurosci. 2021;14:556215. https://doi.org/10.3389/fnmol.2021.556215.
15. Kawamata H, Manfredi G. Mitochondrial dysfunction and intracellular calcium dysregulation in ALS. Mech Ageing Dev. 2010;131(7–8):517. https://doi.org/10.1016/j.mad.2010.05.003.
16. Ivanova MV, Chekanova EO, Belugin BV, Tutykhina IL, Dolzhikova IV, Zakroishchikova IV, et al. Exosomal transport and progression of neurodegeneration in amyotrophic lateral sclerosis. Neurochem J. 2019;13(3):229–39. https://doi.org/10.1134/S1819712419030085.
17. Gomes C, Keller S, Altevogt P, Costa J. Evidence for secretion of Cu,Zn superoxide dismutase via exosomes from a cell model of amyotrophic lateral sclerosis. Neurosci Lett. 2007;428(1): 43–6.
18. Iguchi Y, Eid L, Parent M, Soucy G, Bareil C, Riku Y, et al. Exosome secretion is a key pathway for clearance of pathological TDP-43. Brain. 2016;139(12):3187–201. https://doi.org/10.1093/brain/aww237.
19. Silverman JM, Christy D, Shyu CC, Moon KM, Fernando S, Gidden Z, et al. CNS-derived extracellular vesicles spread toxic proteinopathy and target oligodendrocytes in mice. Sci Transl Med. 2019;11(493):eaaw5075. https://doi.org/10.1126/scitranslmed.aaw5075.
20. Cihankaya H, Bader V, Winklhofer KF, Vorgerd M, Matschke J, Stahlke S, et al. Elevated NLRP3 Inflammasome activation is associated with motor neuron degeneration in ALS. Cells. 2024;13(12) https://doi.org/10.3390/cells13120995.
21. Zhao W, Beers DR, Henkel JS, Zhang W, Urushitani M, Julien JP, et al. Extracellular mutant SOD1 induces microglial-mediated motoneuron injury. Glia. 2010;58(2):231–43. https://doi.org/10.1002/glia.20919.
22. Lall D, Baloh RH. Microglia and C9orf72 in neuroinflammation and ALS and frontotemporal dementia. J Clin Invest. 2017;127(9):3250–8. https://doi.org/10.1172/JCI90607.
23. Beers DR, Henkel JS, Zhao W, Wang J, Huang A, Wen S, et al. Endogenous regulatory T lymphocytes ameliorate amyotrophic lateral sclerosis in mice and correlate with disease progression in patients with amyotrophic lateral sclerosis. Brain. 2011;134(5):1293–314. https://doi.org/10.1093/brain/awr074.
24. Tang Y, Le W. Differential roles of M1 and M2 microglia in neurodegenerative diseases. Mol Neurobiol. 2016;53(2):1181–94.
25. Liao B, Zhao W, Beers DR, Henkel JS, Appel SH. Transformation from a neuroprotective to a neurotoxic microglial phenotype in a mouse model of ALS. Brain. 2012;135(4):1113–25. https://doi.org/10.1093/brain/aws037.
26. Burk K, Pasterkamp RJ. Disrupted neuronal trafficking in amyotrophic lateral sclerosis. Acta Neuropathol. 2019;137(6):859. https://doi.org/10.1007/s00401-019-01964-7.
27. Boivin M, Zhang YJ, Xu YF, Buratti E, Krainer AR, Cashman NR, et al. Reduced C9ORF72 expression and DPR toxicity impair lysosomal function and autophagy. EMBO J. 2020;39: e104068. https://doi.org/10.15252/embj.2020104068.
28. Xie Y, Wang L, Qu C, Liu Z, Shi J. Mutant SOD1 impairs autophagy and endolysosomal trafficking via Rab7 in ALS models. Neurobiol Dis. 2017;106:242–55. https://doi.org/10.1016/j.nbd.2017.07.008.
29. Harding O, Evans CS, Ye J, Cheung J, Maniatis T, Holzbaur ELF. ALS-And FTD-associated missense mutations in TBK1 differentially disrupt mitophagy. Proc Natl Acad Sci USA. 2021;118(24) https://doi.org/10.1073/pnas.2025053118.
30. Chen S, Zhang X, Song L, Le W. Autophagy dysregulation in amyotrophic lateral sclerosis. Brain Pathol. 2011;22(1):110. https://doi.org/10.1111/j.1750-3639.2011.00546.x.
31. Freischmidt A, Wieland T, Richter B, Ruf W, Schaeffer V, Müller K, et al. Haploinsufficiency of TBK1 causes familial ALS and frontotemporal dementia. Nat Neurosci. 2015;18(5):631–6. https://doi.org/10.1038/nn.3990.
32. Richter B, Sliter DA, Herhaus L, Stolz A, Wang C, Beli P, et al. Phosphorylation of OPTN by TBK1 enhances its binding to ubiquitin and promotes selective autophagy. Proc Natl Acad Sci USA. 2016;113(15):4039–44. https://doi.org/10.1073/pnas.1601055113.

33. Chen J, Bassot A, Giuliani F, Simmen T. Amyotrophic Lateral Sclerosis (ALS): stressed by dysfunctional Mitochondria-Endoplasmic Reticulum Contacts (MERCs). Cells. 2021;10(7): 1789. https://doi.org/10.3390/cells10071789.
34. Lau DHW, Hartopp N, Welsh NJ, Mueller S, Glennon EB, Mórotz GM, et al. Disruption of ER−mitochondria signalling in fronto-temporal dementia and related amyotrophic lateral sclerosis. Cell Death Dis. 2018;9(3):327. https://doi.org/10.1038/s41419-017-0022-7.
35. Scheper W, Hoozemans JJM. The unfolded protein response in neurodegenerative diseases: a neuropathological perspective. Acta Neuropathol. 2015;130(3):315–31. https://doi.org/10.1007/s00401-015-1462-8.
36. Coleman MP. Axon biology in ALS: mechanisms of axon degeneration and prospects for therapy. Neurotherapeutics. 2022;19(4):1133. https://doi.org/10.1007/s13311-022-01297-6.
37. Wang Y, Song M, Song F. Neuronal autophagy and axon degeneration. Cell Mol Life Sci. 2018;75(13):2389. https://doi.org/10.1007/s00018-018-2812-1.
38. Zingale VD, Gugliandolo A, Mazzon E. MiR-155: An important regulator of Neuroinflammation. Int J Mol Sci. 2021;23(1):90. https://doi.org/10.3390/ijms23010090.
39. Soubannier V, Rippstein P, McBride HM. Astrocyte-derived exosomes mediate intercellular transfer of miRNAs regulating autophagy in neurons. Front Cell Neurosci. 2020;14:141. https://doi.org/10.3389/fncel.2020.00141.
40. Thome AD, Harms AS, Volpicelli-Daley LA, Standaert DG. MicroRNA-155 regulates autophagy and inflammatory signaling in neurodegenerative models. J Neurosci. 2016;36(8): 2383–96.
41. Jia W, et al. TNF-α inhibits autophagy through NF-κB–mediated suppression of ULK1 in neurons. Cell Death Dis. 2019;10:945.
42. Minchev D, Kazakova M, Sarafian V. Neuroinflammation and autophagy in Parkinson's disease—novel perspectives. Int J Mol Sci. 2022;23(23):14997. https://doi.org/10.3390/ijms232314997.
43. Zhao Y, Wang Z, Wu D, et al. IL-1β impairs autophagy through downregulation of FOXO1 in neuronal cells. Cell Signal. 2020;73:109688. https://doi.org/10.1016/j.cellsig.2020.109688.
44. Scherz-Shouval R, Shvets E, Fass E, Shorer H, Gil L, Elazar Z. Reactive oxygen species are essential for autophagy and specifically regulate the activity of Atg4. EMBO J. 2007;26(7): 1749.
45. López-Fabuel I, et al. Mitochondrial ROS produced by astrocytes modulate autophagy in neurons. Cell Metab. 2016;24:354–67. https://doi.org/10.1016/j.cmet.2016.07.004.
46. Aits S, Jäättelä M. Lysosomal cell death at a glance. J Cell Sci. 2013;126(9):1905–12. https://doi.org/10.1242/jcs.091181.
47. Zhao Y, et al. Neuronal lysosomal disruption induces microglial NLRP3 activation in ALS models. Cell Death Dis. 2021;12:1037. https://doi.org/10.1038/s41419-021-04379-7.
48. Sliter DA, Martinez J, Hao L, Chen X, Sun N, Fischer TD, et al. Parkin and PINK1 mitigate STING-induced inflammation. Nature. 2018;561(7722):258–62. https://doi.org/10.1038/s41586-018-0448-9.
49. Zhang Q, et al. Circulating mitochondrial DAMPs cause inflammatory responses via TLR9. Proc Natl Acad Sci USA. 2010;107(50):20353–8. https://doi.org/10.1073/pnas.1012383107.
50. Arnoux I, Audinat E. Fractalkine signaling and microglia functions in the developing brain. Neural Plast. 2015;2015:689404. https://doi.org/10.15252/embj.2019101812.
51. Cardona AE, et al. Control of microglial neurotoxicity by the fractalkine receptor. Nat Neurosci. 2006;9:917–24. https://doi.org/10.1038/nn1715.
52. Wang T, et al. TFEB-mediated autophagy rescues neuronal toxicity in ALS models. Nat Commun. 2019;10:1–15. https://doi.org/10.1038/s41467-019-12577-0.
53. Zhang X, et al. Trehalose activates TFEB and improves autophagy in neurodegenerative disease models. Autophagy. 2018;14:1–14. https://doi.org/10.1080/15548627.2018.1502495.
54. Guerra F, Bucci C. Multiple roles of the small GTPase Rab7. Cells. 2016;5(3):34. https://doi.org/10.3390/cells5030034.

55. Yan J. Interplay BETWEEN HDAC6 and its interacting partners: essential roles in the Aggresome-autophagy pathway and neurodegenerative diseases. DNA Cell Biol. 2014;33(9): 567–80. https://doi.org/10.1089/dna.2013.2300.
56. d'Ydewalle C, et al. HDAC6 inhibition reverses axonal transport defects in motor neurons from patients with Charcot–Marie–tooth disease. Nat Med. 2011;17:968–74. https://doi.org/10.1038/nm.2397.
57. Trias E, Ibarburu S, Barreto-Núñez R, Babdor J, Maciel TT, Guillo M, et al. Post-paralysis tyrosine kinase inhibition with masitinib abrogates neuroinflammation and slows disease progression in inherited amyotrophic lateral sclerosis. J Neuroinflammation. 2016;13(1):1–12. https://doi.org/10.1186/s12974-016-0620-9.
58. Han J, Chitu V, Stanley ER, Wszolek ZK, Karrenbauer VD, Harris RA. Inhibition of colony stimulating factor-1 receptor (CSF-1R) as a potential therapeutic strategy for neurodegenerative diseases: opportunities and challenges. Cell Mol Life Sci. 2022;79(4):1–15. https://doi.org/10.1007/s00018-022-04225-1.
59. Vargas MR, Johnson DA, Sirkis DW, Messing A, Johnson JA. Nrf2 activation in astrocytes protects against neurodegeneration in mouse models of familial amyotrophic lateral sclerosis. J Neurosci. 2008;28(50):13574. https://doi.org/10.1523/JNEUROSCI.4099-08.2008.
60. Dempsey C, Rubio Araiz A, Bryson KJ, Finucane O, Larkin C, Mills EL, et al. Inhibiting the NLRP3 inflammasome with MCC950 promotes non-phlogistic clearance of amyloid-β and cognitive function in APP/PS1 mice. Brain Behav Immun. 2017;61:306–16. https://doi.org/10.1016/j.bbi.2016.12.014.
61. Liu Z, Chen J, Xu M, Ho S, Wei Y, Ho HP, Yong KT. Engineered multi-domain lipid nanoparticles for targeted delivery. Chem Soc Rev. 2025;54(12):5961–94.
62. Yun Y, An J, Kim HJ, Choi HK, Cho HY. Recent advances in functional lipid-based Nanomedicine as drug carriers for organ-specific delivery. Nanoscale. 2025;17(13):7617–38.
63. Foust KD, Salazar DL, Likhite S, et al. Therapeutic AAV9-mediated suppression of mutant SOD1 slows disease progression and extends survival in models of inherited ALS. Mol Ther. 2013;21(12):2148–59.
64. Bibee KP, Cheng YJ, Ching JK, Marsh JN, Li AJ, Keeling RM, et al. Rapamycin nanoparticles target defective autophagy in muscular dystrophy to enhance both strength and cardiac function. FASEB J. 2014;28(5):2047. https://doi.org/10.1096/fj.13-237388.
65. Putyrski M, Schultz C. Protein translocation as a tool: the current rapamycin story. Org Biomol Chem. 2012;10(31):5907–16. https://doi.org/10.1039/C2OB25355A.
66. MacLean M, López-Díez R, Vasquez C, et al. Neuronal–glial communication perturbations in murine SOD1G93A spinal cord. Commun Biol. 2022;5(1):1–14. https://doi.org/10.1038/s42003-022-03128-y.
67. Rudnick ND, Tsvetkov AS, Hetz C, et al. Distinct roles for motor neuron autophagy early and late in the SOD1G93A mouse model of ALS. Proc Natl Acad Sci. 2017;114(9):E1536–45. https://doi.org/10.1073/pnas.1704294114.
68. Zhang Y, Zhang Y, Zhang Y, et al. Overexpression of autophagy enhancer PACER/RUBCNL in neurons delays disease progression in SOD1G93A ALS mouse model. Biol Res. 2024;57(1): 1–11. https://doi.org/10.1186/s40659-024-00567-1.
69. Debye B, Schmülling L, Zhou L, Rune G, Beyer C, Johann S. Neurodegeneration and NLRP3 inflammasome expression in the anterior thalamus of SOD1(G93A) ALS mice. Brain Pathol. 2018;28(1):14–27. https://doi.org/10.1111/bpa.12467.
70. Krishnan G, Raitcheva D, Bartlett D, Prudencio M, McKenna-Yasek DM, Douthwright C, et al. Poly(GR) and poly(GA) in cerebrospinal fluid as potential biomarkers for C9ORF72-ALS/FTD. Nat Commun. 2022;13(1):1–8. https://doi.org/10.1038/s41467-022-30387-4.
71. Kwon I, Xiang S, Kato M, Wu L, Theodoropoulos P, Wang T, et al. Poly-dipeptides encoded by the C9orf72 repeats bind nucleoli, impede RNA biogenesis, and kill cells. Science. 2014;345 (6201):1139–45. https://doi.org/10.1126/science.1254917.

72. Freibaum BD, Lu Y, Lopez-Gonzalez R, Kim NC, Almeida S, Lee KH, et al. GGGGCC repeat expansion in C9orf72 compromises nucleocytoplasmic transport. Nature. 2015;525(7567): 129–33. https://doi.org/10.1038/nature14974.
73. Loveland AB, Svidritskiy E, Susorov D, Lee S, Park A, Zvornicanin S, et al. Ribosome inhibition by C9ORF72-ALS/FTD-associated poly-PR and poly-GR proteins revealed by cryo-EM. Nat Commun. 2022;13(1):1–13. https://doi.org/10.1038/s41467-022-30418-0.
74. Shoji-Kawata S, Sumpter R, Leveno M, et al. Identification of a candidate therapeutic autophagy-inducing peptide. Nature. 2013;494:201–6. https://doi.org/10.1038/nature11866.
75. Amin A, Perera ND, Tomas D, Cuic B, Radwan M, Hatters DM, Turner BJ, Shabanpoor F. Systemic administration of a novel Beclin 1-derived peptide significantly upregulates autophagy in the spinal motor neurons of autophagy reporter mice. Int J Pharm. 2024;659: 124198.
76. Feng B, Ma Y, Zheng F, Huang X, Feng X, Zhang K, Liu L, Chen F, Zeng W. Development of near-infrared lysosomal pH-activatable fluorescent probe for real-time visualization of autophagy progression. Chem Eng J. 2023;464:142554.
77. Loos B, du Toit A, Hofmeyr JH. Non-invasive monitoring of autophagy. Nat Biomed Eng. 2022;6(9):1015–6.
78. Song H, Dong C, Ren J. In vivo measurement of autophagic flux by fluorescence correlation spectroscopy. Analyst. 2023;148(15):3498–508.
79. Philips T, Rothstein JD. Rodent models of amyotrophic lateral sclerosis. Curr Protoc Pharmacol. 2015;69:5.67.1–5.67.21. https://doi.org/10.1002/0471141755.ph0567s69.
80. Chen YA, Kankel MW, Hana S, Lau SK, Zavodszky MI, McKissick O, Mastrangelo N, Dion J, Wang B, Ferretti D, Koske D. In vivo genome editing using novel AAV-PHP variants rescues motor function deficits and extends survival in a SOD1-ALS mouse model. Gene Ther. 2023;30 (5):443–54.
81. Baird MC, Likhite SB, Vetter TA, Caporale JR, Girard HB, Roussel FS, Howard AE, Schwartz MK, Reed AR, Kaleem A, Zhang X. Combination AAV therapy with galectin-1 and SOD1 downregulation demonstrates superior therapeutic effect in a severe ALS mouse model. Mol Ther Methods Clin Dev. 2024;32(3):101312.
82. Singh G, Mehra A, Arora S, Gugulothu D, Vora LK, Prasad R, Khatri DK. Exosome-mediated delivery and regulation in neurological disease progression. Int J Biol Macromol. 2024;264: 130728.
83. Varcianna A, Myszczynska MA, Castelli LM, O'Neill B, Kim Y, Talbot J, Nyberg S, Nyamali I, Heath PR, Stopford MJ, Hautbergue GM. Micro-RNAs secreted through astrocyte-derived extracellular vesicles cause neuronal network degeneration in C9orf72 ALS. EBioMedicine. 2019;40:626–35.
84. Dehghani S, Ocakcı O, Hatipoglu PT, Özalp VC, Tevlek A. Exosomes as biomarkers and therapeutic agents in neurodegenerative diseases: current insights and future directions. Mol Neurobiol. 2025;62(7):9190–215.
85. Kawada J, Kaneda S, Kirihara T, Maroof A, Levi T, Eggan K, Fujii T, Ikeuchi Y. Generation of a motor nerve organoid with human stem cell-derived neurons. Stem Cell Rep. 2017;9(5): 1441–9.
86. Li W, Liu J, Li W, Sun L, Zhang H, Gao L, Gao C. Patient-derived induced pluripotent stem cell organoids for amyotrophic lateral sclerosis drug discovery. Acta Mater Med. 2025;4(1):36–50.

Protein Aggregation and Neuroinflammation in Huntington's Disease

11

Sagarika Deepthy Tallapragada, Saroj Singhmura, Atreyee Ganguly, Kamalika Mazumder, and Yadu Nandan Dey

Abstract

Huntington's disease is a progressive, autosomal dominant neurodegenerative disorder caused by expanded CAG trinucleotide repeats in the HTT gene, resulting in a mutant huntingtin protein. The mutant protein undergoes misfolding and intraneuronal aggregation, leading to transcriptional dysregulation of neuronal survival genes, mitochondrial dysfunction, and defective autophagic clearance of protein aggregates. These toxic aggregates also activate microglia and astrocytes, triggering chronic neuroinflammation. Despite advances in understanding pathology, there is currently no cure, and treatment is symptomatic. Emerging therapeutic strategies, including gene-silencing approaches such as antisense oligonucleotides, RNA interference, and gene editing, are pursued to lower HTT expression. Autophagy enhancers and novel peptide-brush polymer scaffolds are investigated to promote mutant HTT clearance. Immunomodulatory agents aim to mitigate glial activation and inflammation. These diverse strategies underscore the translational relevance of mechanistic perceptions in cellular pathogenesis. These pave the way to targeted neuroprotective and disease-modifying therapeutic development. This chapter highlights the molecular underpinnings of Huntington's disease and explores emerging therapeutic

S. D. Tallapragada · S. Singhmura · A. Ganguly · Y. N. Dey (✉)
Pharmacology Division, Dr. B. C. Roy College of Pharmacy & Allied Health Sciences, Durgapur, West Bengal, India
e-mail: sagarika.d@bcrcp.org; saroj.singhmura@bcrcp.org; atreyee.ganguly@bcrcp.org; yadunandan.dey@bcrcp.org

K. Mazumder
Department of Pharmaceutical Chemistry, BCDA College of Pharmacy & Technology, Hridaypur, Barasat, West Bengal, India
e-mail: kamalika.mazumder@bcdapt.com

P. Khanal et al. (eds.), *Neuroinflammation and Autophagy in Neurodegeneration*,
https://doi.org/10.1007/978-981-95-3945-1_11

avenues, emphasizing the potential for translating mechanistic insights into targeted, disease-modifying interventions.

Keywords

Cognitive decline · Huntington's disease · Neurodegeneration · Protein aggregation

11.1 Introduction

Huntington's disease (HD) is a rare, inherited neurodegenerative disorder characterized by progressive motor dysfunction, cognitive decline, and psychiatric disturbances. It results from an expanded CAG trinucleotide repeat in the HTT gene, which leads to the production of mutant huntingtin. This mutant protein accumulates in neurons, forms toxic aggregates, disrupts cellular functions, and drives neuroinflammation. Globally, HD affects approximately 4.88 individuals per 100,000, with notably higher prevalence in Europe and North America. Northern Scotland reports rates as high as 15 per 100,000, nearly five times the global average [1, 2]. The disease is invariably fatal, with affected individuals typically dying within 10–20 years after symptom onset due to progressive loss of motor and cognitive function. A vital pathological feature of HD is the polyglutamine tract formed by the expanded CAG repeats in mutant huntingtin. Although normal huntingtin contains 10–35 CAG repeats, pathogenic expansions of $\geq$36 lead to an elongated polyglutamine stretch that promotes misfolding and aggregation. These aggregates form β-sheet-rich inclusion bodies, disturb protein homeostasis, and interfere with transcriptional regulation by binding abnormally to transcription factors and co-regulators. Additionally, mutant huntingtin disrupts mitochondrial function, causing oxidative stress, impaired energy production, and calcium imbalance, and the effects are detrimental to neurons. It hinders axonal transport by destabilizing microtubules, contributing to synaptic dysfunction and cell death. Furthermore, mutant huntingtin aggregates provoke chronic neuroinflammation, activating microglia and astrocytes to release pro-inflammatory cytokines (e.g., IL-6 and TNF-α), which exacerbate neuronal damage. Moreover, these molecular and cellular disruptions underlie the relentless progression of HD [1, 2].

In HD, the age of onset, penetrance, and phenotypic variability are largely determined by the length of the CAG trinucleotide repeat expansion in the HTT gene, along with additional genetic, epigenetic, and environmental modifiers. The typical onset occurs between 30 and 50 years of age, while juvenile HD, associated with $\geq$60 CAG repeats, manifests before age 20. In general, longer CAG expansions correlate with earlier onset, though substantial individual variability exists. HD follows an autosomal dominant inheritance pattern, with full penetrance observed at $\geq$40 CAG repeats. Expansions of 36–39 repeats exhibit reduced penetrance, and individuals with fewer than 36 repeats are typically asymptomatic. However, clinical presentation varies, encompassing motor symptoms (e.g., chorea, dystonia, bradykinesia), cognitive impairments (e.g., executive dysfunction and memory

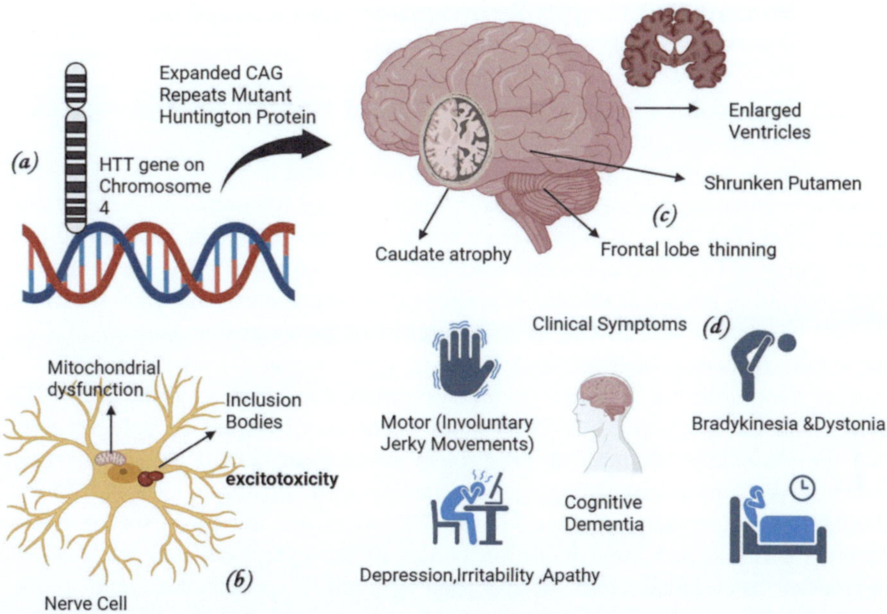

Fig. 11.1 Pathophysiology and clinical manifestations of HD: HD is caused by expanded CAG repeats (**a**) in the HTT gene, producing mutant huntingtin protein. This results in neuronal dysfunction (**b**) through mitochondrial damage, excitotoxicity, and inclusion body formation. (**c**) Brain changes include caudate atrophy, ventricular enlargement, putamen shrinkage, and frontal lobe thinning. (**d**) Clinical symptoms include involuntary motor movements, bradykinesia, dystonia, cognitive decline, and psychiatric disturbances such as depression and apathy. Created using BioRender (https://www.biorender.com/)

loss), and psychiatric disturbances, including depression, irritability, and psychosis. Additionally, modifying factors include genetic variants in DNA repair genes (e.g., MLH1, MSH3), as well as epigenetic and environmental influences [2].

Therapeutic strategies for HD are complex, targeting upstream events like mutant huntingtin expression through gene silencing (e.g., antisense oligonucleotides) and downstream pathologies. These include preventing mutant huntingtin aggregation, restoring autophagic flux, mitigating transcriptional dysregulation (e.g., HDAC inhibitors), combating neuroinflammation (e.g., TNF-α inhibitors), and supporting mitochondrial function to halt or reverse neurodegeneration [3] (Fig. 11.1).

11.2 Mutant Huntingtin Aggregation and Autophagy Dysregulation

11.2.1 Molecular Pathogenesis of Mutant Huntingtin Aggregates

Serine, threonine, glutamic acid, and proline are present at the N-terminal of HTT. Some upregulated proteases, like calpain, caspases, and lysosomal proteases, cleave the N-terminal of HTT. Now, the generated N-terminal fragments, also called exon 1 of mutant HTT, encompass enlarged polyglutamine tracts. This polyglutamine tract containing mutant huntingtin is prone to form aggregates. Abnormal splicing of mutant huntingtin occurs due to the presence of an excessive number of CAG nucleotide repeats, which causes HTT exon 1 generation [4]. The aggregation of mutant huntingtin can generate monomers, oligomers, and amyloid fibrils. The conformational changes of the N-terminus occur due to expanded polyglutamine, and the secondary structure of the protein shifts from α-helix to a β-sheet-rich structure, promoting protein aggregation. Small oligomeric intermediates are formed during this aggregation process [5, 6]. Oligomers are the precursor of mutant huntingtin fibrils and large inclusion bodies in the cytoplasm and nucleus. The aggregates accumulate into fibrils, their growth is accelerated, and finally, nucleation-dependent fibrils are generated, which further form fibril branches. These fibril-rich aggregates are insoluble and accumulate in affected cells, interfering with cellular pathways. The soluble mutant huntingtin causes more toxic interactions than insoluble aggregates [3]. Some studies showed that before disease progression, small mutant huntingtin oligomers and fibrils are formed, and these oligomers play an important role in cell toxicity and neuronal cell death [3, 7]. Moreover, prion-like propagation of misfolded proteins can be implicated in neurodegenerative diseases, including HD. These misfolded aggregates are transmitted through an intercellular path and are responsible for seeding further conversion of toxic disease-related molecules in neighboring cells, and are responsible for disease propagation [8].

11.2.2 Autophagy Dysfunction in HD

As the body's recycling and cleanup mechanism, autophagy is a basic cellular function that breaks down and recycles misfolded proteins and damaged organelles to preserve cellular health. This crucial process is intricately regulated by a precise protein network. The ULK1, along with other proteins like FIP200 and ATG101, plays a non-catalytic role in autophagy initiation and promotes the autophagosome formation. When ULK1 and ATG13 are phosphorylated by mTORC1, the initiation of bulk autophagy is inhibited. The PI3KC3 complex subunit BECN1 is phosphorylated by ULK1 and is responsible for activating PI3KC3. AMPK inactivates mTORC1 and directly phosphorylates ULK1 and stimulates autophagy [9]. mTOR is activated by oxidative stress and is influenced by reactive oxygen species (ROS). Additionally, the inhibition of autophagy is initiated by mTORC1 by targeting ULK1. It reduces the cell's ability to clear damaged components, and overactivation of the mTOR signaling pathway is a negative regulator of autophagy. Moreover, dysregulation of the ULK1–ATG13 complex leads to reduced interactions with autophagy-related proteins and interferes with autophagosome formation [10]. In HD, mutant huntingtin aggregates are responsible for the degradation of autophagy, and wild-type HTT facilitates selective autophagy. HD cells contain empty autophagosomes that lack substrates. There is interference of mutant huntingtin with autophagic cargo recognition and sequestration (SQSTM1/p62). SQSTM1/p62 is a receptor protein for selective autophagy, which coordinates with cargoes and phagophores. Wild and mutant huntingtin both interact with SQSTM1/p62 [11]. However, mutant huntingtin interacts with SQSTM1-mediated cargo recognition, and failure of cargo sequestration by autophagosomes occurs [12]. This showed that mutant huntingtin disrupted SQSTM1–cargo interaction. Furthermore, lysosome controls cell survival and death and are key players in the complete cellular degradation process. To maintain normal lysosomal function, enzyme activity, and organelle biogenesis, the optimal lysosomal acidification of pH 4.0–5.0 is required. Sufficiently acidified lysosomes fuse with autophagosomes, forming acidified autolysosomes, which can effectively clear accumulated cell debris and degrade toxic protein aggregates. Lysosomes with increased pH do not fuse with autophagosomes, which causes no degradation of cellular debris and toxic protein aggregates. Under some conditions with increased pH lysosome, it fuses with autophagosomes and forms low acidified autolysosomes, but it is not sufficient for cellular degradation and induces the presence of toxic protein aggregates [13].

11.2.3 Downstream Proteostatic Collapse

Different studies reported that the elevation of spliced XBP1 mRNA and phosphorylated IRE1 is responsible for neurodegenerative diseases. Proteasomal activity is inhibited by polyglutamine, which induces endoplasmic reticulum stress and activates IRE1 and the TRAF2-ASK1-JNK pathway. When a cell is under stress, it activates an unfolded protein response and directs the UPR signal to apoptosis or cell survival. ATF6, IRE1α–XBP1, PERK–eIF2α–ATF4, and ERAD contribute to restoring endoplasmic reticulum function by activating genes involved in protein refolding, degradation, and stress resolution. However, when endoplasmic reticulum stress is prolonged or severe, it directs unfolded protein response to follow the IRE1-ASK1-JNK or PERK-eIF2α-ATF4-CHOP apoptosis pathway, causing cell death [14, 15]. The HD pathophysiology is associated with mitochondrial dysregulation or impairment in the mitochondrial–lysosomal axis. Decreased ATP production, increased reactive oxygen species production, and a reduction in mitochondrial movement are correlated with HD [16]. Mutant huntingtin exhibits a negative effect on mitophagy machinery, which causes the accumulation of dysfunctional mitochondria and damaged organelles [3]. As a result of mitochondrial damage, oxidative stress increases, which promotes neurodegeneration. Some studies showed that HD cells release mitochondrial DNA and proteins in extracellular vesicles. It is observed that HD patients showed both mitochondrial and autophagy impairment with increased release of extracellular vesicles [17]. Pathological changes at the synapse may affect cognitive changes. Autophagy maintains appropriate synaptic dynamics in the brain. The dysfunction in autophagy impairs pre- and post-synaptic function, leads to HD, and is related to morphologic changes in dendrites and alterations in dendritic spine size, shape, and number. Mutant huntingtin abolishes synaptic function and impairs plasticity by interfering with autophagic turnover of synaptic proteins. It causes loss of regular synaptic input and cognitive impairment, and cell death. Since synaptic input and feedback are essential for neuronal cell survival, their interference with synaptic communication plays an important role in cell death [18]. Synaptophysin is a marker protein of synaptic vesicles, along with calcineurin, and takes part in the release of neurotransmitters at synapses. Synaptophysin loss in the striatum causes the degeneration of medium-sized spiny neurons, which are characteristic of HD [19].

11.3 Microglial Hyperactivation and Cytokine Networks

Microglia play a major role in the pathogenesis of HD. Physiologically, they are resident immune cells of the brain. In the absence of any stimulus, microglia resume a state of surveillance, whereas any inflammatory stimulus leads to their activation and mitigates the stimulus by initiating an inflammatory response. In HD, accumulation of mutant huntingtin in neurons leads to microglia activation [20]. Due to the altered structure of mutant huntingtin, the cell undergoes stress and produces damage-associated molecular patterns. These are recognized by pattern recognition proteins, such as toll-like receptors on microglia. The activation of TLR2/4 further activates NF-κB signaling. Furthermore, there is an interaction between the IκB kinase (IKK) γ subunit of the IKK complex and mutant huntingtin. This results in the activation of the IKK complex composed of IKKα and β subunits. IKKβ kinase then phosphorylates IκBα and releases NFkB, which facilitates gene expression of proinflammatory cytokines such as TNF-α, IL-1β, IL-6, and IL-8 [21]. Moreover, prolonged activation of P2X7 receptors due to the presence of continuous ATP leads to pannexin pore formation, causing cell death [22]. Besides this phenomenon, activation of P2X7 promotes the NFkB signaling, producing NLRP3 inflammasomes that convert procaspase 1 to caspase 1. Moreover, toll-like and P2X7 receptor activation leads to cytokine overproduction, exhibiting neuroinflammation, which is a pathophysiological basis of precipitating HD [23].

Astrocytes are glial cells in the nervous system that play an essential role in neuronal homeostasis by clearing excess excitatory neurotransmitters such as glutamate and preventing excitotoxicity. Physiologically, glutamate is stored in secretory vesicles of presynaptic neurons in the brain and is released into the synaptic cleft, which activates three receptors in postsynaptic neurons, that is, (a) non-NMDA receptors, (b) NMDA receptors, and (c) metabotropic receptors. Upon activation of non-NMDA receptors, Na^+ ions enter the cell, which depolarizes the cell membrane, causing the extrusion of Mg^{2+} from the NMDA receptors. This exposes the binding site of glutamate on the receptor. Glutamate, along with glycine (co-agonist), binds to NMDA receptors and allows Ca^{2+} entry into the cell. Furthermore, pericellular glutamate is tightly regulated by astrocytic Na^+ dependent transporters such as glutamate transporter-1 (GLT1), which takes up the glutamate for recycling. In HD, the mutant huntingtin interferes with GLT1 transcription. This reduces the mRNA and protein glutamate GLT1 expression, and this evidence has been found in HD postmortem tissues. Moreover, GLT1 protein and mRNA reduction have been found in the striatum and cerebral cortex of presymptomatic R/6 mice [24, 25].

Neuroinflammation involves M1 microglia releasing pro-inflammatory mediators that induce neurotoxic A1 astrocytes. These A1 astrocytes upregulate NF-κB, produce complement proteins like C3 (causing synaptic damage), and further release inflammatory cytokines (IL-1β, TNF-α, IL-6). A1 astrocytes also attract immune cells. Although C3 disrupts neuronal $Ca^{2+,}$ leading to excitotoxicity, A1 astrocytes are characteristic of advanced HD, aggravating later-stage pathology but not contributing to early development. Furthermore, HD disrupts neuronal and glial glucose metabolism, affecting the astrocyte-neuronal lactate shuttle. Normally, increased synaptic activity leads to astrocytic lactate production, which fuels neurons. Astrocytes also produce ascorbic acid, promoting neuronal lactate uptake. In HD, downregulated GLT1 and underexpressed SVCT2 transporters impair glutamate uptake and subsequent lactate/ascorbic acid delivery to neurons, resulting in a neuronal energy crisis [26–28].

Mitochondria are highly dynamic organelle that undergoes perpetual fission and fusion to conserve their functional integrity. Mitochondrial fission includes the fragmentation into smaller organelles, and fusion involves the combination of smaller mitochondrial units to produce an integrated network. This process is known as mitochondrial dynamics. Several proteins, such as DRP1, FIS1, MFF, and mitofusins, optic atrophy 1, are responsible for carrying mitochondrial dynamics, contributing to cellular homeostasis [29]. The HD patient shows a marked interaction between mutant huntingtin and DRP1, leading to DRP1 upregulation and MFN1, MFN2, and OPA1 downregulation. This results in the acceleration of the mitochondrial fission process, disrupting mitochondrial dynamics. Furthermore, a study showed that mutant huntingtin produced nitric oxide in cultured neurons, causing s-nitrosylation of DRP1 and GTPase, resulting in their hyperactivity and neuronal death [30].

11.4 Emerging Multimodal Therapeutics

11.4.1 Autophagy-Inducing Agents

One of the HD hallmarks is abnormal mutant huntingtin accumulation; thus, autophagy is considered one of the major pathological processes due to its critical role in aggregated protein degradation and maintaining cellular integrity [31]. Apart from a few compounds like tetrabenazine, olanzapine, and aripiprazole, which are effective in reducing symptoms of HD with potential side effects, these are the only choices, as current agents cannot halt or reverse disease progression. Hence, there is a need to explore new emerging avenues for HD management [32].

Cysteamine is a naturally occurring aminothiol derived from the decarboxylation of cysteine, and its oxidized form is cystamine [33]. It has been approved by the U.S. Food and Drug Administration and European Medicines Agency to treat nephropathic Cystinosis [34, 35]. Cysteamine and cystamine have been explored as a potential drug in the treatment of certain mouse models of neurodegenerative diseases, and their ability to cross blood blood–brain barrier makes them a promising agent for HD. Studies have shown that cystamine has inhibitory effects on transglutaminases [33]. Transglutaminase 2 is an inducible transamidating acyltransferase that catalyzes Ca^{2+}-dependent protein modification. In HD, increased Ca^{2+} levels cause cell stress, catalyze the formation of ε-N-(γ-glutamyl)-lysyl crosslinks between proteins, promote aggregation into insoluble inclusions, and thus reduce clearance and impair autophagy [33, 36]. In HD, increased Ca^{2+} levels activate the enzyme during cellular stress, catalyzing the formation of ε-N-(-γ-glutamyl)-lysyl crosslinks between proteins. This promotes the aggregation of insoluble inclusions, thereby reducing clearance and impairing autophagy. In cells, cystamine is reduced to cysteamine, which acts as a nucleophilic acceptor that attacks the thioester intermediate formed during the creation of ε-N-(γ-glutamyl)-lysyl crosslinks, competitively inhibiting the transamidation reaction [33, 37–39]. Although transglutaminase ablation does not prevent neurodegeneration in HD, transglutaminase 2-deficient mice exhibit improved motor function and extended survival, indicating involvement of additional pathogenic pathways [36]. Cysteamine possesses potent antioxidant properties that mitigate oxidative stress in HD by promoting cysteine uptake for glutathione synthesis. It serves as a hydrogen sulfide precursor with multiple physiological protective effects. Cysteamine increases brain-derived neurotrophic factor (BDNF) levels by enhancing its secretion from the Golgi region. However, this process is disrupted by reduced HSJ1b levels or transglutaminase overexpression. Additionally, the transcription factor TFEB is a master regulator of autophagy–lysosomal gene expression and is impaired in HD. Although TFEB activation is a promising therapeutic approach, its expression must be firmly controlled, as overactivation may worsen disease progression. Cysteamine has been shown to reduce neuropathology and extend lifespan due to its combined effects as a transglutaminase 2 inhibitor and neuroprotective agent by enhancing cysteine, glutathione, and BDNF levels. Its autophagy-inducing effects may involve mTOR inhibition, which is elevated in HD and suppresses TFEB activation. Additionally, clinical trials have shown cysteamine to be safe and well-tolerated, with dose-limiting gastrointestinal side effects such as vomiting, diarrhea, abdominal discomfort, body odor, and halitosis [34, 40–43].

11.4.2 Gene-Targeted Approaches

Recent advances in gene-targeted approaches aim to directly reduce mutant huntingtin protein production. Such approaches are CRISPR/Cas9, antisense oligonucleotide, and RNAi therapies [44]. CRISPR–Cas9 gene-editing technology is a highly advanced tool that can be used to precisely and efficiently correct deleterious base mutations at the genomic level [45]. This approach uses single-guide RNAs to target DNA sequences associated with the CAG-expanded mutant HTT allele in HD. Single-guide RNAs can be designed to target genetic variations that are linked only to the mutant HTT allele. This enables allele-specific inactivation (mutated allele, with too many CAG repeats in HD), reducing mutant huntingtin protein expression without affecting wild-type, which is essential for normal brain homeostasis. The method was used in fibroblasts derived from HD patients and in mouse models, leading to reduced mutant huntingtin RNA and protein levels [46]. Next, base and prime editing are advanced, safer alternatives to CRISPR/Cas9-mediated excision, designed to correct the mutant gene [47]. Antisense oligonucleotides are synthetic DNA that bind to the mRNA of a target gene. They block protein expression, and this approach was approved by the FDA for spinal muscular atrophy. This has to explore potential therapeutics for HD, although trials such as involving tominersen have been halted due to poor tolerability. Despite this, huntingtin remains a promising therapeutic target, and antisense oligonucleotide strategies continue to show potential in addressing the needs of HD research. To achieve effective neuronal expression, improved delivery methods must be adopted, as any therapeutic agent must cross the blood–brain barrier to exert its effects. However, lipid nanoparticles offer a safer and more efficient method for delivering small interfering RNA into brain cells, presenting a promising avenue for gene therapy [48].

11.4.3 Combinatorial and Precision Strategies

The combination of rapamycin and lithium enhances autophagy via both mTOR-dependent and mTOR-independent pathways, as rapamycin inhibits mTOR, whereas lithium suppresses IMPase, reducing inositol and IP_3 levels. In HD fly models, this dual approach provides greater neuroprotection than either drug alone [49]. Other agents tested in HD Drosophila models that induce autophagy by inhibiting inositol synthesis and reducing IP_3 levels include sodium valproate, carbamazepine, and rilmenidine [50]. A newer molecule, SMER28, has also been shown to act through mTOR independently [51]. An IL-1β antagonist, Anakinra, restores proteostasis by coupling autophagy with the proteasome system [52]. Treatments that restore mitochondrial dynamics by promoting fusion, limiting fission, enhancing biogenesis, and increasing mitophagy can help slow disease progression. Several agents targeting disrupted mitochondrial dynamics are under clinical trials, including Mdivi-1, P110, Tubastatin A, MitoQ, and SS-31 [3, 53]. Next, riluzole has been found to block glutamatergic neurotransmission and help in improving HD-associated chorea [54]. Additionally, ^{11}C-CHDI-180R has emerged as a promising PET imaging tool to assess mutant huntingtin aggregate load in visualizing and quantifying the distribution of mutant huntingtin aggregation in the specific HD brain regions [55]. Moreover, it has been observed that cerebrospinal fluid concentrations of mutant huntingtin and neurofilament light are early detectable changes in HD, which can be considered as a biomarker for therapeutic trials [56]. For therapeutic purposes, extracellular vesicles' ability to cross the blood–brain barrier can help in transferring therapeutics to the target area [57].

11.5 Conclusion

In summary, HD progression is driven by self-reinforcing cycles of mutant huntingtin protein aggregation, autophagy dysfunction, and neuroinflammation. Therapeutic breakthroughs likely hinge on disrupting these loops by activating TFEB to restore lysosomal function, inhibiting NLRP3 inflammasome signaling to curb inflammation, and rescuing mitochondrial integrity. However, the impermeability of the blood–brain barrier poses a major challenge for systemic delivery of large biologics like antisense oligonucleotides and CRISPR-based tools. Therapies may be most effective during pre-symptomatic stages before widespread neuronal damage. Advancements in patient-derived brain organoids allow for tailored drug testing to accelerate the promising candidates. These integrated strategies show a novel approach towards more precise, timely, and individualized treatments for HD.

References

1. Medina A, Mahjoub Y, Shaver L, Pringsheim T. Prevalence and incidence of Huntington's disease: an updated systematic review and meta-analysis. Mov Disord. 2022 Dec;37(12): 2327–35. https://doi.org/10.1002/mds.29228.
2. Rawlins MD, Wexler NS, Wexler AR, Tabrizi SJ, Douglas I, Evans SJ, Smeeth L. The prevalence of Huntington's disease. Neuroepidemiology. 2016;46(2):144–53. https://doi.org/10.1159/000443738.
3. Tong H, Yang T, Xu S, Li X, Liu L, Zhou G, et al. Huntington's disease: complex pathogenesis and therapeutic strategies. Int J Mol Sci. 2024;25(7):3845. https://doi.org/10.3390/ijms25073845.
4. Li X, Tong H, Xu S, Zhou G, Yang T, Yin S, et al. Neuroinflammatory proteins in Huntington's disease: insights into mechanisms, diagnosis, and therapeutic implications. Int J Mol Sci. 2024;25(21):11787. https://doi.org/10.3390/ijms252111787.
5. Jarosińska OD, Rüdiger SGD. Molecular strategies to target protein aggregation in Huntington's disease. Front Mol Biosci. 2021;8:769184. https://doi.org/10.3389/fmolb.2021.769184.
6. Pandey NK, Isas JM, Rawat A, Lee RV, Langen J, Pandey P, Langen R. The 17-residue-long N terminus in huntingtin controls stepwise aggregation in solution and on membranes via different mechanisms. J Biol Chem. 2018;293(7):2597–605. https://doi.org/10.1074/jbc.M117.813667. Epub 2017 Dec 27. PMID: 29282287; PMCID: PMC5818184
7. Chen JY, Parekh M, Seliman H, Bakshinskaya D, Dai W, Kwan K, Chen KY, Liu AYC. Heat shock promotes inclusion body formation of mutant huntingtin (mHtt) and alleviates

mHtt-induced transcription factor dysfunction. J Biol Chem. 2018;293(40):15581–93. https://doi.org/10.1074/jbc.RA118.002933.
8. Victoria GS, Zurzolo C. The spread of prion-like proteins by lysosomes and tunneling nanotubes: implications for neurodegenerative diseases. J Cell Biol. 2017;216(9):2633–44. https://doi.org/10.1083/jcb.201701047.
9. Lin MG, Hurley JH. Structure and function of the ULK1 complex in autophagy. Curr Opin Cell Biol. 2016;39:61–8. https://doi.org/10.1016/j.ceb.2016.02.010.
10. Li L, Zou J, Zhou T, Liu X, Tan D, Xiang Q, Yu R. mTOR-mediated nutrient sensing and oxidative stress pathways regulate autophagy: a key mechanism for traditional Chinese medicine to improve diabetic kidney disease. Front Pharmacol. 2025;16 https://doi.org/10.3389/fphar.2025.1578400.
11. Yang J, Chen X, Xu H. SQSTM1/p62 droplet -mediated autophagosome formation: insights into Huntington disease. Autophagy. 2021;17(10):3256–9. https://doi.org/10.1080/15548627.2021.1953820.
12. Martinez-Vicente M, Talloczy Z, Wong E, Tang G, Koga H, et al. Cargo recognition failure is responsible for inefficient autophagy in Huntington's disease. Nat Neurosci. 2010;13(5):567–76. https://doi.org/10.1038/nn.2528.
13. Zhu SY, Yao RQ, Li YX, Zhao PY, Ren C, Du XH, et al. Lysosomal quality control of cell fate: a novel therapeutic target for human diseases. Cell Death Dis. 2020;11(9):817. https://doi.org/10.1038/s41419-020-03032-5.
14. Jiang D, Niwa M, Koong AC. Targeting the IRE1α-XBP1 branch of the unfolded protein response in human diseases. Semin Cancer Biol. 2015;33:48–56. https://doi.org/10.1016/j.semcancer.2015.04.010. Epub 2015 May 16. PMID: 25986851; PMCID: PMC4523453
15. Xu X, Sun B, Zhao C. Poly (ADP-ribose) polymerase 1 and parthanatos in neurological diseases: from pathogenesis to therapeutic opportunities. Neurobiol Dis. 2023;187:106314. https://doi.org/10.1016/j.nbd.2023.106314.
16. Jurcau A, Jurcau CM. Mitochondria in Huntington's disease: implications in pathogenesis and mitochondrial-targeted therapeutic strategies. Neural Regen Res. 2023;18(7):1472–7. https://doi.org/10.4103/1673-5374.360289.
17. Johri A, et al. Mitochondrial dysfunction in neurodegenerative diseases. J Pharmacol Exp Ther. 342(3):619–30. https://doi.org/10.1124/jpet.112.192138.
18. Menzies F, Fleming A, Rubinsztein D. Compromised autophagy and neurodegenerative diseases. Nat Rev Neurosci. 2015;16:345–57. https://doi.org/10.1038/nrn3961.
19. Goto S, Hirano A. Synaptophysin expression in the striatum in Huntington's disease. Acta Neuropathol. 1990;80(1):88–91. https://doi.org/10.1007/bf00294227.
20. Yang HM, Yang S, Huang SS, Tang BS, Guo JF. Microglial activation in the pathogenesis of Huntington's disease. Front Aging Neurosci. 2017;9:193. https://doi.org/10.3389/fnagi.2017.00193.
21. Land WG. The role of damage-associated molecular patterns in human diseases: part I – promoting inflammation and immunity. Sultan Qaboos Univ Med J. 2015;15(1):e9–e21.
22. Lara R, Adinolfi E, Harwood CA, Philpott M, Barden JA, Di Virgilio F, et al. P2X7 in cancer: from molecular mechanisms to therapeutics. Front Pharmacol. 2020;11:793. https://doi.org/10.3389/fphar.2020.00793.
23. Kong H, Zhao H, Chen T, Song Y, Cui Y. Targeted P2X7/NLRP3 signaling pathway against inflammation, apoptosis, and pyroptosis of retinal endothelial cells in diabetic retinopathy. Cell Death Dis. 2022;13(4):336. https://doi.org/10.1038/s41419-022-04786-w.
24. Satarker S, Bojja SL, Gurram PC, Mudgal J, Arora D, Nampoothiri M. Astrocytic glutamatergic transmission and its implications in neurodegenerative disorders. Cells. 2022;11(7):1139. https://doi.org/10.3390/cells11071139.
25. Jimenez-Sanchez M, Licitra F, Underwood BR, Rubinsztein DC. Huntington's disease: mechanisms of pathogenesis and therapeutic strategies. Cold Spring Harb Perspect Med. 2017;7(7):a024240. https://doi.org/10.1101/cshperspect.a024240.

26. Kwon HS, Koh SH. Neuroinflammation in neurodegenerative disorders: the roles of microglia and astrocytes. Transl Neurodegener. 2020;9(1):42. https://doi.org/10.1186/s40035-020-00221-2.
27. Lawrence JM, Schardien K, Wigdahl B, Nonnemacher MR. Roles of neuropathology-associated reactive astrocytes: a systematic review. Acta Neuropathol Commun. 2023;11(1): 42. https://doi.org/10.1186/s40478-023-01526-9.
28. Huang M, Long A, Hao L, Shi Z, Zhang M. Astrocyte in neurological disease: pathogenesis and therapy. MedComm (2020). 2025;6(8):e70299. https://doi.org/10.1002/mco2.70299.
29. Scott I, Youle RJ. Mitochondrial fission and fusion. Essays Biochem. 2010;47:85–98. https://doi.org/10.1042/bse0470085.
30. Reddy PH. Increased mitochondrial fission and neuronal dysfunction in Huntington's disease: implications for molecular inhibitors of excessive mitochondrial fission. Drug Discov Today. 2014;19(7):951–5. https://doi.org/10.1016/j.drudis.2014.03.020.
31. Lei Y, Klionsky DJ. The emerging roles of autophagy in human diseases. Biomedicine. 2021;9 (11):1651. https://doi.org/10.3390/biomedicines9111651.
32. Frank S. Treatment of Huntington's disease. Neurotherapeutics. 2014;11(1):153–60. https://doi.org/10.1007/s13311-013-0244-z.
33. Paul BD, Snyder SH. Therapeutic applications of cysteamine and cystamine in neurodegenerative and neuropsychiatric diseases. Front Neurol. 2019;10:1315. https://doi.org/10.3389/fneur.2019.01315.
34. Borrell-Pagès M, Canals JM, Cordelières FP, Parker JA, Pineda JR, Grange G, et al. Cystamine and cysteamine increase brain levels of BDNF in Huntington disease via HSJ1b and transglutaminase. J Clin Invest. 2006;116(5):1410–24. https://doi.org/10.1172/JCI27607.
35. Dedeoglu A, Kubilus JK, Jeitner TM, Matson SA, Bogdanov M, Kowall NW, et al. Therapeutic effects of cystamine in a murine model of Huntington's disease. J Neurosci. 2002;22(20): 8942–50. https://doi.org/10.1523/JNEUROSCI.22-20-08942.2002.
36. McColgan P, Tabrizi SJ. Huntington's disease: a clinical review. Eur J Neurol. 2018;25(1): 24–34. https://doi.org/10.1111/ene.13413.
37. Karpuj MV, Becher MW, Springer JE, Chabas D, Youssef S, Pedotti R, et al. Prolonged survival and decreased abnormal movements in transgenic model of Huntington disease, with administration of the transglutaminase inhibitor cystamine. Nat Med. 2002;8(2):143–9. https://doi.org/10.1038/nm0202-143.
38. Zuccato C, Cattaneo E. Role of brain-derived neurotrophic factor in Huntington's disease. Prog Neurobiol. 2007;81(5–6):294–330. https://doi.org/10.1016/j.pneurobio.2007.01.003.
39. Karpuj MV, Garren H, Slunt H, Price DL, Gusella J, Becher MW, et al. Transglutaminase aggregates huntingtin into nonamyloidogenic polymers, and its enzymatic activity increases in Huntington's disease brain nuclei. Proc Natl Acad Sci USA. 1999;96:7388–93. https://doi.org/10.1073/pnas.96.13.7388.
40. Klöck C, Khosla C. Regulation of the activities of the mammalian transglutaminase family of enzymes. Protein Sci. 2012;21(12):1781–91. https://doi.org/10.1002/pro.2162.
41. Fox JH, Barber DS, Singh B, Zucker B, Swindell MK, Norflus F, et al. Cystamine increases L-cysteine levels in Huntington's disease transgenic mouse brain and in a PC12 model of polyglutamine aggregation. J Neurochem. 2004;91(2):413–22. https://doi.org/10.1111/j.1471-4159.2004.02726.x.
42. Ojalvo-Pacheco J, Yakhine-Diop SMS, Fuentes JM, Paredes-Barquero M, Niso-Santano M. Role of TFEB in Huntington's disease. Biology (Basel). 2024;13(4):238. https://doi.org/10.3390/biology13040238.
43. Napolitano G, Esposito A, Choi H, Matarese M, Benedetti V, Di Malta C, et al. mTOR-dependent phosphorylation controls TFEB nuclear export. Nat Commun. 2018;9(1):3312. https://doi.org/10.1038/s41467-018-05862-6.
44. Fields E, Vaughan E, Tripu D, Lim I, Shrout K, Conway J, et al. Gene targeting techniques for Huntington's disease. Ageing Res Rev. 2021;70:101385. https://doi.org/10.1016/j.arr.2021.101385.

45. Li T, Yang Y, Qi H, Cui W, Zhang L, Fu X, et al. CRISPR/Cas9 therapeutics: progress and prospects. Signal Transduct Target Ther. 2023;8(1):36. https://doi.org/10.1038/s41392-023-01309-7.
46. Alkanli SS, Alkanli N, Ay A, Albeniz I. CRISPR/Cas9 mediated therapeutic approach in Huntington's disease. Mol Neurobiol. 2023;60(3):1486–98. https://doi.org/10.1007/s12035-022-03150-5.
47. Duan W, Urani E, Mattson MP. The potential of gene editing for Huntington's disease. Trends Neurosci. 2023;46(5):365–76. https://doi.org/10.1016/j.tins.2023.02.005.
48. Rook ME, Southwell AL. Antisense oligonucleotide therapy: from design to the Huntington disease clinic. BioDrugs. 2022;36(2):105–19. https://doi.org/10.1007/s40259-022-00519-9.
49. Sarkar S, Krishna G, Imarisio S, Saiki S, O'Kane CJ, Rubinsztein DC. A rational mechanism for combination treatment of Huntington's disease using lithium and rapamycin. Hum Mol Genet. 2008;17(2):170–8. https://doi.org/10.1093/hmg/ddm294.
50. Hochfeld WE, Lee S, Rubinsztein DC. Therapeutic induction of autophagy to modulate neurodegenerative disease progression. Acta Pharmacol Sin. 2013;34(5):600–4. https://doi.org/10.1038/aps.2012.189.
51. Kirchenwitz M, Stahnke S, Grunau K, Melcher L, van Ham M, et al. The autophagy inducer SMER28 attenuates microtubule dynamics mediating neuroprotection. Sci Rep. 2022;12(1): 17805. https://doi.org/10.1038/s41598-022-20563-3.
52. van de Veerdonk FL, Renga G, Pariano M, Bellet MM, Servillo G, Fallarino F, et al. Anakinra restores cellular proteostasis by coupling mitochondrial redox balance to autophagy. J Clin Invest. 2022;132(2):e144983. https://doi.org/10.1172/JCI144983.
53. Joshi DC, Chavan MB, Gurow K, Gupta M, Dhaliwal JS, Ming LC. The role of mitochondrial dysfunction in Huntington's disease: implications for therapeutic targeting. Biomed Pharmacother. 2025;183:117827. https://doi.org/10.1016/j.biopha.2025.117827.
54. Rosas HD, Koroshetz WJ, Jenkins BG, Chen YI, Hayden DL, Beal MF, et al. Riluzole therapy in Huntington's disease (HD). Mov Disord. 1999 Mar;14(2):326–30. https://doi.org/10.1002/1531-8257(199903)14:2<326::aid-mds1019>3.0.co;2-q.
55. Bertoglio D, Weiss AR, Liguore W, Martin LD, Hobbs T, Templon J, et al. In vivo cerebral imaging of mutant huntingtin aggregates using 11C-CHDI-180R PET in a nonhuman primate model of Huntington disease. J Nucl Med. 2023;64(10):1581–7. https://doi.org/10.2967/jnumed.123.265569.
56. Rodrigues FB, Byrne LM, Tortelli R, Johnson EB, Wijeratne PA, Arridge M, et al. Mutant huntingtin and neurofilament light have distinct longitudinal dynamics in Huntington's disease. Sci Transl Med. 2020 Dec 16;12(574):eabc2888. https://doi.org/10.1126/scitranslmed.abc2888.
57. Ananbeh H, Vodicka P, Kupcova Skalnikova H. Emerging roles of exosomes in Huntington's disease. Int J Mol Sci. 2021;22(8):4085. https://doi.org/10.3390/ijms22084085.

Inflammatory Demyelination and Autophagic Dysfunction in Multiple Sclerosis

12

Praveen Halagali, Salini P. Nair, S. N. Manjula, Himanshu Sharma, and Vamshi Krishna Tippavajhala

Abstract

Multiple sclerosis is a multifaceted and progressive neurological disorder marked by persistent inflammation, demyelination, and axonal injury. Due to autoreactive Th1 and Th17 lymphocytes infiltration into the central nervous system, which disrupts the blood–brain barrier and initiates inflammatory assaults on oligodendrocytes and neuronal axons. Defective autophagy in microglia and neurons impairs the clearance of myelin fragments and damaged organelles, thereby maintaining a chronic inflammatory environment. Mitochondrial impairment and oxidative stress further destabilize neuronal function in repressing autophagy-associated genes and promoting inflammasome activation. Due to the limitations of single-target immunomodulatory therapies in halting disease progression, this chapter proposes combinatorial strategies. The potential of gene therapies in modulating transcription factors, as well as nanoparticle-facilitated delivery of autophagy-enhancing compounds, is also discussed. This

P. Halagali · V. K. Tippavajhala (✉)
Department of Pharmaceutics, Manipal College of Pharmaceutical Sciences, Manipal Academy of Higher Education, Manipal, Karnataka, India
e-mail: praveen.mcopsmpl2023@learner.manipal.edu; vamshi.krishna@manipal.edu

S. P. Nair · S. N. Manjula
Department of Pharmacology, JSS College of Pharmacy, JSS Academy of Higher Education and Research, Mysuru, Karnataka, India
e-mail: snmanjula@jssuni.edu.in

H. Sharma
Teerthanker Mahaveer College of Pharmacy, Teerthanker Mahaveer University, Moradabad, UP, India

P. Khanal et al. (eds.), *Neuroinflammation and Autophagy in Neurodegeneration*,
https://doi.org/10.1007/978-981-95-3945-1_12

chapter emphasizes the importance of targeting both inflammatory and cellular clearance pathways and advocates for a systems-level therapeutic model, guided by biomarkers and precision medicine, to counter neurodegeneration in multiple sclerosis.

Keywords

Multiple sclerosis · Autophagic dysfunction · Inflammatory demyelination · Blood–brain barrier

12.1 Introduction

Multiple sclerosis (MS) is a chronic autoimmune disorder of the central nervous system (CNS), defined by inflammation, demyelination, and ongoing neurodegeneration. It is a leading cause of non-traumatic neurological disability in young adults, affecting over 2.8 million individuals globally. Onset typically occurs between the ages of 20 and 40 and is notably more prevalent in women than in men [1]. Clinically, MS presents in various forms, most frequently as relapsing-remitting MS, which can evolve into secondary progressive MS, characterized by gradual neurological decline independent of relapses [2]. Pathologically, MS begins with active lesions exhibiting perivascular immune infiltration, macrophage activation, and extensive myelin damage [3]. As the disease advances, it shifts toward a neurodegenerative phenotype dominated by chronic microglial activation, mitochondrial failure, axonal loss, and gray matter atrophy, especially in its progressive stages [4].

Although current disease-modifying therapies are effective in reducing relapse frequency and modulating immune responses, their effect on long-term disability and established CNS damage remains limited. Consequently, there is growing interest in therapeutic approaches that enhance cellular resilience, promote neurorepair, and restore metabolic stability. One such emerging aspect of MS pathophysiology is autophagic dysfunction, which is characterized by impaired intracellular recycling and degradation processes. Autophagy is fundamental to the maintenance of cellular health in long-lived CNS cell types like neurons and oligodendrocytes [5]. Disruption of this process leads to myelin debris accumulation, increased oxidative stress, and inflammasome-mediated neuroinflammation [6]. Furthermore, proinflammatory cytokines such as IFN-γ and IL-17 not only exacerbate immune-mediated damage but also compromise autophagic flux, increasing neuronal vulnerability and hindering repair mechanisms [7].

This chapter explores the intricate relationship between inflammatory demyelination and autophagic dysfunction in MS. It examines how proinflammatory mediators, including IFN-γ, TNF-α, and IL-17, not only breach the blood–brain barrier (BBB) and facilitate immune cell infiltration but also disrupt autophagy in glial and neuronal cells. This chapter further discusses how combined autophagy impairment, mitophagy defects, and glial metabolic dysfunction impede remyelination and exacerbate neurodegeneration. By synthesizing insights into mechanisms such as NF-κB-mediated suppression of autophagy genes and mitochondrial reactive oxygen species (ROS) overload, this chapter highlights emerging therapeutic modalities.

12.2 MS Pathophysiology with Inflammation and Autophagy Failure

12.2.1 Inflammatory Demyelination

In MS, the interplay between T-cell-driven CNS infiltration and impaired autophagy significantly contributes to disease pathophysiology [5, 8]. $CD4^+$, autoreactive T-cells play a central role in the pathology of MS; they become activated in peripheral tissues by antigen-presenting cells and subsequently enter the CNS by disrupting the integrity of the blood–brain barrier (BBB) [9, 10]. Once inside the CNS, these T-cells reactivate and proliferate, attack myelin sheaths, and cause neuronal loss [5, 9]. Th1 cells release IL-17 and are closely linked to the development of MS. The combined effects of IFN-γ and TNF-α can disrupt the structural organization of tight junction proteins in the BBB and upregulate molecules involved in cell adhesion, thereby enhancing lymphocyte attachment and their movement into the CNS. IL-17 further amplifies the inflammatory response by inducing endothelial cells of the BBB to produce chemokines that attract lymphocytes [10]. As resident CNS immune cells, microglia become activated and secrete pro-inflammatory cytokines in the absence of the autophagy regulator Atg7, which further augments inflammatory responses [5, 8]. The pro-inflammatory environment, generated by Th1/Th17 cytokines and activated microglia, also promotes oxidative stress and mitochondrial damage in the axons [11]. Mitochondrial dysfunction also results in elevated levels of ROS and nitric oxide, which inflict direct damage on myelin and axons [9, 11]. In progressive MS, oxidative stress and mitochondrial injury contribute to neurodegeneration, with observed mutations in mitochondrial DNA and decreased expression of electron transport chain components in MS lesions [12].

Oligodendrocytes are also highly vulnerable to the inflammatory microenvironment of MS. During active disease stages, there is a significant loss of oligodendrocytes due to immune-mediated mechanisms [10]. Pro-inflammatory cytokines secreted by activated microglia and other immune cells directly induce the destruction of oligodendrocyte progenitor cells and mature oligodendrocytes [13, 14]. TNF-α also triggers oligodendrocyte cell death through necroptosis. The balance of TNF receptor expression shifts toward TNF receptor 1 in progressive MS, promoting apoptosis and necrosis rather than the protective TNFR2 pathway. Furthermore, activated microglia can induce neurotoxic reactive astrocytes, which contribute to oligodendrocyte destruction [14]. This proinflammatory environment significantly hampers the ability to repair myelin. The build-up of myelin remnants severely hinders the differentiation of oligodendrocytes and the remyelination process, which is worsened by ineffective clearance mechanisms in microglia, ultimately leading to reduced myelin regeneration [8]. Defects in microglial autophagy

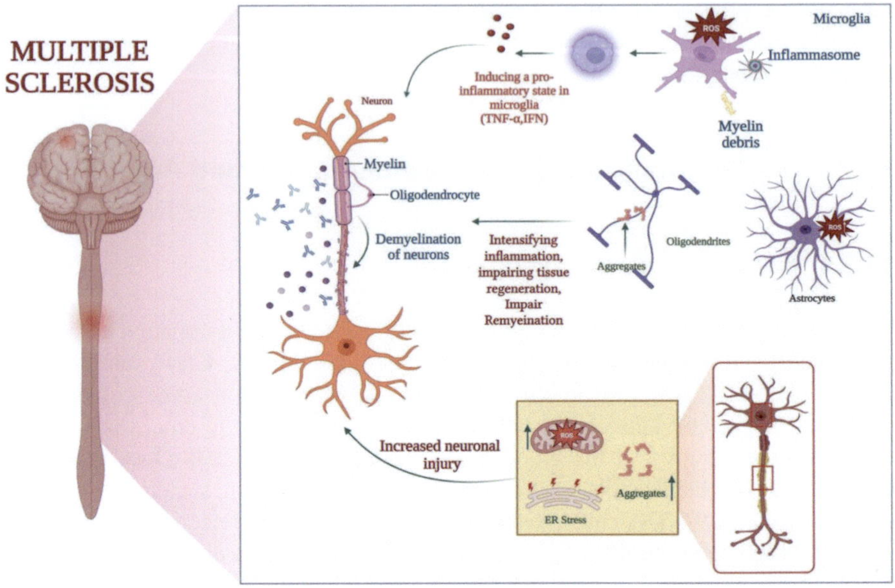

Fig. 12.1 Cellular mechanisms underlying inflammatory demyelination in MS. Activated microglia initiate a pro-inflammatory response and release inflammasome-related molecules, promoting myelin debris accumulation. Astrocytes contribute to oxidative stress by generating ROS, while mitochondrial impairment further damages axons. Protein aggregates within glial cells hinder remyelination and drive progressive neurodegeneration in the CNS. Created using BioRender (https://www.biorender.com/)

machinery, such as Atg7 deficiency, hinder efficient myelin debris clearance [5, 8]. Additionally, pro-inflammatory factors such as IL-17 directly impair the differentiation of oligodendrocyte precursor cells (OPCs) in MS, leading to inefficient remyelination and further axonal injury. Although microglia can promote remyelination by phagocytosing debris and secreting anti-inflammatory factors (Fig. 12.1), inadequate activation or recruitment of these cells can lead to the production of toxic mediators that disrupt the remyelination process [14].

12.2.2 Autophagic Dysfunction in MS Pathology

Oligodendrocytes are susceptible to the inflammatory characteristics of MS, leading to impaired function and survival. A significant aspect of this vulnerability involves the failure to efficiently clear cellular and myelin debris through autophagy-lysosomal pathways. Although microglia are primarily responsible for clearing myelin debris, they exhibit reduced efficiency in clearing myelin, resulting in the myelin fragments, which in turn hinder efficient remyelination and oligodendrocyte differentiation. Evidence of general autophagy impairment in MS lesions includes a reduced LC3-II/LC3-I ratio and elevated levels of p62/SQSTM1, indicative of disrupted protein turnover and degradation [15–17]. This dysfunctional state impacts the overall health and function of CNS cells, including oligodendrocytes. Furthermore, oligodendrocytes experience endoplasmic reticulum (ER) stress and unfolded protein response activation under inflammatory conditions [18]. It acts as a cytoprotective mechanism, and its activation in MS lesions in oligodendrocytes, microglia, and astrocytes is a compensatory response aimed at regulating ER function and promoting myelin synthesis. Specifically, activation of EIF2A (eIF2α) and EIF2AK3 (PERK) has been identified as a potential therapeutic target in MS, demonstrating that p-EIF2A augmentation improves myelination and oligodendrocyte function under inflammatory challenges [19, 20].

Axonal degeneration is a defining feature of MS and plays a major role in the development of permanent neurological impairments and disease progression. Emerging studies underscore the essential function of autophagy in preserving axonal function. In MS, this autophagic process within axons becomes severely compromised, largely due to two interrelated factors: the suppression of autophagy by IFN-γ through an mTOR-independent pathway, and disruption of mitochondrial function caused by elevated ROS levels and impaired mitophagy mechanism. Autophagy is precisely controlled through multiple signaling cascades, among which the mTOR pathway plays a central regulatory role. In MS, however, pro-inflammatory cytokines such as IFN-γ can inhibit autophagy through mechanisms that do not involve mTOR by interfering with the ULK1 complex. This complex is essential for triggering autophagy, and when inhibited, it blocks the development of autophagosomal vesicles, which encapsulate and break down unwanted cellular components. Such mTOR-independent suppression is predominantly harmful to neurons, where autophagy is essential for eliminating misfolded proteins and damaged organelles. When this process is disrupted, harmful protein aggregates accumulate, resulting in neuronal impairment and ultimately axonal degeneration. Additionally, IFN-γ-induced inhibition of autophagy compromises the clearance of myelin debris, which further obstructs remyelination and hampers recovery in MS [6]. Next, ROS overproduction plays a significant role in this process as activated microglia and macrophages, prevalent in MS lesions, are a source of ROS [21]. These ROS, together with nitric oxide, impair the function of complex IV by modifying its catalytic subunits through nitration, altering them post-translationally. The reduction in ATP production in neuronal cell bodies, along with metabolic requirements in axons lacking myelin, results in cellular energy failure and elevated production of ROS in the axon [22]. The accumulation of damaged mitochondria contributes to axonal vulnerability. PINK1 and Parkin are key proteins that work together to selectively remove damaged mitochondria by ubiquitination, targeting them for degradation via autophagy [23]. Genetic alterations affecting PINK1 and Parkin have been associated with neurodegenerative conditions such as Parkinson's disease, indicating that impaired autophagy plays a role in driving disease progression. Studies have shown that axonal mitophagy can be impaired in neurons deficient in Parkin or PINK1, leading to the accumulation of dysfunctional mitochondria. This buildup of malfunctioning organelles, worsened by disrupted retrograde transport of autophagosomes, can cause oxidative stress and trigger apoptosis inside the axon, increasing its vulnerability to degeneration [23, 24].

12.2.3 Bidirectional Crosstalk Between Pathways

The NF-κB functions as a crucial molecular integrator in the inflammatory demyelination in MS. In MS, abnormal and persistent activation of NF-κB contributes significantly to disease pathogenesis [25]. Pro-inflammatory cytokines such as TNF-α and IL-1β act as triggers of the NF-κB pathway in CNS cells, including astrocytes and microglia [26]. Activated NF-κB subunits are present in astrocytes, macrophages, perivascular lymphocytes, oligodendrocytes, and microglia within active MS plaques [27]. This exacerbates inflammation as astroglial NF-κB inactivation has been shown to reduce pro-inflammatory cytokines and improve outcomes in experimental autoimmune encephalomyelitis [28, 29]. Importantly, NF-κB activation can suppress key autophagy genes. A canonical NF-κB subunit, RelA/p65, binds to promoter regions of autophagy genes, suppressing their expression. Though direct evidence in oligodendrocytes is limited, the general upregulation of BCL2 proteins through NF-κB may extend to oligodendrocyte autophagy suppression under inflammatory stress [30]. Prolonged activation of NF-κB signaling has been observed to inhibit the expression of BECN1 and ATG5 genes, repressing autophagy in macrophages [31]. Furthermore, NF-κB plays a direct role in the upregulation of the microglial NLRP3 inflammasome. NF-κB actively triggers NLRP3 inflammasome, leading to increased expression and release of pro-inflammatory cytokines like IL-1β and IL-18 by activated microglia [32]. This NF-κB-driven inflammasome activation with impaired autophagy leads to misfolded protein accumulation and damaged organelles, creating a vicious cycle that further fuels inflammation and neurodegeneration in the CNS.

Conversely, autophagy itself possesses a vital anti-inflammatory role, and its breakdown significantly contributes to MS pathology [33]. A key indicator of impaired autophagic flux is the accumulation of SQSTM1/p62, which typically acts as a cargo receptor for autophagic degradation. When SQSTM1/p62 accumulates due to defective autophagy, it leads to increased inflammasome activation and pro-inflammatory cytokines overproduction [34]. Microglial autophagy-associated phagocytosis is crucial for myelin degradation and clearance, and its impairment leads to increased tissue deposits of myelin debris, hindering remyelination and intensifying neuroinflammation [35]. The dysfunction of autophagy in microglia promotes a pro-inflammatory phenotype that augments inflammatory responses and initiates NLRP3 inflammasome activation. Furthermore, mitophagy loss in neurons promotes mitochondrial dysfunction and oxidative stress. Next, in progressive MS, cortical neurons show mutations in mitochondrial DNA, and MS lesions exhibit decreased expression of electron transport chain components. This accumulation of damaged mitochondria due to impaired mitophagy leads to persistent ROS production, activates inflammasomes, fueling a cyclic inflammatory response and exacerbating neurodegeneration in MS [36].

12.3 Understanding Molecular Drivers and Pathways

12.3.1 NF-κB-Mediated Autophagic Suppression

NF-κB signaling actively contributes to autophagic suppression through transcriptional repression of specific autophagy initiators. The RelA/p65 subunit of NF-κB represses the transcription of key autophagy-related genes [37]. This repression is facilitated by NF-κB's ability to upregulate the transcription of EZH2, which increases repressive H3K27me3 histone marks at the promoters of these autophagy genes. Although this specific mechanism was demonstrated in THP-1 macrophages, it illustrates a broader capacity of NF-κB to transcriptionally suppress autophagy. Furthermore, NF-κB represses the transcription of *BNIP3* by constitutively binding to its promoter under normal conditions. Although p65/RelA is also known to modulate/induce BECN1 transcription, the overall context of NF-κB's role in chronic inflammation and aging relevant to MS suggests a net suppressive effect on autophagy [38]. Complementing this transcriptional control, NF-κB activation also upregulates anti-autophagic BCL2 family proteins in oligodendrocytes. NF-κB is a potent inducer of anti-apoptotic genes by stimulating their transcription [39]. These Bcl-2 family proteins are well-established inhibitors of BECN1-dependent autophagy, thereby preventing the initiation of complex formation [38]. In MS lesions, there is a significant increase in Bcl-2 expression in oligodendrocytes, with the highest numbers observed in remyelinating lesions and a high proportion of oligodendrocytes. This NF-κB-driven upregulation of Bcl-2 family proteins is believed to rescue oligodendrocytes from cell death and support subsequent remyelination in the inflammatory environment [40]. Moreover, NF-κB contributes to autophagic suppression by activating the mTOR pathway. Furthermore, NF-κB activation stimulates the expression of A20/TNFAIP3, thereby inhibiting its autophagic capacity [41]. This multi-pronged strategy by NF-κB, involving transcriptional repression of autophagy initiators and the upregulation of anti-autophagic proteins, contributes to the overall cytoprotective suppression of autophagy for oligodendrocyte survival and function amidst inflammatory stress during MS. The relationship between NF-κB and cathepsins in glial cells is complex and primarily involves NF-κB being activated by, or working in concert with, cathepsins to promote neuroinflammation, rather than NF-κB directly inducing cathepsin inhibition [42]. NF-κB can inhibit autophagy in macrophages through epigenetic mechanisms, which may indirectly impair cellular degradation pathways involving cathepsins. Specifically, pyocyanin stimulation in THP-1 macrophages promotes NF-κB nuclear translocation, which in turn enhances the transcription of EZH2, leading to increased levels of H3K27me3 [30]. The increased level of H3K27me3, through its strengthened association with NF-κB, inhibits the transcription of autophagy-related genes, limiting the autophagy process. Conversely, cathepsins in glia actively promote NF-κB activation and drive neuroinflammation. For example, Cathepsin B maintains NF-κB in a persistently activated state by promoting the autophagic degradation of IκBα in activated microglia, pushing them toward a neurotoxic M1-like phenotype [43]. Another critical pathway involving cathepsin B and NF-κB is NLRP3 inflammasome activation, which occurs following the cytosolic release of cathepsin B due to lysosomal membrane degradation [43]. Activation of the inflammasome results in caspase-1 cleavage, promoting the maturation and secretion of pro-inflammatory cytokines. Further, cathepsin C promotes microglial M1 polarization and aggravates neuroinflammation, primarily through the activation of the Ca2 + -dependent protein kinase C/p38 MAPK/NF-κB pathway. Cathepsin E, secreted from activated microglia, can proteolytically liberate TRAIL from the cell surface, which activates NF-κB through its death receptor DR5, contributing to the M1-M2 phenotypic switch of microglia. Cathepsin L inhibition has been shown to alleviate microglia-mediated neuroinflammatory responses through its interaction with the caspase-8 and NF-κB pathways. Similarly, the impact of Cathepsin X on neurotoxin-mediated molecular pathways is reflected in the NF-κB pathway, where its inhibition reduces NF-κB nuclear translocation and blocks IκBα degradation [44]. Additionally, cellular stress, such as that induced by manganese exposure, can significantly damage lysosomes, leading to increased membrane permeability and the release of intralysosomal components, like cathepsins, into the cytoplasm. This cytosolic release of cathepsins not only contributes to cell death but also enhances the secretion of inflammatory cytokines from microglia. For example, manganese overexposure directly causes such lysosomal damage, leading to Cathepsin B release and subsequent NLRP3-inflammasome activation [45]. Study highlights the importance of these pathways, highlighting that P2X4 receptor activation by ivermectin has been shown to improve lysosomal acidification and promote myelin engulfment and degradation, thus mitigating MS pathology. Furthermore, cathepsin H deficiency has been linked to the early onset of EAE due to an impairment of the toll-like receptor 3 cascade [42].

12.3.2 Cytokine Interference with Neuronal Autophagy

Cytokine interference with neuronal autophagy is a complex, multifaceted process involving direct and indirect modulation of autophagic pathways by inflammatory and regulatory cytokines. Prolonged or dysregulated IFN-γ signaling can impair autophagic flux by blocking autophagosome-lysosome fusion, leading to accumulation of autophagosomes and defective degradation. However, the blockage of autophagosome-lysosome fusion is mediated via JAK-STAT pathway activation, which alters lysosomal function or trafficking machinery in axons, preventing fusion with autophagosomes [7]. Consequently, another pro-inflammatory cytokine, IL-17, activates signaling cascades that synergize with IFN-γ to enhance neuroinflammation and may similarly impair autophagy flux. However, direct evidence of IL-17 blocking fusion through JAK-STAT in axons is less well-characterized. IL-17 is known to promote inflammatory responses that can indirectly disrupt autophagy [46]. The downstream activation of STAT3 transcription factor by various cytokine receptors (e.g., IL-6, IL-10, IL-17) through the JAK-STAT signaling pathway can directly or indirectly repress TFEB transcriptional activity [47]. TFEB is a master regulator of lysosomal biogenesis and autophagy

[48]. STAT3 may induce TFEB translocation by promoting phosphorylation of TFEB at sites that retain it in the cytoplasm or by downregulating TFEB gene expression. Persistent activation of STAT3 by pro-inflammatory cytokines (e.g., IL-6, IL-17) in microglia and neurons suppresses TFEB, contributing to defective lysosomal clearance in neurodegenerative diseases such as MS [49].

Concurrently, mitochondrial-axon dynamics are a vital factor in neuronal health, which is disrupted in neurodegeneration via mitochondrial-axon stress. The stress involves a cascade of dysfunctions primarily driven by impaired mitochondrial dynamics and neuron energy failure, with significant consequences for axonal transport and synaptic integrity. Dynamin-related protein 1 is a key regulator of mitochondrial fission. Under stress, it is recruited to the outer mitochondrial membrane, where it oligomerizes and constricts mitochondria, leading to their division [50], whereas mitophagy relies on fission, segregation, and removal of dysfunctional mitochondrial fragments. Hence, mitophagy failure leads to elongated, dysfunctional mitochondria, impaired mitochondrial quality control, and increased susceptibility to cell death [51]. However, inhibition delays mitochondrial fragmentation, cytochrome c release, and apoptosis, highlighting its pivotal role in neuronal cell death [52]. Consequently, ATP depletion, whether from mitochondrial dysfunction or oxidative stress, rapidly impairs both anterograde and retrograde axonal transport [53]. This depletion concurrently impairs the activity of Na^+/K^+ ATPases and Ca^{2+} exchange, which is essential for action potential propagation and synaptic signaling, and myosin-driven mobilization of synaptic vesicles, leading to vesicle pool depletion during sustained activity and impaired neurotransmission. Experimental evidence shows that restoring ATP levels may rescue synaptic defects due to mitochondrial dysfunction, highlighting the mitochondrial ATP production for synaptic function [54].

12.3.3 Oligodendrocyte-Specific Stress Responses

Furthermore, TREM2 plays a crucial role in facilitating myelin debris clearance in MS through its function in microglia. Elevated TREM2 expression is observed on phagocytes loaded with myelin in areas of active demyelination in human MS. Quantitative studies confirm significantly higher TREM2 mRNA expression in active MS lesions compared to normal appearing white matter or control tissue [55]. TREM2 plays a critical role in microglial function by acting as a lipid sensor, binding to myelin-associated lipids to trigger and sustain phagocytosis. Beyond just uptake, TREM2 is crucial for the intracellular degradation of myelin for regulating microglial cholesterol metabolism upon chronic phagocytic challenge, which is vital for efficient myelin clearance [56]. TREM2 downregulation in microglia leads to impaired myelin debris uptake and clearance. In mouse models of demyelination, TREM2 knockout mice exhibit significant defects in myelin debris clearance, with myelin debris persistently accumulating, leading to chronic failure of remyelination [56]. Even TREM2 haploinsufficiency (Trem2+/− mice) results in defective clearance, indicating sensitivity to TREM2 gene dosage. This impaired function is linked to defective microglial migration and phagocytosis of myelin debris, as well as lipid metabolism defects, where cells are unable to effectively process the engulfed myelin. Furthermore, human macrophages from individuals with genetic TREM2 deficiencies display defects in phagocytic pathways, with genes involved in phagocytosis being specifically downregulated [55]. Although oligodendrocyte precursor cells (OPCs) phagocyte myelin debris, studies have shown that oligodendrocytes (Olig2+ cells) increase myelin basic protein immunoreactivity after inflammatory insult [57]. The detrimental impact of TREM2 deficiency on OPCs observed in models is indirect, resulting from dysfunctional microglia failing to clear residual myelin, which in turn disrupts the recruitment and maturation of OPCs required for effective remyelination. The sustained presence of myelin debris itself acts as an inhibitor of remyelination. Therefore, the functional impairment of TREM2 contributes significantly to persistent demyelination and hinders CNS repair in MS [58]. The complement system also plays a significant role in tagging damaged myelin debris and synapses in CNS lesions for removal. These complement-tagged structures are internalized by phagocytes, which are equipped to degrade them through lysosomal structures [59]. Although this tagging is crucial in the clearance process, if the subsequent degradative processes within microglia are compromised, the increased influx of complement-opsonized material can lead to an exacerbation of autophagy-lysosomal overload. For example, TREM2 downregulation in microglia impairs their ability to degrade myelin debris, resulting in its persistent accumulation within phagolysosomes [60]. This dysfunctional clearance is further complicated by the interplay between mitochondrial damage and complement activation [61]. Such cellular stress, including ER stress and oxidative stress, which are prevalent in MS lesions and affect oligodendrocytes and other glial cells, can compromise the metabolic integrity required for efficient lysosomal function [62]. Inflammatory stress significantly impacts metabolic cooperation between astrocytes and oligodendrocytes. Astrocytes provide a crucial energy substrate to neurons and oligodendrocytes, especially during periods of high demand [63]. However, in MS, astrocytic functions are profoundly modified, including a strong reduction in their metabolic and homeostatic roles. Studies show a downregulation of the astrocyte-neuron lactate shuttle mechanism in MS [64–66]. Reactive astrocytes can display impaired ability to release lactate. This is further aggravated by stress conditions that interfere with communication between astrocytes, predominantly through the downregulation of gap junction proteins like Cx30 and Cx43, which are essential for forming intercellular channels between these cells [67]. The loss of Cx43 in astrocytic networks is linked to rapidly progressive MS and can be associated with inflammatory cytokines via microglia-dependent IL-1β production, disrupting astrocytic intercellular communication. Disruption of gap junction protein function interferes with organized lactate movement between astrocytes, thereby limiting the effective delivery of energy substrates to neurons in need [68]. Oligodendrocytes, which prefer lactate over glucose for myelin production and express high levels of MCT1 to support axons, are affected by shuttling deficits [69]. The downregulation of connexins Cx32 and Cx47 in oligodendrocytes, alongside the upregulation of astrocytic Cx43, is also associated with MS pathogenesis [70].

AMPK plays a central role in regulating autophagy. Under metabolic stress, AMPK generally promotes autophagy to support cell survival [71]. In MS, where energy deficits and mitochondrial dysfunction are linked to disease progression, the impact of AMPK inactivation on autophagic flux in glia is critical [72]. Studies show that oligodendrocytes faced with nutrient deprivation maintain a low glycolytic rate and upregulate autophagy to enhance their survival, despite resulting in process retraction and demyelination [73, 74]. However, if the AMPK role is to promote catabolic processes and maintain energy balance, chronic mitochondrial defects can lead to a decrease in the phosphorylated active form of AMPK [75]. The decreased activity of AMPK could potentially hinder the efficiency of this protective autophagic response under sustained metabolic impairment conditions. Conversely, restoring AMPK activity or supplementing with lactate can counteract negative effects in other neurological models. Therefore, although the direct link of AMPK inactivation impairing autophagic flux in MS specifically is not fully detailed, the general functions of AMPK and autophagy under stress suggest a potential for such impairment if AMPK signaling is compromised [70].

12.4 Multimodal Therapies Connecting Mechanisms to Clinical Practice

12.4.1 Limitations of Current Approaches

Immunomodulatory monotherapies, although effective in modulating specific immune components, face significant limitations that restrict their overall therapeutic impact in chronic neuroinflammatory and neurodegenerative conditions. Anti-CD20 agents such as ocrelizumab effectively deplete B-cells and reduce inflammatory activity; however, they do not address autophagy dysfunction in chronic lesions, a critical mechanism for clearing damaged proteins and organelles that, when impaired, contributes to persistent inflammation and tissue damage. This neglects the fact that autophagy can lead to the accumulation of toxic cellular debris, exacerbating chronic lesion pathology despite B-cell depletion [76, 77]. Next, glucocorticoids, contradictorily, exacerbate mitochondrial dysfunction in neurons. They induce mitochondrial damage by promoting the opening of the mitochondrial permeability transition pore through upregulation of cyclophilin D, leading to impaired mitochondrial quality control, increased oxidative stress, and synaptic loss. Although glucocorticoids suppress inflammation, their detrimental effects on neuronal mitochondria limit their long-term safety and efficacy [78].

12.4.2 Combined Therapeutic Strategies

A well-known antidiabetic drug, metformin, activates AMPK, which bypasses the mTOR-dependent autophagy blockade, thereby promoting autophagic clearance of damaged organelles and proteins. Additionally, preclinical studies suggest that metformin supports oligodendrocyte survival, which is crucial for myelin maintenance [79, 80]. Trehalose is a natural disaccharide and acts as an mTOR-independent lysosomal stabilizer that enhances autophagic flux by promoting lysosomal function. In MS models, trehalose reduces pathological protein aggregation, such as α-synuclein, which contributes to neurodegeneration and chronic inflammation [81]. Next, spermidine induces mitophagy primarily through the inhibition of acetyltransferase EP300, which suppresses autophagy-related gene expression. This positions spermidine as a key component in multimodal therapeutic strategies that simultaneously target mitochondrial dysfunction and neuroinflammation [79, 80].

As discussed earlier, immunomodulatory monotherapies have significant limitations that reduce overall effectiveness in treating chronic neuroinflammatory and neurodegenerative conditions. However, immunomodulatory agents with autophagy benefits are a promising area of research in neurodegenerative diseases. Siponimod is a selective sphingosine-1-phosphate receptor modulator that acts as a functional antagonist at S1P1 receptors on lymphocytes. It prevents lymphocytes from exiting lymph nodes, reducing their entry into the bloodstream and limiting immune cell infiltration into the CNS. This mechanism is crucial in treating MS and other autoimmune diseases [82, 83]. In MS lesions, there is upregulation of S1P1 and S1P3 receptors on reactive astrocytes and S1P5 on oligodendrocytes, precipitating direct modulation of astrocyte activity through inhibiting pro-inflammatory signaling pathways. Furthermore, siponimod has been shown to inhibit microglial inflammasome activation, further reducing CNS inflammation and potentially protecting against neurodegeneration [84, 85]. Further, S1PR1 activation triggers downstream signaling of PI3K/Akt and the mTOR pathway. Although mTOR generally inhibits autophagy, S1PR1 receptor signaling activated through S1P can modulate this inhibition and promote autophagy and mitophagy by regulating the gene expression of key proteins such as Bnip3I, Pink1, and p62, which are directly involved in the mitophagy process [86, 87]. NOD-like receptor protein-3 inflammasome is a multiprotein complex in microglia known to induce the production of pro-inflammatory cytokine IL-1β, directing autophagic arrest. Likewise, MCC950 NLRP-3 Inflammasome inhibitor has been shown to reduce IL-1β signaling and neuroinflammation in microglia through direct inhibition of NLRP-3. MCC950 blocks the assembly of the NOD-like receptor protein-3 inflammasome complex, thereby suppressing the activation of caspase-1 and subsequent maturation and release of IL-1β and IL-18, alleviating autophagic dysfunction [88, 89].

Considering various immunomodulatory therapies for restoring autophagy and neuronal function, the emerging approach has drifted to gene therapy and novel drug delivery systems such as nanoparticle-mediated deliveries. Several previous preclinical studies show TFEB overexpression in oligodendrocytes enhances lysosomal biogenesis and autophagy, promoting pathological protein clearance such as α-synuclein, and preventing neurodegeneration [90]. Concurrently, TFEB also regulated oligodendrocyte differentiation and the length of myelin sheath formation, suggesting a role in maintaining myelin integrity [91]. Consequently, rapamycin and its analogues, such as everolimus, are known as mTOR inhibitors and induce autophagy. However, systemic administration can have limited CNS penetration and off-target effects. Hence, nanoparticle-mediated delivery of rapamycin analogues is a cutting-edge system designed to target demyelinated neurons in the CNS. When administered intravenously, these nanoparticles can accumulate at the site of inflammation and/or demyelinated lesions due to disrupted integrity and increased permeability of affected cells. Once localized, nanoparticles release the encapsulated drug, which enters the target cell through endocytosis or extracellular release, followed by cellular uptake [92].

12.4.3 A Hypothetical Approach with Metformin and Siponimod in Progressive MS

Progressive MS remains a challenging subtype of MS, marked by neurodegeneration and limited response to immunomodulatory therapy alone. A sphingosine-1-phosphate receptor modulator, siponimod, is approved for secondary progressive MS, but its neuroprotective benefits may be limited. Recent preclinical research suggests that metformin, an AMPK activator, can promote autophagy and improve mitochondrial function, offering potential neuroprotective effects in neurodegenerative diseases.

A 52-year-old patient diagnosed with active secondary progressive MS, characterized by a gradual accrual of disability independent of relapses, but with evidence of recent inflammatory activity on MRI. Based on her active disease, the neurologist initially prescribed siponimod, which prevents lymphocytes from egressing lymphoid tissues, assuming it will significantly reduce the risk of disability progression, decrease new MRI lesions, slow brain volume loss, and improve cognitive processing speed as measured by the Symbol Digit Modalities Test. Despite these benefits, the neurologist seeks to address the underlying neurodegeneration. Therefore, an add-on treatment is also prescribed, drawing on its promising investigational role in progressive MS, as realized in the trial MACSiMiSE-BRAIN (NCT05893225). Metformin is hypothesized to offer anti-inflammatory, neuroprotective, and remyelination-promoting properties, primarily through AMPK activation. A prior cohort study in MS patients on metformin showed a significant reduction in new or expanding T2 brain lesions, increased AMPK expression, and beneficial shifts in immune cell populations. In this hypothetical scenario, the combination aims for synergistic effects, that is, siponimod effectively controls overt inflammatory activity and its direct consequences on disability progression and brain atrophy, while metformin concurrently promotes remyelination and exerts broader neuroprotective effects. Over time, neurologists would observe two key outcomes. Firstly, a reduced serum neurofilament light chain is a recognized biomarker for neuro-axonal injury and is analyzed in siponimod trials and explicitly included as a sub study in metformin's MACSiMiSE-BRAIN trial, indicating a potential for this combination to further mitigate neuronal damage. Secondly, the laboratory results would show improved autophagy biomarkers, such as an enhanced LC3-II/p62 ratio, reflecting the positive modulation of cellular waste recycling. This combined approach, leveraging distinct yet complementary mechanisms, hypothetically may lead to a more comprehensive outbreak of MS progression in both reduced neuronal damage markers and enhanced cellular self-renewal processes.

12.5 Conclusion

The interplay between chronic inflammation and autophagy failure in MS forms a self-reinforcing cycle that accelerates neurodegeneration, with autophagy dysfunction not merely a secondary consequence but a central driver of pathology. Impaired microglial autophagy promotes NLRP3 inflammasome activation and myelin debris accumulation, while neuronal mitophagy failure exacerbates oxidative stress and axonal injury. This cycle is perpetuated by NF-κB-mediated repression of key autophagy genes and BCL2 upregulation. To break this cycle, future therapies must be multimodal, targeting glial-axonal metabolic coupling through mitochondrial biogenesis enhancers and modulating inflammasome-autophagy crosstalk through combinations like NLRP3 inhibitors with rapalogs. Oligodendrocyte resilience may be improved by correcting TNFR1/TNFR2 imbalances and alleviating ER stress. Precision medicine efforts will benefit from integrated biomarker strategies, as autophagy flux assays and cytokine profiling can identify patient subgroups likely to respond to autophagy enhancers or anti-inflammatory agents. Advanced modeling using patient-derived iPSC brain organoids allows investigation of oligodendrocyte–neuron interactions and therapeutic responses within simulated lesion environments, although systems biology tools can decode lesion heterogeneity and uncover cell-type-specific drivers. These directions emphasize the need for integrated, personalized, and systems-level approaches to effectively disrupt the inflammation–autophagy axis and improve outcomes in MS.

References

1. Wallin MT, Culpepper WJ, Campbell JD, Nelson LM, Langer-Gould A, Marrie RA, et al. The prevalence of MS in the United States: a population-based estimate using health claims data. Neurology. 2019;92(10):E1029–40.
2. Lublin FD, Reingold SC, Cohen JA, Cutter GR, Sørensen PS, Thompson AJ, et al. Defining the clinical course of multiple sclerosis: the 2013 revisions. Neurology. 2014;83(3):278–86.
3. Lassmann H, Brück W, Lucchinetti CF. The immunopathology of multiple sclerosis: an overview. Brain Pathol. 2007;17(2):210–8.
4. Mahad DH, Trapp BD, Lassmann H. Pathological mechanisms in progressive multiple sclerosis. Lancet Neurol. 2015;14(2):183–93.
5. Misrielal C, Mauthe M, Reggiori F, Eggen BJL. Autophagy in multiple sclerosis: two sides of the same coin. Front Cell Neurosci. 2020;14:14.
6. Shen D, Liu K, Wang H, Wang H. Autophagy modulation in multiple sclerosis and experimental autoimmune encephalomyelitis. Clin Exp Immunol. 2022;209(2):140–50.
7. Chang YP, Tsai CC, Huang WC, Wang CY, Chen CL, Lin YS, et al. Autophagy facilitates IFN-δ-induced Jak2-STAT1 activation and cellular inflammation. J Biol Chem. 2010;285(37):28715–22.

8. Berglund R, Guerreiro-Cacais AO, Adzemovic MZ, Zeitelhofer M, Lund H, Ewing E, et al. Microglial autophagy-associated phagocytosis is essential for recovery from neuroinflammation. Sci Immunol. 2020;5(52):eabb5077.
9. Liang P, Le W. Role of autophagy in the pathogenesis of multiple sclerosis. Neurosci Bull. 2015;31(4):435–44.
10. Mohammed EMA. Understanding multiple sclerosis pathophysiology and current disease-modifying therapies: a review of unaddressed aspects. Front Biosci. 2024;29(11):386.
11. Ashwani, Sharma A, Choudhary MK, Gugulothu D, Pandita D, Verma S, et al. Epigenetic and mitochondrial metabolic dysfunction in multiple sclerosis: a review of herbal drug approaches and current clinical trials. Mol Neurobiol. 2025;62(8):10045–90.
12. Kamma E, Lasisi W, Libner C, Ng HS, Plemel JR. Central nervous system macrophages in progressive multiple sclerosis: relationship to neurodegeneration and therapeutics. J Neuroinflammation. 2022;19(1):45.
13. Papiri G, D'Andreamatteo G, Cacchiò G, Alia S, Silvestrini M, Paci C, et al. Multiple sclerosis: inflammatory and Neuroglial aspects. Curr Issues Mol Biol. 2023;45(2):1443–70.
14. Mado H, Adamczyk-Sowa M, Sowa P. Role of microglial cells in the pathophysiology of MS: synergistic or antagonistic? Int J Mol Sci. 2023;24(3):1861.
15. Feng X, Hou H, Zou Y, Guo L. Defective autophagy is associated with neuronal injury in a mouse model of multiple sclerosis. Bosn J Basic Med Sci. 2017;17(2):95–103.
16. Al-kuraishy HM, Jabir MS, Al-Gareeb AI, Saad HM, Batiha GES, Klionsky DJ. The beneficial role of autophagy in multiple sclerosis: yes or no? Autophagy. 2024;20(2):259–74.
17. Hassanpour M, Hajihassani F, Hiradfar A, Aghamohammadzadeh N, Rahbarghazi R, Safaie N, et al. Real-state of autophagy signaling pathway in neurodegenerative disease; focus on multiple sclerosis. J Inflamm (United Kingdom). 2020;17(1):6.
18. Kapoor A, Sanyal AJ. Endoplasmic reticulum stress and the unfolded protein response. Clin Liver Dis. 2009;13(4):581–90.
19. Chang A, Tourtellotte WW, Rudick R, Trapp BD. Premyelinating oligodendrocytes in chronic lesions of multiple sclerosis. N Engl J Med. 2002;346(3):165–73.
20. Lin W, Bailey SL, Ho H, Harding HP, Ron D, Miller SD, et al. The integrated stress response prevents demyelination by protecting oligodendrocytes against immune-mediated damage. J Clin Invest. 2007;117(2):448–56.
21. Mahad DJ, Ziabreva I, Campbell G, Lax N, White K, Hanson PS, et al. Mitochondrial changes within axons in multiple sclerosis. Brain. 2009;132(5):1161–74.
22. Campbell G, Mahad D. Neurodegeneration in progressive multiple sclerosis. Cold Spring Harb Perspect Med. 2018;8(10):a028985.
23. Maday S. Mechanisms of neuronal homeostasis: autophagy in the axon. Brain Res. 2016;1649: 143–50.
24. van Horssen J, Schreibelt G, Drexhage J, Hazes T, Dijkstra CD, van der Valk P, et al. Severe oxidative damage in multiple sclerosis lesions coincides with enhanced antioxidant enzyme expression. Free Radic Biol Med. 2008;45(12):1729–37.
25. Kopitar-Jerala N. The role of interferons in inflammation and inflammasome activation. Front Immunol. 2017;8:873.
26. Kaltschmidt B, Uherek M, Volk B, Baeuerle PA, Kaltschmidt C. Transcription factor NF-κB is activated in primary neurons by amyloid β peptides and in neurons surrounding early plaques from patients with Alzheimer disease. Proc Natl Acad Sci USA. 1997;94(6):2642–7.
27. Wesselborg S, Bauer MKA, Vogt M, Schmitz ML, Schulze-Osthoff K. Activation of transcription factor NF-κB and p38 mitogen-activated protein kinase is mediated by distinct and separate stress effector pathways. J Biol Chem. 1997;272(19):12422–9.
28. Hilliard B, Samoilova EB, Liu TST, Rostami A, Chen Y. Experimental autoimmune encephalomyelitis in NF-κB- deficient mice: roles of NF-κB in the activation and differentiation of autoreactive T cells. J Immunol. 1999;163(5):2937–43.

29. van Loo G, De Lorenzi R, Schmidt H, Huth M, Mildner A, Schmidt-Supprian M, et al. Inhibition of transcription factor NF-κB in the central nervous system ameliorates autoimmune encephalomyelitis in mice. Nat Immunol. 2006;7(9):954–61.
30. Ashburner BP, Westerheide SD, Baldwin AS. The p65 (RelA) subunit of NF-κB interacts with the histone deacetylase (HDAC) corepressors HDAC1 and HDAC2 to negatively regulate gene expression. Mol Cell Biol. 2001;21(20):7065–77.
31. Djavaheri-Mergny M, Amelotti M, Mathieu J, Besançon F, Bauvy C, Souquère S, et al. NF-κB activation represses tumor necrosis factor-α-induced autophagy. J Biol Chem. 2006;281(41): 30373–82.
32. Bauernfeind FG, Horvath G, Stutz A, Alnemri ES, MacDonald K, Speert D, et al. Cutting edge: NF-κB activating pattern recognition and cytokine receptors license NLRP3 Inflammasome activation by regulating NLRP3 expression. J Immunol. 2009;183(2):787–91.
33. Harris J, Lang T, Thomas JPW, Sukkar MB, Nabar NR, Kehrl JH. Autophagy and inflammasomes. Mol Immunol. 2017;86:10–5.
34. Shi CS, Shenderov K, Huang NN, Kabat J, Abu-Asab M, Fitzgerald KA, et al. Activation of autophagy by inflammatory signals limits IL-1β production by targeting ubiquitinated inflammasomes for destruction. Nat Immunol. 2012;13(3):255–63.
35. Wang X, Sun Y, Yu H, Xue C, Pei X, Chen Y, et al. The regulation of microglia by aging and autophagy in multiple sclerosis. Pharmacol Res. 2025;216:107786.
36. Wager K, Russell C. Mitophagy and neurodegeneration: the zebrafish model system. Autophagy. 2013;9(11):1693–709.
37. Wang J, Bian M, Liang S, Yi X, Du Y. NF-κB-mediated enhancement of H3K27me3 through EZH2: a mechanism to suppress pyocyanin-induced autophagy in macrophages. Eur J Med Res. 2025;30(1):346.
38. Salminen A, Hyttinen JMT, Kauppinen A, Kaarniranta K. Context-dependent regulation of autophagy by IKK-NF- B signaling: impact on the aging process. Int J Cell Biol. 2012;2012: 849541.
39. Stone S, Jamison S, Yue Y, Durose W, Schmidt-Ullrich R, Lin W. NF-γB activation protects oligodendrocytes against inflammation. J Neurosci. 2017;37(38):9332–44.
40. Kuhlmann T, Lucchinetti C, Zettl UK, Bitsch A, Lassmann H, Brück W. Bcl-2-expressing oligodendrocytes in multiple sclerosis lesions. Glia. 1999;28(1):34–9.
41. Lei Z, Yue Y, Stone S, Wu S, Lin W. NF-κB activation accounts for the Cytoprotective effects of PERK activation on oligodendrocytes during EAE. J Neurosci. 2020;40(33):6444–56.
42. Quick JD, Silva C, Wong JH, Lim KL, Reynolds R, Barron AM, et al. Lysosomal acidification dysfunction in microglia: an emerging pathogenic mechanism of neuroinflammation and neurodegeneration. J Neuroinflammation. 2023;20(1):185.
43. Nakanishi H. Cathepsin regulation on microglial function. Biochim Biophys Acta Proteins Proteom. 2020;1868(9):140465.
44. Pišlar A, Bolčina L, Kos J. New insights into the role of cysteine cathepsins in neuroinflammation. Biomolecules. 2021;11(12):1796.
45. Tinkov AA, Paoliello MMB, Mazilina AN, Skalny AV, Martins AC, Voskresenskaya ON, et al. Molecular targets of manganese-induced neurotoxicity: a five-year update. Int J Mol Sci. 2021;22(9):4646.
46. Zheng MY, Luo LZ. The role of IL-17A in mediating inflammatory responses and progression of neurodegenerative diseases. Int J Mol Sci. 2025;26(6):2505.
47. Liu Y, Che X, Yu X, Shang H, Cui P, Fu X, et al. Phosphorylation of STAT3 at Tyr705 contributes to TFEB-mediated autophagy-lysosomal pathway dysfunction and leads to ischemic injury in rats. Cell Mol Life Sci. 2023;80(6):160.
48. IASB. At a glance at a glance. Teachers' Pension Plan Annual Report [Internet]. 2020;12(3): 1–2. Available from: https://www.ippf.org/sites/default/files/2020-07/At a Glance 2019.pdf
49. Zhao Y, Hong Z, Lin Y, Shen W, Yang Y, Zuo Z, et al. Exercise pretreatment alleviates neuroinflammation and oxidative stress by TFEB-mediated autophagic flux in mice with ischemic stroke. Exp Neurol. 2023;364:114380.

50. Zerihun M, Sukumaran S, Qvit N. The Drp1-mediated mitochondrial fission protein Interactome as an emerging Core player in mitochondrial dynamics and cardiovascular disease therapy. Int J Mol Sci. 2023;24(6):5785.
51. Tong M, Zablocki D, Sadoshima J. The role of Drp1 in mitophagy and cell death in the heart. J Mol Cell Cardiol. 2020;142:138–45.
52. Suen DF, Norris KL, Youle RJ. Mitochondrial dynamics and apoptosis. Genes Dev. 2008;22 (12):1577–90.
53. Fang C, Bourdette D, Banker G. Oxidative stress inhibits axonal transport: implications for neurodegenerative diseases. Mol Neurodegener. 2012;7(1):29.
54. Sheng ZH, Cai Q. Mitochondrial transport in neurons: impact on synaptic homeostasis and neurodegeneration. Nat Rev Neurosci. 2012;13(2):77–93.
55. Cignarella F, Filipello F, Bollman B, Cantoni C, Locca A, Mikesell R, et al. TREM2 activation on microglia promotes myelin debris clearance and remyelination in a model of multiple sclerosis. Acta Neuropathol. 2020;140(4):513–34.
56. Wang Y, Kyauk RV, Shen YAA, Xie L, Reichelt M, Lin H, et al. TREM2-dependent microglial function is essential for remyelination and subsequent neuroprotection. Glia. 2023;71(5): 1247–58.
57. Cammarota M, Boscia F. Contribution of oligodendrocytes, microglia, and astrocytes to myelin debris uptake in an explant model of inflammatory demyelination in rats. Cells. 2023;12(17): 2203.
58. You YF, Chen M, Tang Y, Yu WX, Pang XW, Chu YH, et al. TREM2 deficiency inhibits microglial activation and aggravates demyelinating injury in neuromyelitis optica spectrum disorder. J Neuroinflammation. 2023;20(1):89.
59. Michailidou I, Willems JGP, Kooi EJ, Van Eden C, Gold SM, Geurts JJG, et al. Complement C1q-C3-associated synaptic changes in multiple sclerosis hippocampus. Ann Neurol. 2015;77 (6):1007–26.
60. Watkins LM, Neal JW, Loveless S, Michailidou I, Ramaglia V, Rees MI, et al. Complement is activated in progressive multiple sclerosis cortical grey matter lesions. J Neuroinflammation. 2016;13(1):161.
61. Lin J, Hwang S, Luo H, Mohamud Y. Double-edged sword: exploring the mitochondria–complement bidirectional connection in cellular response and disease. Biology (Basel). 2024;13(6):431.
62. Mháille AN, McQuaid S, Windebank A, Cunnea P, McMahon J, Samali A, et al. Increased expression of endoplasmic reticulum stress-related signaling pathway molecules in multiple sclerosis lesions. J Neuropathol Exp Neurol. 2008;67(3):200–11.
63. Chamaa F, Magistretti PJ, Fiumelli H. Astrocyte-derived lactate in stress disorders. Neurobiol Dis. 2024;192:106417.
64. Mason S. Lactate shuttles in neuroenergetics-homeostasis, allostasis and beyond. Front Neurosci. 2017;11:43.
65. Zeis T, Allaman I, Gentner M, Schroder K, Tschopp J, Magistretti PJ, et al. Metabolic gene expression changes in astrocytes in multiple sclerosis cerebral cortex are indicative of immune-mediated signaling. Brain Behav Immun. 2015;48:313–25.
66. Srivastava R, Aslam M, Kalluri SR, Schirmer L, Buck D, Tackenberg B, et al. Potassium Channel KIR4.1 as an immune target in multiple sclerosis. N Engl J Med. 2012;367(2):115–23.
67. Barmpagiannos K, Theotokis P, Petratos S, Pagnin M, Einstein O, Kesidou E, et al. The diversity of astrocyte activation during multiple sclerosis: potential cellular targets for novel disease modifying therapeutics. Healthcare (Switzerland). 2023;11(11):1585.
68. Kunkl M, Amormino C, Tedeschi V, Fiorillo MT, Tuosto L. Astrocytes and inflammatory T Helper Cells: a dangerous liaison in multiple sclerosis. Front Immunol. 2022;13:824411.
69. Hu X, Yu G, Liao X, Xiao L. Interactions between astrocytes and oligodendroglia in myelin development and related brain diseases. Neurosci Bull. 2023;39(3):541–52.
70. Liu Z, Guo Y, Zhang Y, Gao Y, Ning B. Metabolic reprogramming of astrocytes: emerging roles of lactate. Neural Regen Res. 2026;21(2):421–32.

71. Segatto M, Rosso P, Fioramonti M, Fracassi A, Marangoni M, Taglietti V, et al. AMPK in the central nervous system: physiological roles and pathological implications. Res Rep Biol. 2016;7:1–13.
72. Tepavčević V. Oligodendroglial energy metabolism and (Re)myelination. Life. 2021;11(3): 238.
73. Rone MB, Cui QL, Fang J, Wang LC, Zhang J, Khan D, et al. Oligodendrogliopathy in multiple sclerosis: low glycolytic metabolic rate promotes oligodendrocyte survival. J Neurosci. 2016;36 (17):4698–707.
74. Fernandes MGF, Luo JXX, Cui QL, Perlman K, Pernin F, Yaqubi M, et al. Age-related injury responses of human oligodendrocytes to metabolic insults: link to BCL-2 and autophagy pathways. Commun Biol. 2021;4(1):20.
75. Agostini F, Bisaglia M, Plotegher N. Linking ROS levels to autophagy: the key role of AMPK. Antioxidants. 2023;12(7):1406.
76. White E, Karp C, Strohecker AM, Guo Y, Mathew R. Role of autophagy in suppression of inflammation and cancer. Curr Opin Cell Biol. 2010;22(2):212–7.
77. Tallantyre EC, Whittam DH, Jolles S, Paling D, Constantinesecu C, Robertson NP, et al. Secondary antibody deficiency: a complication of anti-CD20 therapy for neuroinflammation. J Neurol. 2018;265(5):1115–22.
78. Du F, Yu Q, Swerdlow RH, Waites CL. Glucocorticoid-driven mitochondrial damage stimulates tau pathology. Brain. 2023;146(10):4378–94.
79. Rahman MA, Rhim H. Therapeutic implication of autophagy in neurodegenerative diseases. BMB Rep. 2017;50(7):345–54.
80. Fujikake N, Shin M, Shimizu S. Association between autophagy and neurodegenerative diseases. Front Neurosci. 2018;12:255.
81. Lin CH, Wu YR, Yang JM, Chen WL, Chao CY, Chen IC, et al. Novel lactulose and Melibiose targeting autophagy to reduce PolyQ aggregation in cell models of spinocerebellar ataxia 3. CNS Neurol Disord Drug Targets. 2016;15(3):351–9.
82. Behrangi N, Fischbach F, Kipp M. Mechanism of siponimod: anti-inflammatory and neuroprotective mode of action. Cells. 2019;8(1):24.
83. Spampinato SF, Merlo S, Sano Y, Kanda T, Sortino MA. Protective effect of the sphingosine-1 phosphate receptor agonist siponimodon disrupted blood brain barrier function. Biochem Pharmacol. 2021;186:114465.
84. Montarolo F, Martire S, Marnetto F, Valentino P, Valverde S, Capobianco MA, et al. The selective agonist for Sphingosine-1-phosphate receptors Siponimod increases the expression level of NR4A genes in microglia cell line. Curr Issues Mol Biol. 2022;44(3):1247–56.
85. Komiya H, Takeuchi H, Ogasawara A, Ogawa Y, Kubota S, Hashiguchi S, et al. Siponimod inhibits microglial inflammasome activation. Neurosci Res. 2025;213:138–45.
86. Xiao S, Peng K, Li C, Long Y, Yu Q. The role of sphingosine-1-phosphate in autophagy and related disorders. Cell Death Discov. 2023;9(1):380.
87. Meacci E, Garcia-Gil M. S1P/S1P receptor signaling in neuromuscular disorders. Int J Mol Sci. 2019;20(24):6364.
88. Xu W, Huang Y, Rongbin Z. NLRP3 inflammasome in neuroinflammation and central nervous system diseases. Cell Mol Immunol. 2025;22(4):341–55.
89. Liu Q, Zhang MM, Guo MX, Zhang QP, Li NZ, Cheng J, et al. Inhibition of microglial NLRP3 with MCC950 attenuates microglial morphology and NLRP3/Caspase-1/IL-1β signaling in stress-induced mice. J Neuroimmune Pharmacol. 2022;17(3–4):503–14.
90. Arotcarena ML, Bourdenx M, Dutheil N, Thiolat ML, Doudnikoff E, Dovero S, et al. Transcription factor EB overexpression prevents neurodegeneration in experimental synucleinopathies. JCI Insight. 2019;4(16):e129719.
91. Bouchard EL, Meireles AM, Talbot WS. Oligodendrocyte development and myelin sheath formation are regulated by the antagonistic interaction between the Rag-Ragulator complex and TFEB. Glia. 2024;72(2):289–99.
92. Shirasu T, Koyama H, Miura Y, Hoshina K, Kataoka K, Watanabe T. Nanoparticles effectively target rapamycin delivery to sites of experimental aortic aneurysm in rats. PLoS One. 2016;11 (6):e0157813.

Disrupted Cellular Clearance and Inflammatory Loops in Frontotemporal Dementia

13

Praveen Halagali, Vamshi Krishna Tippavajhala, Himanshu Sharma, Adil Abalkhail, and Wesam Taher Almagharbeh

Abstract

Frontotemporal dementia (FTD) is a neurodegenerative disorder marked by progressive neuronal loss in the frontal and temporal lobes, driven by distinct molecular and cellular pathologies. This chapter explores the critical intersection of neuroinflammation and autophagy–lysosomal dysfunction as central mechanisms in FTD pathogenesis. Key genetic contributors MAPT, GRN, and C9ORF72 initiate diverse pathological cascades, including tau aggregation, TDP-43 mislocalization, and impaired vesicular trafficking to disrupt proteostasis and immune regulation. Autophagy failure leads to the accumulation of toxic proteins and organelles, which activate microglia and amplify pro-inflammatory signaling through pathways such as nuclear factor kappa B (NF-κB), NLRP3 inflammasome, and cGAS-STING. This self-reinforcing cycle exacerbates neuronal vulnerability and accelerates disease progression. Additionally, stress

P. Halagali · V. K. Tippavajhala
Department of Pharmaceutics, Manipal College of Pharmaceutical Sciences, Manipal Academy of Higher Education, Manipal, Karnataka, India
e-mail: praveen.mcopsmpl2023@learner.manipal.edu; vamshi.krishna@manipal.edu

H. Sharma (✉)
Teerthanker Mahaveer College of Pharmacy, Teerthanker Mahaveer University, Moradabad, UP, India

A. Abalkhail
Department of Public Health, College of Applied Medical Sciences, Qassim University, Buraydah, Saudi Arabia
e-mail: abalkhail@qu.edu.sa

W. T. Almagharbeh
Medical and Surgical Nursing Department, Faculty of Nursing, University of Tabuk, Tabuk, Saudi Arabia
e-mail: Walmagharbeh@ut.edu.sa

P. Khanal et al. (eds.), *Neuroinflammation and Autophagy in Neurodegeneration*,
https://doi.org/10.1007/978-981-95-3945-1_13

granule persistence, mitochondrial dysfunction, and lysosomal stress further entrench this pathological feedback loop. Genetic modifiers like transmembrane protein 106B (TMEM106B) intensify these effects, influencing disease severity and onset. Understanding the mechanistic convergence of inflammation and proteostatic failure is essential for developing effective interventions aimed at halting or reversing neurodegeneration in FTD.

Keywords

Autophagy–lysosomal dysfunction · Frontotemporal dementia · Neuroinflammation · TDP-43 aggregation

13.1 Introduction

Frontotemporal dementia (FTD) is a variant of dementia that causes selective atrophy in the frontal and temporal lobes of the brain [1]. It is the second most common neurodegenerative dementing illness, typically occurring in individuals aged 45 and 65. Between 15 and 22 cases per 100,000 people is the estimated global prevalence, with significant regional and ethnic variations. Moreover, primary progressive aphasia is characterized by a progressive loss of language skills, and behavioral variant FTD is characterized by changes in executive functioning, personality, and social conduct, two main clinical subtypes of FTD [2]. These subtypes reflect the focal degeneration of specific brain regions, where primary progressive aphasia involves the temporal lobe and FTD is linked to frontal lobe atrophy. Pathologically, FTD is typified by severe neuronal loss, microglial activation, and gliosis, mostly in the frontal and temporal cortices [3]. A defining characteristic is the presence of either hyperphosphorylated tau or TAR DNA-binding protein 43 (TDP-43), depending on the genetic and molecular subtype. Unlike Alzheimer's disease, FTD does not have amyloid-β plaques and frequently exhibits distinct tauopathy profiles [4]. Clinical neuropsychological tests, imaging biomarkers, and cerebrospinal fluid (CSF) markers are all part of the diagnostic evaluation. Moreover, its unique histopathological and molecular characteristics make it a unique clinical entity that necessitates targeted research and treatment approaches, even though careful differential diagnosis is necessary due to the clinical overlap with other dementias [5].

13.2 Fundamental Genetic Factors in Familial FTD

About 30–40% of FTD cases have a positive family history, and the majority of inherited cases are brought on by mutations in GRN, C9ORF72, and MAPT [6]. In familial FTD, MAPT mutations lead to abnormal tau isoform production and aggregation, causing tauopathy with neuronal cytoskeletal disruption [7]. GRN mutations that cause haploinsufficiency of progranulin are essential regulators of neuroinflammation and lysosomal function, which are often the cause of TDP-43 pathology [8]. The most common genetic cause of C9ORF72 repeat expansions also leads to toxic dipeptide repeat proteins and RNA foci that cause widespread neurodegeneration and impede autophagy and nucleocytoplasmic transport. Traumatic brain injury (TBI) and persistent psychological stress have been associated with increased susceptibility by promoting neuroinflammation and protein aggregation [9]. Recent findings demonstrate that, without altering the underlying DNA sequence, epigenetic modifications such as DNA methylation and histone acetylation can regulate the gene expression linked to FTD-associated proteins [10]. These changes can be triggered by environmental factors, which contribute to the disease onset and/or advancement. Concurrently, another vital factor influencing C9ORF72-linked FTD is oxidative stress, which exacerbates deficiencies in nucleocytoplasmic transport and promotes toxic species accumulation [11]. This stress-driven cellular dysfunction, mainly when environmental factors like toxins lead to progressive motor and cognitive decline in affected individuals [12]. Moreover, variations in lysosomal-associated transmembrane protein 106B (TMEM106B) have been identified as genetic modifiers that impact the severity and disease progression in GRN mutation carriers, and this effect is observed through lysosomal function and progranulin trafficking. Consequently, TMEM106B polymorphisms alter the severity of TDP-43 pathology [13]. Considering these facts, TMEM106B is a significant therapeutic target in genetically susceptible individuals, as protective alleles are associated with milder and delayed onset phenotypes [14].

13.3 Genetic Factors Contributing to Autophagy Failure and Neuroinflammation

It is recognized that in FTD, genetic mutations impair fundamental cellular processes, autophagy–lysosomal pathways, and neuroinflammation [15]. The three most studied genetic contributors, MAPT, GRN, and C9ORF72, contribute to disease pathogenesis through distinct but connected pathways that result in neuronal dysfunction and death. Mutations in the MAPT gene are primarily responsible for tauopathies in FTD as its pathogenic variants lead to aberrant splicing, hyperphosphorylation, and aggregation into insoluble filaments [16]. These aggregates disrupt microtubule stability and axonal transport, but recent studies associate tau with autophagic flux alterations [17]. These aggregates prevent autophagosome maturation and lysosome fusion, causing damaged proteins and organelles to accumulate. Furthermore, mutant tau may indirectly activate microglia and astrocytes, creating a chronically pro-inflammatory environment that exacerbates neurodegeneration [18]. Haploinsufficiency and reduced levels of multifunctional protein progranulin are caused by GRN mutations, whereas microglial homeostasis, neuroinflammation, and lysosomal activity are regulated by progranulin. Thus, in the absence of this protein, microglia are hyperactivated and pro-inflammatory cytokines like tumor necrosis factor-α (TNF-α) and interleukin-6 (IL-6) are released in excess. Concurrently, lysosomal acidification and enzyme trafficking are impacted by progranulin deficiency, which impairs autophagic degradation. Furthermore, neuronal damage is sustained in a feed-forward loop due to the combined effects of neuroinflammation and impaired cellular clearance. Importantly, TDP-43 proteinopathy links autophagy failure to this RNA-binding protein mislocalization and aggregation and is closely associated with GRN mutations [19]. The most common genetic cause of familial FTD and amyotrophic lateral sclerosis is influenced by a variety of pathogenic pathways. These include gain-of-function toxicity from dipeptide repeat proteins, loss-of-function mechanisms caused by decreased expression of the C9ORF72 protein, and RNA foci formation. The C9ORF72 influences vesicular trafficking and autophagy initiation by interacting with SMCR8 and the Rab GTPase family [20]. The absence of C9ORF72 in humans inhibits autophagosome formation and cargo degradation, leading to cytoplasmic misfolded protein accumulation. The lack of C9ORF72 dysfunction of innate immune signaling pathways drives constitutive microglial activation and further enhances IFN-1 responses. In turn, this altered vantage point promotes neuroinflammatory factors that increase neuronal vulnerability through defective autophagy. Moreover, genetic modifiers, including TMEM106B, dysregulate lysosomal function, increasing the severity of GRN and C9ORF72-associated diseases [21]. Risk alleles in TMEM106B point reinforce the link between dysregulated autophagy and inflammation in FTD, explaining their increased potential for lysosomal dysfunction, TDP-43 accumulation, and microglial activation [22].

13.3.1 Tau-Driven Pathogenesis and MAPT Mutations

The MAPT gene is among the most common genetic causes of FTD, especially in cases described as FTDP-17 [23]. The MAPT gene encodes the tau protein, which is essential for maintaining cytoskeletal integrity and stabilizing microtubules in neuronal axons during axonal transport. Alternative splicing of exon 10 of the MAPT gene causes MAPT-associated FTD through an imbalance of the 3- and 4-repeat (3R/4R) tau isoforms [24]. The balance between 3R and 4R tau isoforms is essential for neuronal microtubule stabilization in healthy adults. Some splice-site mutations in MAPT can disrupt this balance, resulting in an excess of 4R tau, leading to increased tau aggregation and decreased microtubule stability. Features of neurodegeneration affected by the 4R/3R ratio exhibit cytoskeleton instability, impaired axonal transport, and early synaptic dysfunction. The isoform shift creates a pro-aggregatory environment in neurons and microglia that promotes tau misfolding and insoluble tau filaments aggregation [25]. Hyperphosphorylated tau is released into the extracellular space, and it may be taken up by neighboring neurons and microglia. When these tau filaments are taken up by cells, they act as a template to facilitate the aggregation and misfolding of native tau. This is a vital part of tau pathology that spreads through transcellular transmission in connected brain regions [26]. This transcellular feature holds relevance in understanding the progressive nature of MAPT-related FTD, which shares similarities with prion diseases. Microglia may also play a role in transcellular tau spread and dispersal of pathogenic species through operational ingestion of aggregated tau, then subsequent exosome release. Furthermore, extracellular tau aggregates simultaneously promote neuroinflammation by activating the NLRP3 inflammasome in microglia [27]. When extracellular tau fibrils interact with pattern recognition receptors, such as TLRs, an innate immune response is activated to assemble and activate the NLRP3 inflammasome complex, leading to pro-inflammatory cytokines like IL-1β and IL-18 release. Prolonged NLRP3 activation leads to neuroinflammation, which promotes tau hyperphosphorylation and neuron damage [28]. Crucially, the inflammatory feedback loop increases tau toxicity and provides a glial cell primed response to other pathogenic stimuli, and tau aggregation adds to dysfunctional autophagy effect. The autophagy–lysosomal pathway is responsible for degrading misfolded proteins and damaged organelles in maintaining cellular homeostasis. Tau aggregates impair an important chaperone protein, HSPB8, which impairs the autophagic clearance of cellular protein aggregates. Furthermore, tau aggregates physically block the formation of autophagosomes fusion with lysosomes, impeding the clearance of aggregated proteins [29]. This blockage results in raised autophagic vacuoles, and further tau aggregation leads to a devastating cycle of proteostasis failure and neuronal dysfunction. Currently, numerous therapeutic approaches are being developed to target tau pathology in FTDs associated with MAPT. The use of antisense oligonucleotides targeting MAPT transcripts is an exciting approach to reduce toxic isoforms, as they reduce tau expression mechanistically as opposed to simply targeting the toxic isoforms [30]. Similarly, to address tau-induced cytoskeletal disassembly and dysfunction, including normal axonal transport restoration, there will be a need to explore microtubule stabilizers to stop the mechanism of modification or damage that tau invokes, which suggests a precision approach to address tau-based FTDs [31].

13.3.2 Progranulin-Linked Mechanisms and GRN Deficiency

Haploinsufficiency of lysosomal protein progranulin, due to mutations in the GRN gene, is the primary driver of FTD with TDP-43 proteinopathy. Progranulin plays a role in neuronal survival, lysosomal homeostasis, and immunology. The loss of progranulin affects autophagy through lysosomal acidification impairment and enzyme sorting, resulting in poor autophagic degradation and misfolded protein accumulation [32]. Paradoxically, loss of progranulin leads to dysregulation of microglia with resultant chronic neuroinflammation and increased microglial activation and pro-inflammatory cytokines promoting synaptic pruning and neuronal loss. Progranulin-deficient neurons are also more susceptible to mitochondrial disruption and oxidative stress. Mutations like TMEM106B worsen lysosomal dysfunction in GRN mutation carriers and are suspected genetic modifiers that influence the disease severity. Given that progranulin deficiency creates a pathological loop that links lysosomal failure and autophagic dysfunction to neuroinflammatory damage, it presents as a possible targeted therapy for atypical FTD [33]. Progranulin, also known as GRN, is a multifunctional glycoprotein that is important for lysosomal hydrolase trafficking, as well as modulating signaling pathways associated with inflammation and neurodevelopment. GRN haploinsufficiency, due to heterozygous loss-of-function mutations, results in GRN reduction in FTD, affecting a cascade of molecular events in neuroinflammation, autophagy dysfunction, and lysosomal dysfunction interlink [34]. The primary cell biological consequence associated with progranulin deficiency is lysosomal stress, which is relevant because cathepsin D is perturbed. Importantly, progranulin is a co-chaperone establishing proper cathepsin D trafficking and maturation, and therefore, in one sense, the functional efficacy of cathepsin D catalytic activity is reduced [35]. Certainly, impaired autophagic flux, coupled with lysosomal dysfunction, is synonymous with GRN-FTD pathology and results in pathological accumulation of TDP-43 and other misfolded proteins. Although stress and lysosome dysfunction can initiate compensatory signals toward transcriptional events, such as TFEB transcriptional activity upregulation, these compensatory signals may be largely effective but will often not fully restore lysosomal function in the diseased brain. Progranulin plays an essential anti-inflammatory role in the immune environment in the central nervous system [36]. It impacts microglial function, and consequently, a lack of progranulin will result in microglial priming and increased sensitivity to inflammatory stimuli. Moreover, the progranulin loss leads to decreased TREM2 expression. Since TREM2 is crucial for microglial phagocytosis and lipid sensing, its loss prevents degenerating myelin and apoptotic cells clearance, resulting in synaptic and neuronal loss. Microglia from progranulin-deficient individuals exhibit a shift toward a disease-associated phenotype, increasing cytokine release. These cytokines promote a negative extracellular environment and enhance neurotoxicity, promoting a hostile neurotoxic environment. Furthermore, the progranulin-deficient state associates with abnormal lipid metabolism, altering cellular pathways involved in lysosomal lipid degradation [37]. Moreover, neurons with GRN mutations accumulate lipofuscin that is auto fluorescent and contains oxidized proteins, lipids, and ceramides, inducing inflammation and apoptosis [38]. The accumulation of lipids results in a slow loss of neuronal function, generates reactive oxygen species (ROS), and alters lysosomal function. Gene therapy approaches based on adeno-associated virus vectors have demonstrated potential in addressing the molecular consequences of GRN haploinsufficiency [39].

13.4 Mechanistic Interaction of Autophagy Failure and Neuroinflammation in FTD

In FTD, autophagy–lysosomal dysfunction and neuroinflammation pathologically combine to produce a self-amplifying loop accelerating neuronal degeneration. These two processes are closely related, and issues in one process often exacerbate in the other. The genetic subtypes of FTD involving mutations, impaired cellular clearance, and chronic immune activation coexist and reinforce one another [40]. Autophagy is a crucial intracellular degradation process that eliminates harmful aggregates, misfolded proteins, and damaged organelles. Since neurons are post-mitotic cells that are highly susceptible to proteostatic imbalance, autophagy serves as a first-line defense against oxidative stress and proteinopathy. However, in FTD, this system is often compromised. For example, MAPT mutations result in hyperphosphorylated tau aggregates that obstruct autophagosome–lysosome fusion, whereas C9ORF72 expansions and GRN mutations impair autophagy initiation and lysosomal acidification, respectively [41]. Accumulation of unbroken cellular waste sets off intracellular danger signals, and damage-associated molecular patterns (DAMPs) such as mitochondrial DNA, ATP, and misfolded proteins are released. These DAMPs cause strong neuroinflammatory responses by activating the innate immune receptors of microglia. Further, microglial activation hinders neuronal autophagy through a variety of mechanisms. Pro-inflammatory cytokines such as IL-1β, TNF-α, and IL-6 are released by activated microglia and inhibit neuronal autophagy through nuclear factor kappa B (NF-κB) signaling and autophagy-initiating kinases, such as ULK1. The already weakened autophagic machinery is overloaded by chronic neuroinflammation that causes mitochondrial dysfunction, oxidative stress, and an increase in ROS [42]. Thus, microglial-derived inflammation not only contributes to synaptic pruning and neuronal injury but also directly prevents the neuron from removing pathological proteins. However, impaired autophagy in microglia themselves results in a persistently pro-inflammatory phenotype. Lysosomal stress and inflammasome activation through NLRP3 complex are caused by autophagic dysfunction in GRN and C9ORF72-deficient microglia [43]. Once inflammasome is activated, IL-1β and IL-18 mature, reinforcing the cycle of inflammation (Fig. 13.1). The C9ORF72-linked impairment in nucleocytoplasmic transport and autophagy in microglia and neurons triggers the cGAS-STING pathway, enhancing IFN-1 responses. These inflammatory signals reinforce tau hyperphosphorylation and TDP-43 mislocalization while maintaining the protein aggregation cycle. Therefore, rather than interacting linearly, autophagy failure and neuroinflammation interact synergistically and self-reinforcingly. Either pathway malfunction alone can disrupt neural homeostasis, but when both occur together, a toxic feedback loop is produced that accelerates disease progression. This vicious cycle may be broken by treatment strategies that simultaneously target both components, such as autophagy inducers and anti-inflammatory medications. Focusing on the intersection of these two processes rather than treating each mechanism independently may be more effective, predominantly in genetically driven forms of FTD, where these pathways are severely disrupted [44].

13.4.1 Stress Granule Dynamics and TDP-43 Proteinopathy

The TDP-43 pathology is present in the majority of FTD cases, predominantly those associated with GRN and C9ORF72 mutations. TDP-43 normally plays a role in RNA metabolism, which includes splicing, transport, and microRNA processing. However, in FTD, TDP-43 is mislocalized from nucleus to cytoplasm, where it clumps together into phosphorylated, insoluble forms, and this mislocalization is

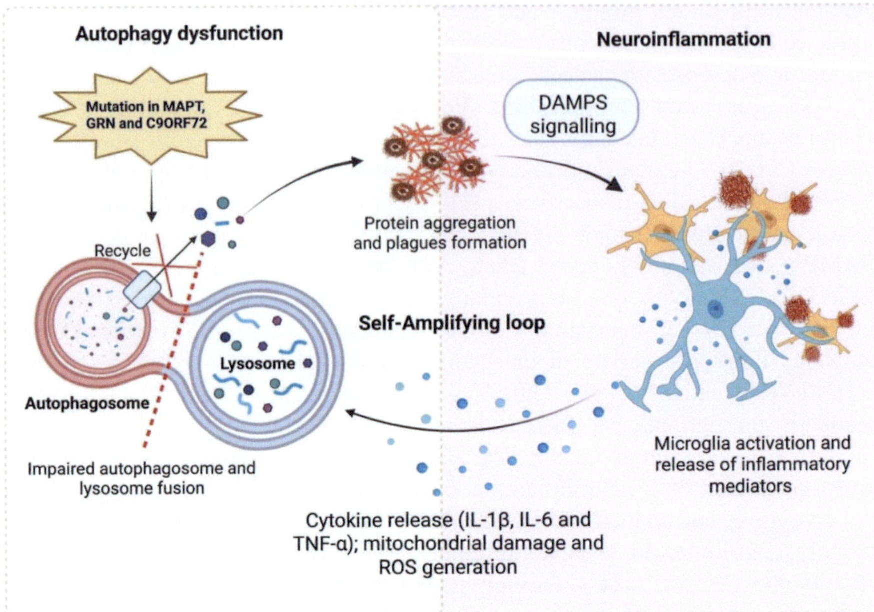

Fig. 13.1 Interplay between autophagy failure and neuroinflammation in FTD. Accumulation of cellular waste due to impaired autophagy activates microglia through DAMPs and induces pro-inflammatory cytokine release. This inflammatory response continues to inhibit autophagy, resulting in a self-amplifying loop that exacerbates neuronal degeneration. (Created using BioRender (https://www.biorender.com/))

closely associated with disturbances in the stress granule dynamics [45]. Under stress, cells and transient stress granules that prevent mRNA translation and protect transcripts. Since the normal disassembly of these granules is inhibited by chronic stress or impaired autophagy, TDP-43 remains in disease states. C9ORF72-FTD involves the arginine-rich dipeptide repeat proteins that interfere with stress granule clearance by modifying liquid–liquid phase separation and promoting abnormal protein–protein interactions. The formation of TDP-43 aggregates in C9ORF72-FTD feeds into stress granule dysfunction as it disrupts RNA metabolism and nucleocytoplasmic transport, leading to downstream neurotoxicity. Furthermore, TDP-43 pathology is an important feature of most FTD cases, predominantly associated with GRN and C9ORF72. In FTD, TDP-43 is mislocalized to the cytoplasm, where it is hyperphosphorylated, ubiquitinated, and forms insoluble aggregates [46]. Next, a significant pathological mechanism of cytoplasmic TDP-43 aggregates is the persistently retained stress granules. When chronic stress or impaired autophagy is present, these granules cannot disassemble, trapping TDP-43 and enabling it to form irreversible aggregates [47]. Additionally, essential autophagy receptors such as p62 and optineurin are blocked by stress granules, preventing the cell from employing selective autophagy to get rid of dangerous aggregates. When TDP-43 pathology causes endoplasmic reticulum stress, the unfolded protein response exacerbates protein misfolding and, if it continues, it initiates apoptotic signaling. Furthermore, mislocalized TDP-43 causes mitochondrial dysfunction by impairing bioenergetics and RNA processing in mitochondria and increasing ROS generation. Since TDP-43 toxicity causes oxidative stress that damages cellular components and activates the NLRP3 inflammasome in microglia, chronic neuroinflammation is directly related to TDP-43 toxicity. This inflammatory cascade concurrently exacerbates neuronal damage. Because of these interconnected mechanisms, treatment strategies that target TDP-43 aggregation are under investigation. One promising alternative is anle138b that restores mitochondrial function and dissolves pathological protein oligomers like TDP-43 aggregates. By treating the proteinopathy and its aftereffects, such targeted therapies offer hope for altering the trajectory of TDP-43-driven FTD [48].

13.4.2 Cytokine Networks and Microglial Activation

Microglial activation plays a key role in FTD pathophysiology, driving and enhancing neurodegeneration through complex cytokine-mediated signaling networks. For example, intrinsic changes in FTD microglial function are brought on by genetic mutations that prime resident immune cells for a pro-inflammatory state [49]. Progranulin deficiency results in hyperresponsive microglia that release more pro-inflammatory cytokines and have reduced efficacy in phagocytose. C9ORF72 promotes the activation of the STING pathway and IFN-1 signaling and chronic neuroinflammation persists when its function is diminished. Additionally, cytokines and chemokines released by activated microglia interact with neurons and astrocytes, exacerbating synaptic pruning, oxidative stress, and neuronal death [50]. The microglial ability to eliminate debris and regulate inflammation is hampered by TREM2 downregulation and other crucial regulatory receptors. This dysregulation creates a feed-forward loop in which cytokine release sustains glial activation and neurotoxicity. Moreover, inflammatory mediators generated by microglia that inhibit autophagy, promote tau phosphorylation, and TDP-43 aggregation further contributes to the disease progression [51]. Understanding the intricate relationships between cytokine networks and microglial activation is essential to developing targeted immunomodulatory therapies that restore FTD immune homeostasis. Here, a major contributing factor is progranulin deficiency, which reduces microglial lysosomal function and interferes with immune homeostasis and CX3CR1 signaling, which is crucial for neuron-microglia communication. Loss of CX3CR1-mediated signaling results in microglial dystrophy and pro-inflammatory phenotype, which is unable to adequately eliminate neuronal debris. This maladaptive state of microglia is made worse by NLRP3 inflammasome activation in response to lysosomal stress and mitochondrial damage [52]. Strong pro-inflammatory cytokines are cleaved and secreted as a result of inflammasome priming, creating a chronically inflammatory environment that promotes synaptic dysfunction and neuronal death. Next, complement-mediated synaptic stripping is a destructive process in which reactive microglia mark functional synapses for removal through C1q and C3 deposition and can result from chronic microglial activation. This inappropriate synapse pruning turns pathological in FTD, leading to cognitive decline and behavioral abnormalities. This is made worse by the incapacity of microglia to regulate lipid-rich debris due to disruptions in the autophagy–microglia axis through LC3-associated phagocytosis. When LC3-associated phagocytosis is under functioning, undigested lipids and damaged organelles accumulate, exacerbating inflammatory signaling and microglial activation, as it is essential for efficient myelin and apoptotic material removal [53]. This decreased clearance is evident in GRN and C9ORF72-deficient models, where lysosomal and phagocytic abnormalities combine with cytokine dysregulation, producing a vicious cycle of neuroinflammation. The use of therapeutic strategies that address both microglial inflammation and autophagic failure is a growing research interest. A Food and Drug Administration (FDA)-approved anti-inflammatory drug, Colchicine, has been used traditionally to treat gout and has shown promise in dual-targeting key FTD mechanisms. Colchicine reduces neuroinflammation and IL-1β secretion by inhibiting NLRP3 inflammasome assembly, likely through its disruption of microtubule dynamics. Additionally, colchicine has been shown to promote autophagy, facilitating the clearance of protein aggregates and damaged organelles. This makes it appealing as a repurposing drug in case of neurodegenerative disease, such as FTD, does inflammation and proteostasis failures seem to be intertwined [54]. Very broadly speaking, what defines the key axis of pathogenesis in FTD is the combination of cellular dysfunction arising from compromised autophagic and/or inflammasome activity at the systemic levels. By delineating the specific abnormalities, including CX3CR1 loss, NLRP3 overactivity, synaptic stripping through C1q/C3, and LC3-associated phagocytosis impairment, we can better identify novel FTD clinical mediators. In this, drugs like colchicine may offer a pragmatic and translationally accessible option for drug repurposing. The theorized potential to prevent and/or even reverse neurodegeneration in FTD is in therapeutics that target both inflammatory and degradative pathways simultaneously [55].

13.5 Conclusion

FTD is driven by complex, interwoven mechanisms involving genetic mutations, autophagy–lysosomal dysfunction, and chronic neuroinflammation. Mutations in MAPT, GRN, and C9ORF72 genes each initiate distinct but converging cellular disruptions that compromise protein clearance and immune regulation. Tau pathology, TDP-43 aggregation, and microglial activation interact through self-reinforcing loops, where defective autophagy perpetuates inflammation, and inflammation, in turn, impairs proteostasis. Lysosomal stress, impaired nucleocytoplasmic transport, and altered stress granule dynamics further amplify neuronal damage. Key modulators such as TMEM106B and disrupted pathways like NLRP3 inflammasome and cGAS-STING may exacerbate disease progression. These interactions show the importance of integrated therapeutic approaches, but targeting only a single pathological process may be effective. Hence, concurrent modulation of both autophagic clearance and immune responses holds greater promise. Agents that enhance lysosomal function, stabilize microtubules, or inhibit inflammasome activation need to be explored. A deeper understanding of these intersecting mechanisms not only clarifies FTD pathogenesis but also enables precision-targeted strategies to mitigate neurodegeneration.

References

1. Snowden JS. Changing perspectives on frontotemporal dementia: a review. J Neuropsychol. 2023;17:211–34.
2. Jiménez-García AM, Tortorella ME, Nishimura AL, Arias N. The differential effects of genetic mutations in ALS and FTD genes on behavioural and cognitive changes: a systematic review and meta-analysis. IJMS. 2025;26:6199.
3. Malpetti M, Cope TE, Street D, Jones PS, Hezemans FH, Mak E, et al. Microglial activation in the frontal cortex predicts cognitive decline in frontotemporal dementia. Brain. 2023;146:3221–31.
4. Tomé SO, Vandenberghe R, Ospitalieri S, Van Schoor E, Tousseyn T, Otto M, et al. Distinct molecular patterns of TDP-43 pathology in Alzheimer's disease: relationship with clinical phenotypes. Acta Neuropathol Commun [Internet] 2020 [cited 2025 July 18];8. Available from: https://actaneurocomms.biomedcentral.com/articles/10.1186/s40478-020-00934-5.
5. Lall D, Baloh RH. Microglia and C9orf72 in neuroinflammation and ALS and frontotemporal dementia. J Clin Invest. 2017;127:3250–8.
6. Genetics of frontotemporal dementia. Factors affecting neurological aging [Internet]. Elsevier; 2021 [cited 2025 July 18]. p. 3–16. Available from: https://linkinghub.elsevier.com/retrieve/pii/B9780128179901000019.
7. Corsi A, Bombieri C, Valenti MT, Romanelli MG. Tau isoforms: gaining insight into MAPT alternative splicing. IJMS. 2022;23:15383.
8. Lysosomal dysfunction and other Pathomechanisms in FTLD: evidence from Progranulin genetics and biology. Advances in experimental medicine and biology [Internet]. Cham: Springer International Publishing; 2021 [cited 2025 July 18]. p. 219–42. Available from: http://link.springer.com/10.1007/978-3-030-51140-1_14.
9. Barker S, Paul B, Pieper A. Increased risk of aging-related neurodegenerative disease after traumatic brain injury. Biomedicine. 2023;11:1154.
10. Yamaguchi M, Omori K, Asada S, Yoshida H. Epigenetic regulation of ALS and CMT: a lesson from drosophila models. IJMS. 2021;22:491.
11. Diab R, Pilotto F, Saxena S. Autophagy and neurodegeneration: unraveling the role of C9ORF72 in the regulation of autophagy and its relationship to ALS-FTD pathology. Front Cell Neurosci [Internet] 2023 [cited 2025 July 18];17. Available from: https://www.frontiersin.org/articles/10.3389/fncel.2023.1086895/full.
12. De Marchi F, Franjkic T, Schito P, Russo T, Nimac J, Chami AA, et al. Emerging trends in the field of inflammation and proteinopathy in ALS/FTD spectrum disorder. Biomedicine (Taipei). 2023;11:1599.
13. Llibre-Guerra JJ, Lee SE, Suemoto CK, Ehrenberg AJ, Kovacs GG, Karydas A, et al. A novel temporal-predominant neuro-astroglial tauopathy associated with *TMEM106B* gene polymorphism in FTLD/ALS-TDP. Brain Pathol. 2021;31:267–82.
14. Jiao H-S, Yuan P, Yu J-T. TMEM106B aggregation in neurodegenerative diseases: linking genetics to function. Mol Neurodegener [Internet]. 2023 [cited 2025 July 18];18. Available from: https://molecularneurodegeneration.biomedcentral.com/articles/10.1186/s13024-023-00644-1.

15. Kim Y, Ha T-Y, Lee M-S, Chang K-A. Regulatory mechanisms and therapeutic implications of lysosomal dysfunction in Alzheimer's disease. Int J Biol Sci. 2025;21:1014–31.
16. Ruiz-Gabarre D, Carnero-Espejo A, Ávila J, García-Escudero V. What's in a gene? The outstanding diversity of MAPT. Cells. 2022;11:840.
17. Ahmad MA, Kareem O, Khushtar M, Akbar M, Haque MR, Iqubal A, et al. Neuroinflammation: a potential risk for dementia. IJMS. 2022;23:616.
18. Rawat P, Sehar U, Bisht J, Selman A, Culberson J, Reddy PH. Phosphorylated Tau in Alzheimer's disease and other Tauopathies. IJMS. 2022;23:12841.
19. Beckers J, Van Damme P. The role of autophagy in the pathogenesis and treatment of amyotrophic lateral sclerosis (ALS) and frontotemporal dementia (FTD). Autophagy Rep [Internet] 2025 [cited 2025 July 18];4. Available from: https://www.tandfonline.com/doi/full/10.1080/27694127.2025.2474796.
20. Pang W, Hu F. Cellular and physiological functions of C9ORF72 and implications for ALS/FTD. J Neurochem. 2021;157:334–50.
21. Todd TW, Shao W, Zhang Y, Petrucelli L. The endolysosomal pathway and ALS/FTD. Trends Neurosci. 2023;46:1025–41.
22. Beckers J, Tharkeshwar AK, Van Damme P. C9orf72 ALS-FTD: recent evidence for dysregulation of the autophagy-lysosome pathway at multiple levels. Autophagy. 2021;17:3306–22.
23. Wang J, Wang B, Zhou T. The advance on frontotemporal dementia (FTD)'s neuropathology and molecular genetics. Zhang F, editor. Mediat Inflamm. 2022;2022:1–6.
24. Giannini LAA, Ohm DT, Rozemuller AJM, Dratch L, Suh E, Van Deerlin VM, et al. Isoform-specific patterns of tau burden and neuronal degeneration in MAPT-associated frontotemporal lobar degeneration. Acta Neuropathol. 2022;144:1065–84.
25. Zhang Y, Wu K-M, Yang L, Dong Q, Yu J-T. Tauopathies: new perspectives and challenges. Mol Neurodegener [Internet]. 2022 [cited 2025 July 18];17. Available from: https://molecularneurodegeneration.biomedcentral.com/articles/10.1186/s13024-022-00533-z.
26. Olfati N, Shoeibi A, Litvan I. Clinical spectrum of tauopathies. Front Neurol [Internet] 2022 [cited 2025 July 18];13. Available from: https://www.frontiersin.org/articles/10.3389/fneur.2022.944806/full.
27. Pampuscenko K, Morkuniene R, Krasauskas L, Smirnovas V, Brown GC, Borutaite V. Extracellular tau stimulates phagocytosis of living neurons by activated microglia via toll-like 4 receptor–NLRP3 inflammasome–caspase-1 signalling axis. Sci Rep [Internet] 2023 [cited 2025 July 18];13. Available from: https://www.nature.com/articles/s41598-023-37887-3.
28. Van Zeller M, Dias D, Sebastião AM, Valente CA. NLRP3 Inflammasome: a starring role in amyloid-β- and tau-driven pathological events in Alzheimer's disease. Wu Z, editor. JAD. 2021;83:939–61.
29. Hamano T, Enomoto S, Shirafuji N, Ikawa M, Yamamura O, Yen S-H, et al. Autophagy and Tau protein. IJMS. 2021;22:7475.
30. Easton A, Jensen ML, Wang C, Hagedorn PH, Li Y, Weed M, et al. Identification and characterization of a MAPT-targeting locked nucleic acid antisense oligonucleotide therapeutic for tauopathies. Mol Ther Nucleic Acids. 2022;29:625–42.
31. Chang MC, Srinivasan K, Friedman BA, Suto E, Modrusan Z, Lee WP, et al. Progranulin deficiency causes impairment of autophagy and TDP-43 accumulation. J Exp Med. 2017;214:2611–28.
32. Sung W, Noh M-Y, Nahm M, Kim YS, Ki C-S, Kim Y-E, et al. Progranulin haploinsufficiency mediates cytoplasmic TDP-43 aggregation with lysosomal abnormalities in human microglia. J Neuroinflammation [Internet] 2024 [cited 2025 July 18];21. Available from: https://jneuroinflammation.biomedcentral.com/articles/10.1186/s12974-024-03039-1.
33. Amin S, Carling G, Gan L. New insights and therapeutic opportunities for progranulin-deficient frontotemporal dementia. Curr Opin Neurobiol. 2022;72:131–9.

34. Huin V, Barbier M, Bottani A, Lobrinus JA, Clot F, Lamari F, et al. Homozygous GRN mutations: new phenotypes and new insights into pathological and molecular mechanisms. Brain. 2020;143:303–19.
35. Logan T, Simon MJ, Rana A, Cherf GM, Srivastava A, Davis SS, et al. Rescue of a lysosomal storage disorder caused by Grn loss of function with a brain penetrant progranulin biologic. Cell. 2021;184:4651–4668.e25.
36. Chitramuthu BP, Bennett HPJ, Bateman A. Progranulin: a new avenue towards the understanding and treatment of neurodegenerative disease. Brain. 2017;140:3081–104.
37. Simon MJ, Logan T, DeVos SL, Di Paolo G. Lysosomal functions of progranulin and implications for treatment of frontotemporal dementia. Trends Cell Biol. 2023;33:324–39.
38. Götzl JK, Mori K, Damme M, Fellerer K, Tahirovic S, Kleinberger G, et al. Common pathobiochemical hallmarks of progranulin-associated frontotemporal lobar degeneration and neuronal ceroid lipofuscinosis. Acta Neuropathol [Internet] 2014 [cited 2025 July 19]. Available from: http://link.springer.com/10.1007/s00401-014-1262-6.
39. Hinderer C, Miller R, Dyer C, Johansson J, Bell P, Buza E, et al. Adeno-associated virus serotype 1-based gene therapy for FTD caused by *GRN* mutations. Ann Clin Transl Neurol. 2020;7:1843–53.
40. Bright F, Werry EL, Dobson-Stone C, Piguet O, Ittner LM, Halliday GM, et al. Neuroinflammation in frontotemporal dementia. Nat Rev Neurol. 2019;15:540–55.
41. Darios F, Stevanin G. Impairment of lysosome function and autophagy in rare neurodegenerative diseases. J Mol Biol. 2020;432:2714–34.
42. Pereira TMC, Côco LZ, Ton AMM, Meyrelles SS, Campos-Toimil M, Campagnaro BP, et al. The emerging scenario of the gut–brain axis: the therapeutic actions of the new actor kefir against neurodegenerative diseases. Antioxidants. 2021;10:1845.
43. Ghasemi M, Keyhanian K, Douthwright C. Glial cell dysfunction in C9orf72-related amyotrophic lateral sclerosis and frontotemporal dementia. Cells. 2021;10:249.
44. Norevik CS, Huuha AM, Røsbjørgen RN, Hildegard Bergersen L, Jacobsen K, Miguel-dos-Santos R, et al. Exercised blood plasma promotes hippocampal neurogenesis in the Alzheimer's disease rat brain. J Sport Health Sci. 2024;13:245–55.
45. Suk TR, Rousseaux MWC. The role of TDP-43 mislocalization in amyotrophic lateral sclerosis. Mol Neurodegener [Internet] 2020 [cited 2025 July 19];15. Available from: https://molecularneurodegeneration.biomedcentral.com/articles/10.1186/s13024-020-00397-1.
46. Cracco L, Doud EH, Hallinan GI, Garringer HJ, Jacobsen MH, Richardson RM, et al. Distinguishing post-translational modifications in dominantly inherited frontotemporal dementias: FTLD-TDP Type A (*GRN*) vs Type B (*C9orf72*). Neuropathol Appl Neurobiol [Internet] 2022 [cited 2025 July 19];48. Available from: https://onlinelibrary.wiley.com/doi/10.1111/nan.12836.
47. Zhang Y, Gu J, Sun Q. Aberrant stress granule dynamics and Aggrephagy in ALS pathogenesis. Cells. 2021;10:2247.
48. Yu C-H, Davidson S, Harapas CR, Hilton JB, Mlodzianoski MJ, Laohamonthonkul P, et al. TDP-43 triggers mitochondrial DNA release via mPTP to activate cGAS/STING in ALS. Cell. 2020;183:636–649.e18.
49. Fenoglio C, Serpente M, Arcaro M, Carandini T, Sacchi L, Pintus M, et al. Inflammatory plasma profile in genetic symptomatic and presymptomatic frontotemporal dementia – a GENFI study. Brain Behav Immun. 2024;122:231–40.
50. García-García R, Martín-Herrero L, Blanca-Pariente L, Pérez-Cabello J, Roodveldt C. Immune signaling kinases in amyotrophic lateral sclerosis (ALS) and frontotemporal dementia (FTD). IJMS. 2021;22:13280.
51. Xie M, Liu YU, Zhao S, Zhang L, Bosco DB, Pang Y-P, et al. TREM2 interacts with TDP-43 and mediates microglial neuroprotection against TDP-43-related neurodegeneration. Nat Neurosci. 2022;25:26–38.

52. Lin M, Yu H, Xie Q, Xu Z, Shang P. Role of microglia autophagy and mitophagy in age-related neurodegenerative diseases. Front Aging Neurosci [Internet] 2023 [cited 2025 July 19];14. Available from: https://www.frontiersin.org/articles/10.3389/fnagi.2022.1100133/full.
53. Chen X, Su Q, Gong R, Ling X, Xu R, Feng Q, et al. LC3-associated phagocytosis and human diseases: insights from mechanisms to therapeutic potential. FASEB J [Internet] 2024 [cited 2025 July 19];38. Available from: https://faseb.onlinelibrary.wiley.com/doi/10.1096/fj.202402126R.
54. Kakoti BB, Bezbaruah R, Ahmed N. Therapeutic drug repositioning with special emphasis on neurodegenerative diseases: threats and issues. Front Pharmacol [Internet] 2022 [cited 2025 July 19];13. Available from: https://www.frontiersin.org/articles/10.3389/fphar.2022.1007315/full.
55. Panza F, Lozupone M, Seripa D, Daniele A, Watling M, Giannelli G, et al. Development of disease-modifying drugs for frontotemporal dementia spectrum disorders. Nat Rev Neurol. 2020;16:213–28.

Part III

Therapeutic Strategies and Emerging Research

Neuroinflammation and Autophagy in Neurodegenerative Therapy

14

Anuroopa G. Nadh, Shuvechha Chakraborty, Rajesh Raju, Arshad Rizvi, and Raushan Kumar Chaudhary

Abstract

Neurodegenerative diseases such as Alzheimer's disease, Parkinson's disease, and amyotrophic lateral sclerosis are driven by intertwined pathological mechanisms. Since traditional therapeutic strategies targeting single pathways have limited clinical success, this chapter underscores the necessity of dual-target approaches that simultaneously modulate neuroinflammatory cascades and restore autophagic flux to halt disease progression. By dissecting the mechanistic crosstalk between inflammasome activation and autophagic failure, this chapter provides a roadmap for developing advanced therapeutics. It critically evaluates current strategies, including small-molecule inhibitors, biologic agents, and emerging multimodal interventions such as nanoparticle-based drug delivery, gene therapies, and repurposed drugs. This chapter also addresses translational challenges like blood–brain barrier penetration, off-target effects, and combinatorial therapy optimization while highlighting preclinical and clinical advancements. The urgency of this dual targeting is underscored by the limitations of monotherapies in addressing the multifactorial nature of diseases. Chronic neuroinflammation exacerbates autophagy impairment by creating a self-sustaining cycle of neuronal damage, whereas defective autophagy amplifies inflammatory signaling through protein aggregate accumulation. Next, it also advocates personalized, multimodal regimens tailored to disease-specific

A. G. Nadh · S. Chakraborty · R. Raju · R. K. Chaudhary (✉)
Centre for Integrative Omics Data Science (CIODS), Yenepoya (Deemed to be University), Mangalore, Karnataka, India
e-mail: rajeshraju@yenepoya.edu.in

A. Rizvi
Department of Chemistry and Biochemistry, University of Oklahoma, Norman, OK, USA
e-mail: arshad.rizvi-1@ou.edu

P. Khanal et al. (eds.), *Neuroinflammation and Autophagy in Neurodegeneration*,
https://doi.org/10.1007/978-981-95-3945-1_14

neuroinflammatory and autophagy signatures, emphasizing the need for biomarker development and ethical considerations in gene-editing trials.

Keywords

Autophagy · Neurodegeneration · Neuroinflammation · Pharmacotherapy

14.1 Introduction

The majority of neurodegenerative diseases are marked by multiple etiologies, where complex interactions between genetic, environmental, molecular, and cellular factors contribute to disease progression. Emerging research highlights the significance of molecular and cellular crosstalk among key neurodegenerative pathways and the association of certain molecules that trigger central nervous system (CNS) disorders. Chronic CNS disorders such as Alzheimer's disease (AD), Parkinson's disease (PD), multiple sclerosis (MS), and Huntington's disease (HD) are multifactorial (Fig. 14.1), arising from diverse pathological factors where multiple triggering factors induce the disease pathology. It is important to recognize that synchronized

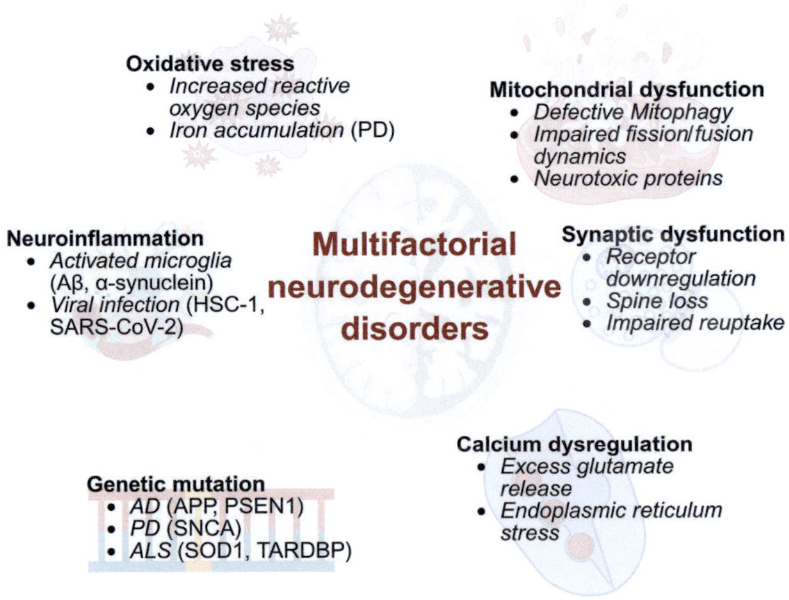

Fig. 14.1 Key mechanisms underlying multifactorial neurodegenerative disorders. Neurodegenerative diseases are contributed to by converging mechanisms including oxidative stress, mitochondrial dysfunction, calcium and synaptic dysregulation, neuroinflammation, and genetic mutations. (Created using BioRender (https://www.biorender.com/))

instruction of signaling molecules is crucial for maintaining effective cellular communication and response.

In CNS, interrupted signaling triggers chronic neuroinflammation, driven by sustained microglial activation and elevated interleukin-1β (IL-1β) and tumor necrosis factor-α (TNF-α), contributing to neuronal damage [1]. Simultaneously, autophagy impairment is marked by p62/SQSTM1 accumulation, which blocks damaged proteins, leading to toxicity [2]. The NLRP3 inflammasome links these processes, as impaired autophagy fails to degrade NLRP3 and mitochondrial damage-associated molecular patterns (DAMPs) (e.g., mtDNA, reactive oxygen species [ROS]), creating a self-perpetuating cycle of neurodegeneration. Common regulators like nuclear factor kappa B (NF-κB) and mammalian target of rapamycin complex (mTOR) integrate inflammatory and survival signaling, and their dysregulation strengthens neuronal stress [3]. This complex crosstalk limits single-target therapies in neurodegeneration. Multimodal strategies targeting inflammation, autophagy, and metabolism may better modify disease trajectories and improve outcomes. To better capture the complexity of human neurodegenerative diseases, preclinical models move beyond crude systems by incorporating genetic diversity, brain region specificity, and developmental dynamics. Advances in single-cell and spatial omics enable refined modeling of brain development and disease-relevant processes. Patient-derived induced pluripotent stem cells (iPSCs) offer human-specific models by recapitulating genetic circumstances and differentiating into neurons or organoids [4]. Although mice remain standard, nonhuman primates and rats provide more relevant cognitive and behavioral parallels to human neurodegenerative diseases. Gene-editing therapies show promise in correcting monogenic neurodegenerative diseases like Rett or Kabuki syndrome, but raise ethical challenges. Responsible application demands balancing therapeutic potential with ethical safeguards and public consensus. CRISPR-based gene editing and single-cell sequencing further dissect genetic and environmental interactions [5]. Overall, gene-editing for neurodegenerative diseases entails complex ethical trade-offs between therapeutic promise and risks to individual rights and future generations.

14.2 Current Therapeutic Strategies

14.2.1 Small-Molecule Interventions

The NLRP3 inflammasome contributes to neurodegenerative diseases by driving chronic inflammation, neuronal pyroptosis, and impaired neuroplasticity, making it a promising therapeutic and diagnostic target [6]. To mitigate its harmful effects, several NLRP3 inflammasome blockers have been developed and studied. MCC950 is one of the most widely studied NLRP3-specific inhibitors, preventing protein complex formation necessary for inflammasome assembly. Specifically, it impairs the interaction between NLRP3 and ASC, which is a key adaptor protein that oligomerizes into large signaling platforms called "ASC specks" [7]. Without ASC oligomerization, caspase-1 activation is reduced, which is essential for processing the pro-inflammatory cytokine IL-1β into its active form. In amyotrophic lateral sclerosis (ALS) models like SOD1^G93A mice, this leads to a decrease in IL-1β secretion from activated microglia [8]. Dapansutrile (OLT1177) is an orally active, selective inhibitor of the NLRP3 inflammasome that has shown promise in preclinical models of AD. In AD mouse models, OLT1177 significantly reduced amyloid-β (Aβ) plaque load in the brain. This was accompanied by memory and learning tasks, suggesting functional recovery in addition to pathological improvement [9]. Another compound, CY-09, specifically inhibits NLRP3 inflammasome activation by directly targeting the ATP-binding site in NACHT domain of NLRP3. By blocking NLRP3 ATPase activity, CY-09 prevents assembly and activation of the inflammasome complex [10]. Several limitations of these inhibitors hinder their clinical translation, especially related to CNS delivery and immune safety. Many inflammasome inhibitors are large or hydrophilic molecules that cannot efficiently cross the blood–brain barrier (BBB), limiting their therapeutic effect in brain [11]. Since inflammasomes play a critical role in pathogen defense and sterile inflammation, their inhibition could impair immune surveillance, increasing susceptibility to infections [10]. Given these limitations, current research focuses on the development of new therapeutic modalities for NLRP3 degradation, like proteolysis-targeting chimera (PROTACs) and other covalent inhibitors. PROTACs reduce protein levels by targeting them for ubiquitin-mediated degradation [12]. Additionally, covalent inhibitors utilize covalent interactions for modern drug design, offering sustained pharmacological activity and enhanced target selectivity [13]. RRx-001 is an anticancer agent that serves as a specific and potent inhibitor of NLRP3. It covalently binds to cysteine 409 of NLRP3 through its bromoacetyl group, preventing essential interaction between NLRP3 and NEK7, which is obligatory for assembly and activation of NLRP3 inflammasome [14].

mTOR modulators are also emerging as promising agents in HD. One approach for HD promotes mutant huntingtin clearance through macroautophagy. Everolimus is an mTORC1 inhibitor that enhances autophagy by blocking mTOR activity through FKBP-12 binding, thereby relieving mTOR suppression [15, 16]. However, clinical management with everolimus may lead to undesirable metabolic disturbances such as dyslipidemia and hyperglycemia [17]. More potent ATP-competitive inhibitors such as Torin1 offer greater suppression of mTOR activity but are limited by hepatotoxicity and alter cellular metabolism significantly [18, 19]. Dual mTORC1/2 inhibitors provide a balanced approach by promoting autophagy, though concurrently downregulating PI3K-Akt pathway, which is associated with cell survival, enhancing neuroprotective effects [20, 21]. Natural compounds like resveratrol activate autophagy through SIRT1, offering a less toxic endogenous route of mTOR modulation [22]. Combination strategies that couple mTOR inhibitors with nonsteroidal anti-inflammatory drugs (NSAIDs) have shown potential in mitigating neuroinflammation [23]. Autophagy enhancers are gaining attention as therapeutic agents in neurodegenerative diseases due to their ability to promote the clearance of toxic protein aggregates and improve cellular homeostasis. A natural polyamine, spermidine, extends health span through the inhibition of EP300 acetyltransferase, which normally suppresses autophagy [24]. A disaccharide sugar, trehalose, also enhances autophagy through mTOR-independent mechanisms by stabilizing lysosomes and has shown efficacy in clearing α-synuclein and mutant huntingtin aggregates [25]. An L-type calcium channel blocker, felodipine, has been identified for repurposing as an autophagy inducer, although detailed mechanistic studies are still emerging. Next, a well-known AMPK activator, metformin, also reduces neuroinflammation and promotes autophagy, showing neuroprotective effects in PD models [26]. Last, a STAT3 inhibitor, niclosamide, has been reported to enhance mitophagy, potentially benefiting ALS, though some evidence also suggests it may impair lysosomal function, highlighting the need for further study [27].

14.2.2 Biologic Agents

Certain anti-inflammatory cytokines have emerged as promising therapeutic options for neuroprotection due to their ability to modulate immune responses and downregulate neuroinflammation. For example, IL-10 gene therapy in APP+PS1 transgenic mice through adeno-associated virus (AAV) delivery systems has shown efficacy in enhancing hippocampal neurogenesis and improving spatial learning, without directly reducing brain Aβ deposition [28]. Similarly, TfR fusion proteins are engineered to cross the BBB by targeting transferrin receptors, which demonstrate targeted anti-inflammatory effects in the CNS [29]. However, the pleiotropic nature of cytokines and the risk of inflammation during post-treatment period pose significant challenges to the treatment, as systemic administration often results in off-target effects and toxicity due to widespread cytokine activity and limited penetration of the blood–brain barrier (BBB) [30, 31]. To overcome the inability to cross the BBB, and to enhance therapeutic efficacy, innovative approaches are to be designed to offer an alternate way of controlled and localized release of therapeutic agents in CNS [30]. Encapsulated cytokine-releasing hydrogels have emerged as promising biomaterial-based delivery systems capable of controlled, localized release within the CNS [32, 33]. Such platforms not only bypass the BBB but also maintain sustained therapeutic cytokine levels while minimizing systemic exposure and inflammation. Direct in vivo evidence supports this strategy: IL-10-loaded collagen hydrogels implanted into the striatum of Parkinsonian rat models achieved prolonged cytokine retention, significantly reduced microglial activation, and preserved graft survival compared with bolus delivery [34]. Similarly, an IL-10–eluting dual-functional gelatin hydrogel with DAMP-scavenging capacity attenuated neuroinflammation and promoted neural regeneration in a mouse spinal cord injury model [33]. Collectively, these studies demonstrate that cytokine-releasing hydrogels represent an innovative and clinically relevant approach for localized CNS therapy.

Monoclonal antibodies are increasingly recognized as potent biological agents for neuroprotection, due to their ability to target inflammation associated with neurodegenerative diseases. An anti-Aβ antibody, aducanumab, not only facilitates Aβ clearance but also regulates microglial activation, reducing neuroinflammation.

Gantenerumab similarly targets Aβ plaques while influencing TREM2-dependent autophagy pathways, potentially enhancing cellular clearance mechanisms in the brain [35]. Amantadine was originally developed as an antiviral, is being investigated as repurposed drug for ALS, reflecting its broader neuroprotective potential [36]. To overcome the challenge of BBB penetration, bispecific antibody designs, such as those combining anti-transferrin receptor and anti-TNF-α specificities, are under development to enable effective CNS delivery [37]. However, systemic administration of monoclonal antibodies carries risks, including cytokine release syndrome, highlighting the need to address dosing strategies and delivery optimization.

14.3 Challenges in Therapeutic Development

14.3.1 Pharmacological Barriers

Pharmacological barriers related to the BBB present significant challenges for CNS therapeutics. The selective permeability of BBB limits drug access to brain, necessitating alternate strategies for effective delivery. Prodrugs such as esterified rapamycin are chemically modified to enhance lipophilicity and BBB permeability, subsequently undergoing enzymatic activation in CNS. It has to be noted that nanocarriers like liposomes offer a more controlled delivery platform, protecting drug payloads, enhancing transport through receptor-mediated or absorptive transcytosis [38]. Another method is using the endogenous transport systems. For example, conjugating the drugs to glucose to exploit GLUT1 transporters offers a promising route for targeted delivery. Also, the intranasal administration method is being used as a non-invasive method to bypass the BBB entirely, exemplified by insulin-loaded chitosan nanoparticles, which show effectiveness in CNS treatment models [39]. BBB-targeted approaches face further complications due to the physiological and biological variability. The changes in cellular, physico-chemical, and molecular events due to aging significantly alter BBB function. The controversial output in experimental assays lacks clarity and results in inconclusive outcomes. The species differences between rodent and human BBB physiology pose translational challenges. Rodent models often overestimate drug permeability, which may not replicate in clinical trials. These hurdles highlight the complexity of BBB-targeted drug design and underscore the necessity for advanced delivery systems and more predictive preclinical models.

Drug stability is a critical determinant of therapeutic efficacy in CNS disorders, where delivery challenges are sensitive. Biologics are vulnerable to proteolytic degradation in brain interstitial fluid, limiting their bioavailability and therapeutic window [40]. One widely adopted strategy to enhance the stability and circulatory half-life of such therapeutics is PEGylation, which shields molecules from enzymatic breakdown and reduces renal clearance. For example, PEGylated peptides show prolonged half-life and can be tailored for controlled release using protease-cleavable linkers [41]. However, PEGylated protein formulations are sensitive to environmental conditions and typically require strict cold-chain logistics to prevent degradation due to heat, humidity, or light exposure. In parallel, small-molecule drugs also face stability challenges. For example, in inflamed CNS environments, drugs like carfilzomib undergo rapid oxidation and hydrolytic degradation, diminishing their therapeutic value [42]. To improve targeting and intracellular delivery, pH-sensitive formulations have been developed to release drugs specifically in acidic compartments like lysosomes, enhancing drug accumulation at the intended site while preserving systemic stability [41, 43]. These strategies underscore the importance of considering degradation pathways and delivery environment in drug development to ensure therapeutic effectiveness and safety.

14.3.2 Safety and Efficacy Concerns

In drug development and clinical efficacy testing, ensuring safety and efficacy remains a critical challenge. The off-target adverse effects caused by unintended interactions of drugs with biological systems outside their intended targets are a key concern, as these effects can significantly compromise the desired therapeutic outcomes. For example, rapamycin has been shown to induce hyperglycemia and insulin resistance, showing its disruptive effect on metabolic regulation [44]. Similarly, NLRP3 inflammasome inhibitors, though effective in regulating inflammation, can disrupt the gut microbiome–immune axis, leading to long-term immune dysregulation or gastrointestinal complications, which may compromise their therapeutic benefits in patients [45].

Moreover, JAK inhibitors, which are used in autoimmune and inflammatory diseases, have been known to be associated with increased risk of herpes zoster reactivation as they impair antiviral immune surveillance [46]. In cancer and neurodegenerative diseases, autophagy enhancers that are effective in clearing cellular debris have been monitored for their tumor-enhancing property in preclinical models, likely due to the protective role of autophagy in established tumors [47]. Also, the antibody-based therapies, chiefly anti-Aβ monoclonal antibodies used in AD, have been observed to have antibody-related imaging abnormalities, such as edema, transient ischemic attack, seizures, or acute cognitive decline, raising serious concerns about neurotoxicity [48]. These examples underscore the complex interplay between therapeutic targets and systemic biology, necessitating rigorous preclinical and clinical evaluation to overcome off-target risks while preserving therapeutic efficacy.

Overactivation of autophagy can also lead to several pathological consequences in brain. One key issue is lysosomal membrane permeabilization, which releases cathepsins into the cytosol and initiates cell death pathways. In particular, cathepsin L contributes to mitochondrial damage and apoptosis following excessive autophagic flux [49]. Nutrient depletion is another risk as prolonged autophagy can lead to excessive self-digestion in neurons, resulting in energy collapse and cell death [50]. Counterintuitively, overactive autophagy can result in paradoxical protein aggregation when autophagosome–lysosome fusion is impaired, causing toxic buildup of undigested material despite elevated autophagy initiation [51]. Furthermore, excessive mitophagy can deplete healthy mitochondria, leading to bioenergetic failure and neuronal death [52]. To mitigate these risks, titrated strategies such as pulsatile mTOR inhibition can be explored to cyclically modulate autophagy activity rather than sustaining prolonged induction, thereby preserving lysosome health and cellular balance [53, 54].

14.3.3 Combinatorial Therapy Complexity

Combinatorial therapies that utilize multiple therapeutic agents simultaneously are employed to target chronic and multifactorial diseases, such as chronic inflammation and neurodegenerative disorders. However, off-target interactions and pharmacodynamic incompatibilities remain a significant concern in this approach. In addition, combining drugs unavoidably complicates both development procedures and clinical translation [55]. An example to list is the combination of rapamycin with NSAIDs. Although this pairing can reduce neuroinflammation by targeting both immune signaling and cellular growth pathways, it simultaneously increases gastrointestinal toxicity, which is a common side effect of NSAIDs [56]. This trade-off between central benefit and peripheral mistreatment highlights the difficulty of balancing efficacy with tolerability in combination regimens. In other cases, combinations can produce unexpected or controversial pharmacological effects. For example, NLRP3 inflammasome inhibitors combined with an autophagy enhancer, trehalose, have shown additive benefits in AD models, reducing both neuroinflammation and protein aggregation [57]. However, a combination of anti-TNF-α agents with autophagy enhancers can yield conflicting outcomes on microglial polarization, including impaired immune clearance. Additionally, pairing mTORC1 inhibitors with growth promoters in balancing cellular atrophy and survival, since growth factors promote anabolic processes that mTORC1 inhibition suppresses. A notable clinical failure illustrating these complexities is the combination of β-secretase inhibitors and anti-Aβ monoclonal antibodies in AD treatment. Even though the approach is individually promising, the combination may not produce significant cognitive improvement and exhibit severe adverse effects, likely due to over-suppression of amyloid pathways and immune overactivation. These examples highlight how mechanistic overlap, tissue specificity, and patient variability can make combinatorial therapy development a delicate approach with unpredictable output.

Dosing optimization plays a crucial role in enhancing therapeutic outcomes in CNS-targeted treatments. Chronotherapy aligns drug administration with body circadian rhythms and shows promise in improving both efficacy and tolerability by timing doses to match biological peaks in autophagy and drug metabolism [58]. This principle is relevant for autophagy-modulating agents, where circadian variations significantly influence cellular responses [59]. Additionally, staggered dosing is used to reduce drug–drug interactions in complex regimens, allowing to reach therapeutic levels without interfering metabolism or absorption of co-administered drugs. For

CNS drugs, pharmacokinetic/pharmacodynamic modeling is essential to ensure adequate brain exposure by evaluating cerebrospinal fluid to plasma ratios for drugs with limited BBB permeability. This is vital after tailoring treatments across populations, as both pediatric and geriatric patients require dose adjustments due to differences in metabolism, organ function, and drug sensitivity. In clinical trials, especially involving combination regimens, adaptive trial designs such as the I-SPY2 (NCT01042379) provide dynamic, real-time strategies to optimize dosing based on patient responses and biomarker profiles. These innovations collectively support a precision medicine approach that maximizes efficacy while minimizing risks in vulnerable CNS populations.

14.4 Emerging Multimodal Approaches

14.4.1 Advanced Delivery Systems

Advanced drug delivery systems using nanoparticles are transforming treatment approaches for neurodegenerative diseases by enabling targeted, efficient, and responsive therapeutic interventions. PLGA nanoparticles have been successfully employed to co-deliver JAK inhibitors and trehalose through immune modulation and autophagy enhancement, with promising outcomes in autoimmune and neurodegenerative diseases [60]. Similarly, gold nanoparticles allow localized delivery in PD models, enhanced safety, and efficacy through light-activated precision [61]. Lipid-based nanoparticles demonstrated effective delivery of siRNAs targeting pro-inflammatory microRNAs to modulate neuroinflammation in CNS disorders [62]. In parallel, quantum dots are emerging for real-time imaging of autophagosome dynamics, providing novel insights into autophagy process and drug responses at the subcellular level [63]. Furthermore, dual-functional nanoparticles that combine BBB penetration with stimuli-responsive release advance precision neurotherapies by ensuring both effective delivery and on-demand drug activation [64]. Together, these innovations represent a leap forward in CNS drug delivery, enabling multifaceted interventions with higher efficacy and reduced systemic toxicity.

Engineered exosomes are emerging as a highly promising platform for targeted therapeutic delivery. For example, TFEB mRNA-loaded exosomes have been proposed in preclinical studies to enhance lysosomal biogenesis and support autophagy in neurodegeneration, although in vivo evidence is still limited. To increase specificity, exosome surface modification with rabies virus glycoprotein (RVG) peptides has been experimentally shown to enable neuronal targeting and systemic delivery of RNA cargos such as miRNAs and neuroprotective peptides to the brain [65, 66]. Exosomes are being engineered to deliver CRISPR-Cas9 components as a non-viral, low-immunogenicity approach to gene editing. Proof-of-concept studies have demonstrated intracellular delivery of Cas9 RNP-loaded exosomes and preliminary targeting of genes such as P2X7 in preclinical models [67, 68]. Another exciting direction is to develop exosome-based vaccines to train microglia against neurotoxic agents like Aβ plaques. These vaccines can elicit robust cytotoxic T-cell responses and have demonstrated efficacy in preclinical models [69]. Despite these advances, significant scalability and good manufacturing practice challenges remain. Producing clinical-grade exosomes with consistent quality, yield, and functional cargo requires standardized protocols and robust bioprocessing techniques, which are still in development [70].

14.4.2 Gene Therapy Innovations

Gene therapy innovations are rapidly reshaping the therapeutic landscape for neurodegenerative diseases by modulating autophagy and inflammation. One approach involves AAV9-mediated TFEB delivery, which showed potential in rescuing α-synucleinopathy in PD models [71]. Combining TFEB-based therapies with NSAIDs demonstrated synergistic effects in reducing neuroinflammation [72]. However, concerns remain with TFEB overactivation leading to oncogenic risk in peripheral tissues. To mitigate this, regulatable promoters like Tet-On systems can fine-tune TFEB expression, and TFEB-TFE3 fusion proteins are being explored to enhance lysosomal and autophagy gene activation. Beyond TFEB, precision gene-editing strategies, such as base editing approaches aimed at regulators like CARD8, are emerging as promising strategies for suppressing NLRP3-driven neuroinflammation in PD, although in vivo validation is still limited [73]. In parallel, prime editing holds potential to correct mutations in autophagy genes such as PARK2. In vivo CRISPR screens also identify novel regulators of neuroinflammation, facilitating target discovery for future therapies [74].

14.4.3 Drug Repurposing

Drug repurposing offers a promising strategy for treating neurodegenerative diseases by leveraging existing therapeutics with established safety profiles. Several drugs originally developed for other disorders have been observed to have efficacy in modulating key neurodegenerative pathways. For instance, a JAK inhibitor, baricitinib, commonly used to treat moderate-to-severe rheumatoid arthritis, is noted for its regulatory effect on microglial IL-6 expression in ALS [75]. Similarly, ruxolitinib, used to treat myelofibrosis, has shown preliminary evidence of attenuating neuroinflammation in MS models, whereas the JAK inhibitor tofacitinib may also modulate PI3K signaling and enhance autophagy flux in preclinical studies. The antidiabetic agent metformin promotes neuroprotection through AMPK-dependent autophagy and NF-κB inhibition. In AD, the GLP-1 receptor agonist exenatide reduces Aβ accumulation and tau hyperphosphorylation, and the PPARγ agonist pioglitazone suppresses NLRP3 inflammasome activation. Finally, SGLT2 inhibitors are under early investigation for their potential to enhance mitophagy in PD models [76]. However, some limitations persist for these drugs, including increased infection risk in elderly populations due to immunomodulatory effects, and possibility of hypoglycemia also poses a significant concern when used in non-diabetic patients with neurodegenerative diseases.

14.5 Translational Progress and Clinical Trials

14.5.1 Preclinical Success Stories

Translational progress in AD clinical research and drug development has been marked by promising preclinical success stories, although significant hurdles stand in their way of clinical applications. Targeted genetic and pharmacological interventions yielded notable improvements in experimental animal models. For example, NLRP3 knockout mice exhibited a 50% reduction in Aβ plaques and enhanced spatial memory in the Morris Water Maze [77]. This points out the critical role of neuroinflammation in AD pathology and indicates that inflammasome modulation may attenuate AD pathology. Similarly, overexpression of TFEB showed enhanced microglial phagocytosis, highlighting the therapeutic potential of lysosomal biogenesis and autophagic clearance [78]. Furthermore, a combination therapy of rapamycin and melatonin in SAMP8 mice demonstrated synergistic effects on lifespan extension, indicating the role of mTOR inhibition and circadian regulation in neurodegeneration. Despite these advances, research gaps persist, notably the limited applicability of transgenic mouse models to the heterogeneous and predominantly sporadic nature of human diseases. Emerging humanized models incorporating patient-derived iPSC microglia offer a more physiologically relevant platform, enabling more predictive assessments of therapeutic efficacy in diverse AD subtypes.

Next, PD organoids are revolutionizing in vitro modeling of disease mechanisms and therapeutic testing. Human midbrain organoids derived from patients with SNCA triplication reliably replicate protein aggregation and dopaminergic neuron loss, whereas AAV-mediated TFEB delivery in these models rescues α-synuclein degradation by boosting lysosomal function [79]. Additionally, CRISPR-edited PINK1 organoids demonstrate restored mitophagy, showing gene correction utility in reversing mitochondrial deficits associated with familial PD mutations [80, 81]. In terms of pharmacological interventions, NLRP3 inhibitor MCC950 significantly reduces dopaminergic neuron loss and motor deficits in α-synuclein-overexpressing models [72]. Despite these breakthroughs, modeling Braak staging in vitro remains a major challenge. Braak staging involves progressive, region-specific pathology that is challenging to replicate in the static spatial architecture of organoids. However, the use of high-content screening platforms enables quantitative analysis of autophagy flux, α-synuclein clearance, and neuronal viability, making organoids a valuable tool for drug screening and mechanistic studies [82]. As these technologies evolve, PD organoids are poised to bridge the gap between genetic modeling and translational drug discovery.

14.5.2 Possible Clinical Trials

Several Phase II and III clinical trials are currently exploring novel therapies for neurodegenerative diseases. A selective NLRP3 inflammasome inhibitor, inzomelid, is under investigation in patients with mild cognitive impairment to explore the Clinical Dementia Rating–Sum of Boxes as the primary endpoint. Next, sirolimus (rapamycin) inhibits mTORC1 signaling, and it is being tested (NCT04629495) to evaluate its effects on autophagy and neuronal survival [83]. In ALS, baricitinib is being evaluated (NCT05189106) for its potential to slow functional decline, which is measured by changes in the ALS Functional Rating Scale-Revised [84]. However, compound targeting glutamatergic signaling and autophagy enhancement troriluzole in an AD Phase III trial (NCT03605667) did not meet efficacy expectations and failed a futility analysis [85]. Meanwhile, a GLP-1 receptor agonist, exenatide, showed promising motor function improvements at 60 weeks in a Phase III PD study (NCT04232969), supporting its potential as a neuroprotective agent [86]. These trials highlight the diversity of mechanisms being targeted in neurodegeneration, including inflammation, autophagy, and metabolic modulation.

Biomarker development trials face significant challenges in patient stratification, monitoring, and trial design. A key issue is the comparison between cerebrospinal fluid and blood biomarkers: while p-Tau/Aβ42 remains the gold standard for AD diagnosis, plasma neurofilament light chain provides a less invasive and promising alternative for detecting neurodegeneration and tracking disease progression [87]. However, differences in diagnostic accuracy and inter-platform variability limit their interchangeability. Molecular autophagy markers such as LC3B remain largely experimental, with limited clinical validation to date [88, 89]. Notably, emerging clinical data from early-stage PD patients demonstrate that CSF LC3B levels can discriminate PD with high sensitivity and specificity, and correlate with disease severity and imaging findings [90]. However, broader clinical validation across other neurodegenerative disorders is still needed. Another barrier is in the quantification of autophagy through autophagosome content in neuron-derived extracellular vesicles, which presents technical difficulties in consistency and standardization. In parallel, integration of multi-omics data by combining proteomics, metabolomics, and transcriptomics holds potential for better disease subtyping, yet challenges persist in harmonizing heterogeneous data and translating findings into actionable trial endpoints [91]. Next, adaptive trial designs using Bayesian frameworks are gaining traction for testing combination therapies, but their complexity requires precise biomarker-based subgroup identification to be effective [92].

14.6 Future Directions

Targeting neuroinflammation and autophagy is rapidly evolving in a precision medicine framework, which is driven by advances in multi-omics technologies and computational modelling. Genetic analysis of patients for inflammasome-related variants, such as the NLRP3 Q705K polymorphism, may enable effective stratification for anti-inflammatory interventions. Considering the critical role of autophagy in disease progression, assessing autophagy activity using biomarkers like the LC3-II/LC3-I ratio in peripheral blood mononuclear cells can serve as a dynamic indicator of treatment efficacy. Also, detailed microglial transcriptional profiling distinguishing between TREM2- and TYROBP-associated subtypes could further update targeted immunomodulation strategies. In parallel, multitudes of diagnostic methods are emerging as a promising way to detect therapeutic efficacy of alternative approaches and treatment strategies. Gut microbiome signatures are emerging as predictive tools to identify individuals likely to benefit from anti-inflammatory therapeutic agents. It must be noted that employing computational tools and techniques could offer unprecedented capabilities to model the biological systems and to predict the drug-target interactions at the molecular level [93]. Also, the advancement of generative artificial intelligence technologies is being utilized to design BBB-penetrant dual inhibitors, whereas machine learning-driven analysis of high-throughput screening data in patient-derived brain organoids is observed to be beneficial in rapid validation of therapeutic candidates [94]. These innovations aiming to manage neuroinflammation hold promise for developing more precise and effective therapies for chronic neurological disorders. Emerging neurotherapeutics, such as engineered probiotics delivering IL-10 and autophagy-enhancing peptides, and DNA origami nanobots, enable targeted drug release in hippocampal regions are pushing the frontiers of precision medicine. However, these innovations introduce significant ethical, regulatory, and ecological concerns due to informed consent complexities in dementia-related gene therapy trials and disparities in global access to expensive biologics. Long-term safety, including CRISPR off-target effects in post-mitotic neurons, necessitates prolonged surveillance, whereas the environmental burden of nanoparticle waste in clinical settings remains largely under-addressed.

14.7 Conclusion

In conclusion, effectively halting neurodegenerative diseases' progression requires a strategic focus on dual modulation of neuroinflammation and autophagy. However, the success of such approaches depends heavily on overcoming critical translational barriers, including limited delivery across the BBB, potential toxicity of therapeutic agents, and need for sensitive, scalable biomarkers. Multimodal strategies that combine pharmacological agents, gene editing technologies, and innovative delivery platforms are likely to define the next generation of therapies. Mechanistic insights into autophagy-inflammation crosstalk must guide adaptive clinical trial designs, with patient stratification and longitudinal biomarker validation serving as cornerstones for maximizing therapeutic effect. Equally important is the formation of robust public–private partnerships to de-risk development pipelines and accelerate clinical translation. Bridging these scientific, logistical, and economic gaps is essential to bring next-generation neurotherapies from bench to bedside.

References

1. Carlson NG, Wieggel WA, Chen J, Bacchi A, Rogers SW, Gahring LC. Inflammatory cytokines IL-1 alpha, IL-1 beta, IL-6, and TNF-alpha impart neuroprotection to an excitotoxin through distinct pathways. J Immunol. 1999;163(7):3963–8.
2. Ma S, Attarwala IY, Xie XQ. SQSTM1/p62: a potential target for neurodegenerative disease. ACS Chem Neurosci. 2019;10(5):2094–114.
3. Xu T, Liu J, Li XR, Yu Y, Luo X, Zheng X, Cheng Y, Yu PQ, Liu Y. The mTOR/NF-κB pathway mediates neuroinflammation and synaptic plasticity in diabetic encephalopathy. Mol Neurobiol. 2021;58(8):3848–62.
4. Wen Z. Modeling neurodevelopmental and psychiatric diseases with human iPSCs. J Neurosci Res. 2017;95(5):1097–109.
5. Sun J, Noss S, Banerjee D, Das M, Girirajan S. Strategies for dissecting the complexity of neurodevelopmental disorders. Trends Genet. 2024;40(2):187–202.
6. Guan Y, Han F. Key mechanisms and potential targets of the NLRP3 Inflammasome in neurodegenerative diseases. Front Integr Neurosci. 2020;14:37.
7. Lu HF, Zhou YC, Hu TY, Yang DH, Wang XJ, Luo DD, Qiu SQ, Cheng BH, Zeng XH. Unraveling the role of NLRP3 inflammasome in allergic inflammation: implications for novel therapies. Front Immunol. 2024;15:1435892.
8. Deora V, Lee JD, Albornoz EA, McAlary L, Jagaraj CJ, Robertson AAB, Atkin JD, Cooper MA, Schroder K, Yerbury JJ, Gordon R, Woodruff TM. The microglial NLRP3 inflammasome is activated by amyotrophic lateral sclerosis proteins. Glia. 2020;68(2):407–21.
9. Lonnemann N, Hosseini S, Marchetti C, Skouras DB, Stefanoni D, D'Alessandro A, Dinarello CA, Korte M. The NLRP3 inflammasome inhibitor OLT1177 rescues cognitive impairment in a mouse model of Alzheimer's disease. Proc Natl Acad Sci U S A. 2020;117(50):32145–54.
10. Jiang H, He H, Chen Y, Huang W, Cheng J, Ye J, Wang A, Tao J, Wang C, Liu Q, Jin T, Jiang W, Deng X, Zhou R. Identification of a selective and direct NLRP3 inhibitor to treat inflammatory disorders. J Exp Med. 2017;214(11):3219–38.
11. Kaur B, Biby S, Namme JN, More S, Xu Y, Zhang S. Biological and therapeutic significance of targeting NLRP3 inflammasome in the brain and the current efforts to develop brain-penetrant inhibitors. Adv Pharmacol. 2025;102:103–57.
12. Sakamoto KM, Kim KB, Kumagai A, Mercurio F, Crews CM, Deshaies RJ. Protacs: chimeric molecules that target proteins to the Skp1-Cullin-F box complex for ubiquitination and degradation. Proc Natl Acad Sci USA. 2001;98(15):8554–9.
13. Hameed MS, Cao H, Guo L, Zeng L, Ren Y. Advancements, challenges, and future frontiers in covalent inhibitors and covalent drugs: a review. Eur J Med Chem Rep. 2024;12:100217.
14. Chen Y, He H, Lin B, Chen Y, Deng X, Jiang W, Zhou R. RRx-001 ameliorates inflammatory diseases by acting as a potent covalent NLRP3 inhibitor. Cell Mol Immunol. 2021;18(6): 1425–36.
15. Guertin DA, Sabatini DM. The pharmacology of mTOR inhibition. Sci Signal. 2009;2(67): pe24.
16. Alonso MM, Jiang H, Yokoyama T, Xu J, Bekele NB, Lang FF, Kondo S, Gomez-Manzano C, Fueyo J. Delta-24-RGD in combination with RAD001 induces enhanced anti-glioma effect via autophagic cell death. Mol Ther. 2008;16(3):487–93.
17. Lombard-Bohas C, Cariou B, Vergès B, Coriat R, N'guyen T, François E, Hammel P, Niccoli P, Hentic O. Management of metabolic disorders induced by everolimus in patients with differentiated neuroendocrine tumors: expert proposals. Bull Cancer. 2014;101(2):175–83.
18. Thoreen CC, Kang SA, Chang JW, Liu Q, Zhang J, Gao Y, Reichling LJ, Sim T, Sabatini DM, Gray NS. An ATP-competitive mammalian target of rapamycin inhibitor reveals rapamycin-resistant functions of mTORC1. J Biol Chem. 2009;284(12):8023–32.
19. Robeyns R, Sisto A, Iturrospe E, da Silva KM, van de Lavoir M, Timmerman V, Covaci A, Stroobants S, van Nuijs ALN. The metabolic and Lipidomic fingerprint of Torin1 exposure in mouse embryonic fibroblasts using untargeted metabolomics. Meta. 2024;14(5):248.

20. Hadley G, Beard DJ, Alexopoulou Z, Sutherland BA, Buchan AM. Investigation of the novel mTOR inhibitor AZD2014 in neuronal ischemia. Neurosci Lett. 2019;706:223–30.
21. Varin J, Poulain L, Hivelin M, Nusbaum P, Hubas A, Laurendeau I, Lantieri L, Wolkenstein P, Vidaud M, Pasmant E, Chapuis N, Parfait B. Dual mTORC1/2 inhibition induces anti-proliferative effect in NF1-associated plexiform neurofibroma and malignant peripheral nerve sheath tumor cells. Oncotarget. 2016;7(24):35753–67.
22. Ma X, Sun Z, Han X, Li S, Jiang X, Chen S, Zhang J, Lu H. Neuroprotective effect of resveratrol via activation of Sirt1 signaling in a rat model of combined diabetes and Alzheimer's disease. Front Neurosci. 2020;13:1400.
23. Tyagi A, Kamal MA, Poddar NK. Integrated pathways of COX-2 and mTOR: roles in cell sensing and Alzheimer's disease. Front Neurosci. 2020;14:693.
24. Pietrocola F, Lachkar S, Enot DP, Niso-Santano M, Bravo-San Pedro JM, Sica V, Izzo V, Maiuri MC, Madeo F, Mariño G, Kroemer G. Spermidine induces autophagy by inhibiting the acetyltransferase EP300. Cell Death Differ. 2015;22(3):509–16.
25. Sarkar S, Davies JE, Huang Z, Tunnacliffe A, Rubinsztein DC. Trehalose, a novel mTOR-independent autophagy enhancer, accelerates the clearance of mutant huntingtin and alpha-synuclein. J Biol Chem. 2007;282(8):5641–52.
26. Gopar-Cuevas Y, Saucedo-Cardenas O, Loera-Arias MJ, Montes-de-Oca-Luna R, Rodriguez-Rocha H, Garcia-Garcia A. Metformin and Trehalose-modulated autophagy exerts a neurotherapeutic effect on Parkinson's disease. Mol Neurobiol. 2023;60(12):7253–73.
27. Gao Y, Liang A, Kang JH, Liu P, Yin XM, Li M. Niclosamide blocks autophagy via lysosomal dysfunction (663.18). FASEB J. 2014;28:663–18.
28. Kiyota T, Ingraham KL, Swan RJ, Jacobsen MT, Andrews SJ, Ikezu T. AAV serotype 2/1-mediated gene delivery of anti-inflammatory interleukin-10 enhances neurogenesis and cognitive function in APP+PS1 mice. Gene Ther. 2012;19(7):724–33.
29. Zhou QH, Sumbria R, Hui EK, Lu JZ, Boado RJ, Pardridge WM. Neuroprotection with a brain-penetrating biologic tumor necrosis factor inhibitor. J Pharmacol Exp Ther. 2011;339(2): 618–23.
30. Moon JJ, Huang B, Irvine DJ. Engineering nano- and microparticles to tune immunity. Adv Mater. 2012;24(28):3724–46.
31. Pakulska MM, Ballios BG, Shoichet MS. Injectable hydrogels for central nervous system therapy. Biomed Mater. 2012;7(2):024101.
32. Kim J, Choi Y, Kim DH, Yoon HY, Kim K. Injectable hydrogel-based combination cancer immunotherapy for overcoming localized therapeutic efficacy. Pharmaceutics. 2022;14(9): 1908.
33. Shen H, Xu B, Yang C, Xue W, You Z, Wu X, Ma D, Shao D, Leong K, Dai J. A DAMP-scavenging, IL-10-releasing hydrogel promotes neural regeneration and motor function recovery after spinal cord injury. Biomaterials. 2022;280:121279.
34. Cabré S, Alamilla V, Moriarty N, Pandit A, Dowd E. Anti-inflammatory cytokine-eluting collagen hydrogel reduces the host immune response to dopaminergic cell transplants in a rat model of Parkinson's disease. Neuronal Signals. 2021;5(3):NS20210028.
35. Bateman RJ, Smith J, Donohue MC, Delmar P, Abbas R, Salloway S, Wojtowicz J, Blennow K, Bittner T, Black SE, Klein G, Boada M, Grimmer T, Tamaoka A, Perry RJ, Turner RS, Watson D, Woodward M, Thanasopoulou A, Lane C, Baudler M, Fox NC, Cummings JL, Fontoura P, Doody RS, GRADUATE I and II Investigators and the Gantenerumab Study Group. Two phase 3 trials of gantenerumab in early Alzheimer's disease. N Engl J Med. 2023;389(20):1862–76.
36. Carroll E, Scaber J, Huber KVM, Brennan PE, Thompson AG, Turner MR, Talbot K. Drug repurposing in amyotrophic lateral sclerosis (ALS). Expert Opin Drug Discov. 2025;20(4):447–64.
37. Singh Gautam A, Pandey SK, Lasure V, Dubey S, Singh RK. Monoclonal antibodies for the management of central nervous system diseases: clinical success and future strategies. Expert Opin Biol Ther. 2023;23(7):603–18.

38. Ahlawat J, Guillama Barroso G, Masoudi Asil S, Alvarado M, Armendariz I, Bernal J, Carabaza X, Chavez S, Cruz P, Escalante V, Estorga S, Fernandez D, Lozano C, Marrufo M, Ahmad N, Negrete S, Olvera K, Parada X, Portillo B, Ramirez A, Ramos R, Rodriguez V, Rojas P, Romero J, Suarez D, Urueta G, Viel S, Narayan M. Nanocarriers as potential drug delivery candidates for overcoming the blood-brain barrier: challenges and possibilities. ACS Omega. 2020;5(22):12583–95.
39. Wu D, Chen Q, Chen X, Han F, Chen Z, Wang Y. The blood-brain barrier: structure, regulation, and drug delivery. Signal Transduct Target Ther. 2023;8(1):217.
40. van den Oetelaar PJ, Jansen PS, Melgers PA, Wagenaars GN, ten Kortenaar PB. Stability assessment of peptide and protein drugs. J Control Release. 1992;21(1–3):11–21.
41. Böttger R, Knappe D, Hoffmann R. Readily adaptable release kinetics of prodrugs using protease-dependent reversible PEGylation. J Control Release. 2016;230:88–94.
42. Sestak V, Roh J, Klepalova L, Kovarikova P. A UHPLC-UV-QTOF study on the stability of carfilzomib, a novel proteasome inhibitor. J Pharm Biomed Anal. 2016;124:365–73.
43. Zhuo S, Zhang F, Yu J, Zhang X, Yang G, Liu X. pH-Sensitive biomaterials for drug delivery. Molecules. 2020;25(23):5649.
44. Blagosklonny MV. Once again on rapamycin-induced insulin resistance and longevity: despite of or owing to. Aging (Albany NY). 2012;4(5):350–8.
45. Yang D, Wang Z, Chen Y, Guo Q, Dong Y. Interactions between gut microbes and NLRP3 inflammasome in the gut-brain axis. Comput Struct Biotechnol J. 2023;21:2215–27.
46. Xu Q, He L, Yin Y. Risk of herpes zoster associated with JAK inhibitors in immune-mediated inflammatory diseases: a systematic review and network meta-analysis. Front Pharmacol. 2023;14:1241954.
47. Xia H, Green DR, Zou W. Autophagy in tumour immunity and therapy. Nat Rev Cancer. 2021;21(5):281–97.
48. Jeremic D, Jiménez-Díaz L, Navarro-López JD. Anti-amyloid antibodies in Alzheimer's disease: what did clinical trials teach us? Neural Regen Res. 2025;20(4):1092–3.
49. Li L, Gao L, Song Y, Qin ZH, Liang Z. Activated cathepsin L is associated with the switch from autophagy to apoptotic death of SH-SY5Y cells exposed to 6-hydroxydopamine. Biochem Biophys Res Commun. 2016;470(3):579–85.
50. Ghavami S, Shojaei S, Yeganeh B, Ande SR, Jangamreddy JR, Mehrpour M, Christoffersson J, Chaabane W, Moghadam AR, Kashani HH, Hashemi M, Owji AA, Łos MJ. Autophagy and apoptosis dysfunction in neurodegenerative disorders. Prog Neurobiol. 2014;112:24–49.
51. Liu C, Gao Y, Barrett J, Hu B. Autophagy and protein aggregation after brain ischemia. J Neurochem. 2010;115(1):68–78.
52. Zhu Y, Zhang J, Deng Q, Chen X. Mitophagy-associated programmed neuronal death and neuroinflammation. Front Immunol. 2024;15:1460286.
53. Jiang T, Du J, Raynald, Wang J, Li C. Presurgical administration of mTOR inhibitors in patients with large subependymal giant cell astrocytoma associated with tuberous sclerosis complex. World Neurosurg. 2017;107:1053.e1–6.
54. Wu M, Arcaro A, Varga Z, Vogetseder A, Le Hir M, Wüthrich RP, Serra AL. Pulse mTOR inhibitor treatment effectively controls cyst growth but leads to severe parenchymal and glomerular hypertrophy in rat polycystic kidney disease. Am J Physiol Renal Physiol. 2009;297(6):F1597–605.
55. Wang B, Warden AR, Ding X. The optimization of combinatorial drug therapies: strategies and laboratorial platforms. Drug Discov Today. 2021;26(11):2646–59.
56. Kaplan MJ, Ellis CN, Bata-Csorgo Z, Kaplan RS, Endres JL, Fox DA. Systemic toxicity following administration of sirolimus (formerly rapamycin) for psoriasis: association of capillary leak syndrome with apoptosis of lesional lymphocytes. Arch Dermatol. 1999;135(5): 553–7.
57. Friedman LG, Qureshi YH, Yu WH. Promoting autophagic clearance: viable therapeutic targets in Alzheimer's disease. Neurotherapeutics. 2015;12(1):94–108.
58. Lévi F. From circadian rhythms to cancer chronotherapeutics. Chronobiol Int. 2002;19(1):1–19.

59. Rabinovich-Nikitin I, Rasouli M, Reitz CJ, Posen I, Margulets V, Dhingra R, Khatua TN, Thliveris JA, Martino TA, Kirshenbaum LA. Mitochondrial autophagy and cell survival is regulated by the circadian *Clock* gene in cardiac myocytes during ischemic stress. Autophagy. 2021;17(11):3794–812.
60. Chen K, Yang Y, Wu Y, Cao W, Zhao Y, Wang S, Wang K. PLGA nanoparticles encapsulating TSHR-A and rapamycin enhance the induction of dendritic cell-specific immune tolerance in mice with Graves' disease. Biomed Mater. 2025;20(2):025045.
61. Mandal S, Debnath K, Jana NR, Jana NR. Trehalose-functionalized gold nanoparticle for inhibiting intracellular protein aggregation. Langmuir. 2017;33(49):13996–4003.
62. Tryphena KP, Singh G, Jain N, Famta P, Srivastava S, Singh SB, Khatri DK. Integration of miRNA's theranostic potential with nanotechnology: promises and challenges for Parkinson's disease therapeutics. Mech Ageing Dev. 2023;211:111800.
63. Aminabad ED, Hasanzadeh M, Saadati A, Feizi MA, Safaralizadeh R, Mobed A. An innovative biodevice towards monitoring of miR-153 using specific DNA immobilized on the surface of poly (chitosan) decorated AgNPrs/GQDs-CysA conductive nano-ink: early-stage diagnosis of Parkinson's disease using biosensor technology. Mater Sci Eng B Solid-State Mater. 2022;286: 116017.
64. Cheng G, Liu Z, Yan Z, Wu J, Li Z, Gao S, Zheng C, Guo S, Pan Y, Chen X, Lin G, Zhou J, Chen T. Minocycline nanoplatform penetrates the BBB and enables the targeted treatment of Parkinson's disease with cognitive impairment. J Control Release. 2025;377:591–605.
65. Yu Y, Li W, Mao L, Peng W, Long D, Li D, Zhou R, Dang X. Genetically engineered exosomes display RVG peptide and selectively enrich a neprilysin variant: a potential formulation for the treatment of Alzheimer's disease. J Drug Target. 2021;29(10):1128–38.
66. Jiang P, Xiao Y, Hu X, Wang C, Gao H, Huang H, Lv J, Qi Z, Wang Z. RVG29 peptide-modified exosomes loaded with Mir-133b mediate the RhoA-ROCK pathway to improve motor and neurological symptoms in Parkinson's disease. ACS Biomater Sci Eng. 2024;10(5): 3069–85.
67. Ye Y, Zhang X, Xie F, Xu B, Xie P, Yang T, Shi Q, Zhang CY, Zhang Y, Chen J, Jiang X, Li J. An engineered exosome for delivering sgRNA:Cas9 ribonucleoprotein complex and genome editing in recipient cells. Biomater Sci. 2020;8:2966.
68. Balaraman AK, Babu MA, Moglad E, Mandaliya V, Rekha MM, Gupta S, Prasad GVS, Kumari M, Chauhan AS, Ali H, Goyal K. Exosome-mediated delivery of CRISPR-Cas9: a revolutionary approach to cancer gene editing. Pathol Res Pract. 2025;266:155785.
69. Chaput N, Schartz NE, André F, Taïeb J, Novault S, Bonnaventure P, Aubert N, Bernard J, Lemonnier F, Merad M, Adema G, Adams M, Ferrantini M, Carpentier AF, Escudier B, Tursz T, Angevin E, Zitvogel L. Exosomes as potent cell-free peptide-based vaccine. II. Exosomes in CpG adjuvants efficiently prime naive Tc1 lymphocytes leading to tumor rejection. J Immunol. 2004;172(4):2137–46.
70. Huda MN, Nurunnabi M. Potential application of exosomes in vaccine development and delivery. Pharm Res. 2022;39(11):2635–71.
71. Chen J, Mao K, Yu H, Wen Y, She H, Zhang H, Liu L, Li M, Li W, Zou F. p38-TFEB pathways promote microglia activation through inhibiting CMA-mediated NLRP3 degradation in Parkinson's disease. J Neuroinflammation. 2021;18(1):295.
72. Grotemeyer A, Fischer J, Koprich JB, Brotchie JM, Volkmann J, Ip CW. P 5 MCC950 treatment reduces neuroinflammation, rescues neuronal cell death and ameliorates motor deficits in the AAV1/2-A53T-αSyn mouse model of Parkinson's disease. Clin Neurophysiol. 2022;137:e17.
73. Panicker N, Kam TI, Wang H, Neifert S, Chou SC, Kumar M, Brahmachari S, Jhaldiyal A, Hinkle JT, Akkentli F, Mao X, Xu E, Karuppagounder SS, Hsu ET, Kang SU, Pletnikova O, Troncoso J, Dawson VL, Dawson TM. Neuronal NLRP3 is a parkin substrate that drives neurodegeneration in Parkinson's disease. Neuron. 2022;110(15):2422–2437.e9.
74. Li H, Wu S, Ma X, Li X, Cheng T, Chen Z, Wu J, Lv L, Li L, Xu L, Wang W, Hu Y, Jiang H, Yin Y, Qiu Z, Hu X. Co-editing PINK1 and DJ-1 genes via adeno-associated virus-delivered

CRISPR/Cas9 system in adult monkey brain elicits classical parkinsonian phenotype. Neurosci Bull. 2021;37(9):1271–88.
75. Saber-Ayad M, Hammoudeh S, Abu-Gharbieh E, Hamoudi R, Tarazi H, Al-Tel TH, Hamid Q. Current status of Baricitinib as a repurposed therapy for COVID-19. Pharmaceuticals (Basel). 2021;14(7):680.
76. Liu J, Ting JP, Al-Azzam S, Ding Y, Afshar S. Therapeutic advances in diabetes, autoimmune, and neurological diseases. Int J Mol Sci. 2021;22(6):2805.
77. Heneka MT, Kummer MP, Stutz A, Delekate A, Schwartz S, Vieira-Saecker A, Griep A, Axt D, Remus A, Tzeng TC, Gelpi E, Halle A, Korte M, Latz E, Golenbock DT. NLRP3 is activated in Alzheimer's disease and contributes to pathology in APP/PS1 mice. Nature. 2013;493(7434): 674–8.
78. Arotcarena ML, Bourdenx M, Dutheil N, Thiolat ML, Doudnikoff E, Dovero S, Ballabio A, Fernagut PO, Meissner WG, Bezard E, Dehay B. Transcription factor EB overexpression prevents neurodegeneration in experimental synucleinopathies. JCI Insight. 2019;4(16): e129719.
79. Decressac M, Mattsson B, Weikop P, Lundblad M, Jakobsson J, Björklund A. TFEB-mediated autophagy rescues midbrain dopamine neurons from α-synuclein toxicity. Proc Natl Acad Sci USA. 2013;110(19):E1817–26.
80. Chen CX, You Z, Abdian N, Sirois J, Shlaifer I, Tabatabaei M, Boivin MN, Gaborieau L, Karamchandani J, Beitel LK, Fon EA, Durcan TM. Generation of homozygous PRKN, PINK1 and double PINK1/PRKN knockout cell lines from healthy induced pluripotent stem cells using CRISPR/Cas9 editing. Stem Cell Res. 2022;62:102806.
81. Rahman MU, Bilal M, Shah JA, Kaushik A, Teissedre PL, Kujawska M. CRISPR-Cas9-based technology and its relevance to gene editing in Parkinson's disease. Pharmaceutics. 2022;14(6): 1252.
82. Antoniou N, Prodromidou K, Kouroupi G, Boumpoureka I, Samiotaki M, Panayotou G, Xilouri M, Kloukina I, Stefanis L, Grailhe R, Taoufik E, Matsas R. High content screening and proteomic analysis identify a kinase inhibitor that rescues pathological phenotypes in a patient-derived model of Parkinson's disease. NPJ Parkinsons Dis. 2022;8(1):15.
83. Estevez-Fraga C, Tabrizi SJ, Wild EJ. Huntington's disease clinical trials corner: august 2023. J Huntingtons Dis. 2023;12(2):169–85.
84. Richardson PJ, Smith DP, de Giorgio A, Snetkov X, Almond-Thynne J, Cronin S, Mead RJ, McDermott CJ, Shaw PJ. Janus kinase inhibitors are potential therapeutics for amyotrophic lateral sclerosis. Transl Neurodegener. 2023;12(1):47.
85. Abdul Manap AS, Almadodi R, Sultana S, Sebastian MG, Kavani KS, Lyenouq VE, Shankar A. Alzheimer's disease: a review on the current trends of the effective diagnosis and therapeutics. Front Aging Neurosci. 2024;16:1429211.
86. Vijiaratnam N, Girges C, Auld G, McComish R, King A, Skene SS, Hibbert S, Wong A, Melander S, Gibson R, Matthews H, Dickson J, Carroll C, Patrick A, Inches J, Silverdale M, Blackledge B, Whiston J, Hu M, Welch J, Duncan G, Power K, Gallen S, Kerr J, Chaudhuri KR, Batzu L, Rota S, Jabbari E, Morris H, Limousin P, Greig N, Li Y, Libri V, Gandhi S, Athauda D, Chowdhury K, Foltynie T. Exenatide once a week versus placebo as a potential disease-modifying treatment for people with Parkinson's disease in the UK: a phase 3, multicentre, double-blind, parallel-group, randomised, placebo-controlled trial. Lancet. 2025;405(10479):627–36.
87. Tang R, Buchholz E, Dale AM, Rissman RA, Fennema-Notestine C, Gillespie NA, Hagler DJ Jr, Lyons MJ, Neale MC, Panizzon MS, Puckett OK, Reynolds CA, Franz CE, Kremen WS, Elman JA. Associations of plasma neurofilament light chain with cognition and neuroimaging measures in community-dwelling early old age men. Alzheimer's Res Ther. 2024;16(1):90.
88. Sivapackiam J, Kabra S, Speidel S, Sharma M, Laforest R, Salter A, Rettig MP, Sharma V. 68Ga-Galmydar: a PET imaging tracer for noninvasive detection of doxorubicin-induced cardiotoxicity. PLoS One. 2019;14(5):e0215579.

89. Liśkiewicz D, Liśkiewicz A, Nowacka-Chmielewska MM, Grabowski M, Pondel N, Grabowska K, Student S, Barski JJ, Małecki A. Differential response of hippocampal and cerebrocortical autophagy and ketone body metabolism to the ketogenic diet. Front Cell Neurosci. 2021;15:733607.
90. Youn J, Lee SB, Lee HS, Yang HO, Park J, Kim JS, Oh E, Park S, Jang W. Cerebrospinal fluid levels of autophagy-related proteins represent potentially novel biomarkers of early-stage Parkinson's disease. Sci Rep. 2018;8(1):16866.
91. Mohr AE, Ortega-Santos CP, Whisner CM, Klein-Seetharaman J, Jasbi P. Navigating challenges and opportunities in multi-omics integration for personalized healthcare. Biomedicine. 2024;12(7):1496.
92. Su L, Chen X, Zhang J, Yan F. MIDAS-2: an enhanced Bayesian platform design for immunotherapy combinations with subgroup efficacy exploration. J Biopharm Stat. 2025;35 (1):37–57.
93. Nadh AG, Amjesh R, Ahmed M, Kumar TKM, Sudhakaran PR, Raju R. Identification of high-affinity molecules from Sesbania grandiflora as potential therapeutic agents for oral ulcers-a multi-target-directed in silico approach. J Indian Chem Soc. 2025;102(5):101674.
94. Zhang H, Huang N, Bian S, Sawan M. Brain organoids-on-chip for neural diseases modeling: history, challenges and trends. J Pharm Anal. 2025:101323. https://doi.org/10.1016/j.jpha.2025.101323.

Advanced Delivery Systems and Gene Editing in Neurotherapeutics

15

Prabhat Upadhyay, Aamir Suhail, and Sudhir Kumar

Abstract

Neurodegenerative diseases require innovative therapies targeting autophagy dysfunction, neuroinflammation, and genetic defects. This chapter examines integrated neurotherapeutic strategies, including nanoparticle delivery, gene editing, and optogenetics, designed to cross biological barriers and correct disease mechanisms. Engineered nanoparticles can deliver autophagy enhancers and anti-inflammatory agents across the blood–brain barrier, whereas avian-associated virus-mediated TFEB overexpression can enhance lysosomal function and suppress inflammasomes. CRISPR corrects genetic defects, such as PARK2 for mitophagy restoration and C9ORF72 repeat excision to reduce toxic dipeptides. Moreover, optogenetics modulates microglial function through TMEM119-driven opsins, reducing neuroinflammation while preserving phagocytosis in neurodegenerative conditions. Challenges remain with nanoparticle endosomal entrapment, immunogenicity, CRISPR off-target effects, and optogenetic

P. Upadhyay
Vincent Center for Reproductive Biology, Department of Obstetrics and Gynaecology, Massachusetts General Hospital, Harvard Medical School, Boston, MA, USA

A. Suhail
The Gene Lay Institute of Immunology and Inflammation, Brigham and Women's Hospital, Mass General Hospital, and Harvard Medical School, Boston, MA, USA

Ann Romney Center for Neurologic Diseases, Brigham and Women's Hospital, Boston, MA, USA

Broad Institute of MIT and Harvard, Cambridge, MA, USA
e-mail: asuhail@bwh.harvard.edu

S. Kumar (✉)
Department of Microbiology and Immunology, School of Medicine, Emory University, Atlanta, GA, USA
e-mail: sudhir.kumar@emory.edu

P. Khanal et al. (eds.), *Neuroinflammation and Autophagy in Neurodegeneration*,
https://doi.org/10.1007/978-981-95-3945-1_15

invasiveness. Emerging solutions include artificial intelligence-optimized blood–brain barrier carriers, sonogenetics, and miRNA-targeted vectors. This chapter outlines a translational framework, balancing molecular insights with clinical hurdles, advancing precision neurotherapeutics to modify disease progression.

Keywords

Autophagy · CRISPR · Nanoparticles · Neuroinflammation · Neurodegenerative diseases · Optogenetics

15.1 Introduction

Neurodegenerative diseases such as Alzheimer's disease (AD) and Parkinson's disease (PD) share key pathological features, including impaired autophagy, toxic protein accumulation, and persistent neuroinflammation, yet effective treatments remain elusive [1, 2]. Despite increasing recognition of the critical role autophagy and inflammation play in disease progression, effective therapeutic development remains a significant challenge and an urgent priority. The blood–brain barrier (BBB) and genetic variability among patients limit the effectiveness of conventional approaches, as many small molecules either fail to reach the central nervous system (CNS) or lack the precision to selectively target disease-relevant pathways without unintended effects [3, 4]. These challenges call for a paradigm shift toward next-generation delivery platforms and gene-editing technologies that can bypass the brain's protective barriers and address the underlying causes of disease, rather than merely managing symptoms [5].

This chapter discusses how emerging technologies are mechanistically addressing these barriers. For example, engineered nanoparticles designed for receptor-mediated transcytosis across the BBB enable dual cargo delivery and allow simultaneous suppression of inflammatory cytokines and autophagy activation in affected neurons and glial cells [6, 7]. Concurrently, viral vectors such as avian-associated virus (AAV) and CRISPR-based systems are being harnessed to rectify autophagy–lysosomal dysfunction at their genetic level, such as reactivating PINK1/Parkin-mediated mitophagy in PD or excising C9ORF72 repeat expansions in amyotrophic lateral sclerosis (ALS)/frontotemporal dementia (FTD) [8–11]. Next, optogenetic tools enable real-time control of microglial activation, offering dynamic regulation of neuroimmune responses in diseased neural circuits [12, 13]. Although each approach brings distinct advantages, their greatest therapeutic potential may emerge through strategic combination. However, translating these approaches into the clinic remains challenging. Multimodal therapies such as combining AAV-CRISPR systems with nanoparticle-delivered autophagy inducers face hurdles in scalability and safety, which require precise optimization to minimize immune responses and off-target effects [14–16]. Furthermore, patient stratification is also essential, as genetic subtypes may exhibit divergent responses to autophagy-targeted therapies. For example, although TFEB overexpression could be beneficial in

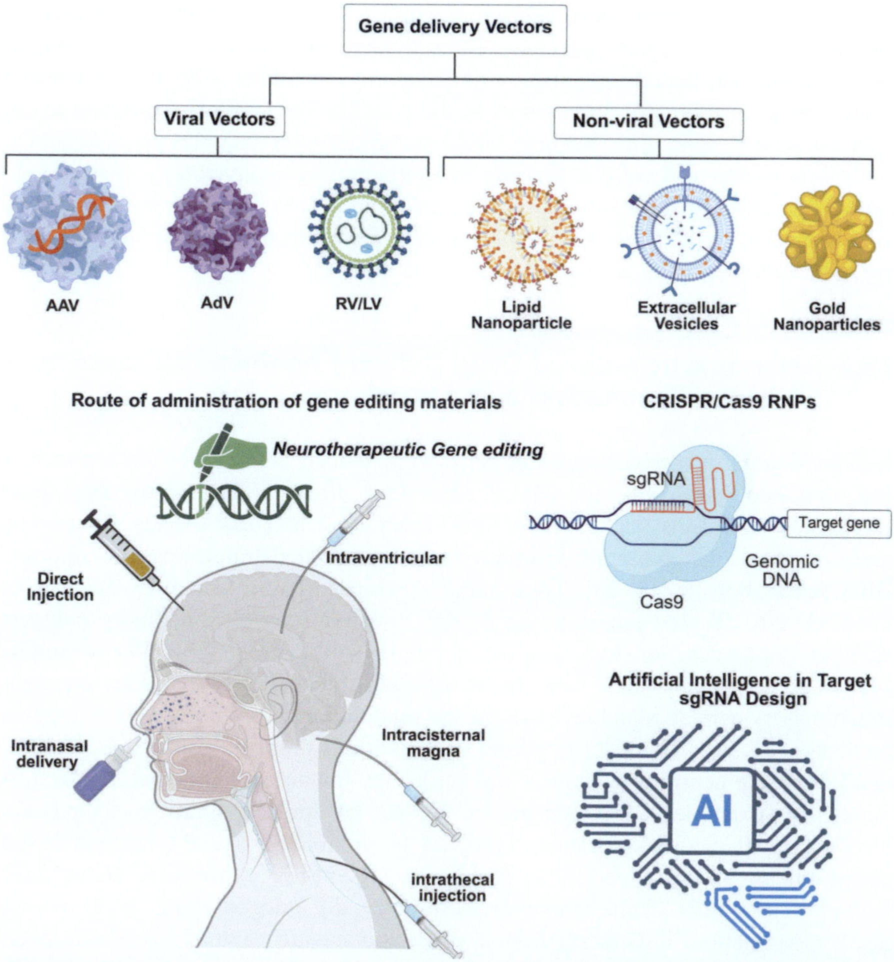

Fig. 15.1 Neurotherapeutic gene editing and gene delivery approaches. Gene delivery vectors include viral and non-viral methods enabling delivery through various administration routes, including intranasal, direct injection, intraventricular, intracisternal magna, or intracochlear injection. CRISPR-Cas9 complexes serve as the primary gene-editing machinery for precise genetic modifications. Artificial intelligence-assisted neurotherapeutic design leverages computational tools to optimize gene-editing strategies for enhanced efficacy and specificity. (Created using BioRender (https://www.biorender.com/))

tauopathies, its effect in C9ORF72-linked ALS may vary, with potential risks of worsening dipeptide repeat (DPR)-related toxicity depending on disease [17–22]. Similarly, advancing optogenetics toward clinical use also depends on suitable non-invasive human-compatible delivery systems (Fig. 15.1) development for long-term application [23].

Hence, by exploring these emerging cutting-edge technologies, this chapter attempts to highlight how precision neurotherapeutics such as nanotechnology, gene editing, and neuromodulation will transform the future of neurodegenerative disease pharmacotherapy. Integration of these technologies offers the potential not only for symptomatic relief but also for disease-modifying therapies personalized to unique molecular and cellular features of individual pathophysiology. As the field shifts beyond one-size-fits-all approaches, integrating advances in delivery technology with genetic and immune understanding may ultimately pave the way for breakthrough treatments in neurodegenerative diseases.

15.2 Nanoparticle-Based Drug Delivery Against Neuroinflammation and Autophagy

Nanoparticle-based delivery systems can simultaneously address neuroinflammation and impaired autophagy in neurodegenerative diseases. By engineering their surfaces with ligands such as transferrin receptor 1 and low-density lipoprotein receptor, these nanoparticles leverage receptor-mediated transcytosis to improve BBB penetration [6, 24, 25]. Their cargo combines anti-inflammatory agents like interleukin-1β (IL-1β) inhibitors or NLRP3 antagonists with autophagy inducers such as rapamycin or trehalose in a single formulation [26]. For example, PEG-PLGA nanoparticles can co-encapsulate rapamycin alongside an anti-inflammatory peptide to achieve synergistic modulation of proteostasis and immune pathways [27, 28]. This dual targeting strategy holds promise for diseases like AD and PD, where neuroinflammation and autophagy impairment are closely linked. A compelling example of this approach is the use of lipid nanoparticles to enhance lysosomal Aβ plaque clearance, followed by inhibiting NLRP3 inflammasome activation in AD models [29]. By increasing TFEB expression, these lipid nanoparticles may simultaneously promote Aβ degradation and reduce neuroinflammation, illustrating how a single therapeutic payload agent can target multiple disease mechanisms. Moreover, preclinical studies demonstrate that TFEB-lipid nanoparticles lower Aβ accumulation and improve cognitive function in transgenic AD mice [30]. This approach underscores the potential of nucleic acid-based nanotherapeutics for durable restoration of autophagy, coupled with immunomodulation.

Despite their potential, nanoparticles face significant challenges. In inflamed brain regions, altered vascular permeability and extracellular matrix remodeling can cause particle aggregation, limiting uniform distribution and cellular uptake. Moreover, efficient endosomal escape also remains a major challenge for nucleic acid cargo like mRNA and siRNA, which require cytosolic delivery to exert their effects [31]. Although strategies such as pH-responsive polymers and fusogenic peptides have been investigated to improve endosomal rupture, consistently achieving release across diverse cell types remains challenging [32, 33]. However, next-generation nanoparticles need to be designed with stimuli-responsive elements such as redox-sensitive linkers and ultrasound-triggered release to achieve precise spatiotemporal control. Integrating lipid nanoparticles with exosome-based delivery may further boost biocompatibility and targeting accuracy. Additionally, artificial intelligence-driven designs of BBB-permeable peptides hold promise to optimize ligand selection and enhance CNS uptake. As these technologies advance, nanoparticle systems could enable personalized combination therapies, such as co-delivering CRISPR gene-editing components alongside autophagy modulators to provide a novel strategy in the pharmacotherapy of complex neurodegenerative diseases.

15.3 AAV-Mediated Gene Therapy to Enhance Autophagy

AAV-mediated TFEB overexpression can be advanced to boost autophagy–lysosomal function in neurodegenerative diseases. As a master transcriptional regulator, TFEB regulates lysosomal biogenesis and autophagy-related gene networks and promotes toxic protein aggregates such as α-synuclein, tau, and TDP-43 [20, 34–36]. Beyond maintaining proteostasis, TFEB also exerts anti-inflammatory effects by suppressing pro-inflammatory cytokines such as tumor necrosis factor-α (TNF-α) and IL-6 through both mammalian target of rapamycin complex 1 (mTORC1)-dependent and independent pathways. Importantly, TFEB activation functions synergistically with mitophagy signaling, especially PINK1/PARKIN, to restore mitochondrial activity in PD disease models, which highlights its broad significance across multiple proteinopathies [37, 38]. This dual functionality positions TFEB not only as an enhancer of autophagy but also as a multimodal neuroprotective factor that integrates proteostatic and immunomodulatory mechanisms. Furthermore, the efficacy of AAV-TFEB depends on vector engineering to optimize CNS delivery and minimize off-target effects. Capsid modifications and their variants can improve BBB penetration, enhancing neuronal tropism [39, 40]. Meanwhile, cell-specific promoters such as SYN1 for neurons and CD68 for microglia confine TFEB expression to disease-associated cells, minimizing the side effects from widespread overexpression [41].

Moreover, in P301S tauopathy mice, AAV-mediated TFEB expression enhances autophagic flux, reducing tau hyperphosphorylation and fewer neurofibrillary tangles, including microglial activation suppression [42, 43]. Similar neuroprotective effects have been reported in α-synucleinopathy models in which TFEB overexpression preserved dopaminergic neurons [20, 34]. These findings highlight TFEB dual efficacy to address both upstream proteotoxicity and downstream immune activation, positioning it as a promising therapeutic candidate for complex neurodegenerative pathologies. However, dose-dependent risks, such as lysosomal overload or aberrant lipid metabolism, also underscore the need for precise dosing strategies in clinical settings. Although TFEB remains a central player in autophagy-targeted gene therapy, emerging strategies aim to enhance its effects or overcome its limitations. Combinatorial approaches, such as co-expressing TFEB with PARKIN to augment mitophagy or with anti-inflammatory transgenes like IL-10, can offer the potential to simultaneously target multiple pathological pathways. Alternatively, CRISPR-activated endogenous TFEB regulation or its downstream targets in the CLEAR network may provide more finely tuned physiologically applicable instructions. Next, with continued advancements in AAV capsid engineering and delivery protocols, the prospect of clinically effective, autophagy-boosting gene therapies is increasingly tangible in neurodegenerative conditions where classical symptomatic treatment has shown limited success.

15.4 Correcting Autophagy-Defective Mutations Using CRISPR/Cas9 Systems

The CRISPR-Cas9 system offers a groundbreaking strategy to tackle the underlying mitochondrial dysfunction in PD by enabling precise correction of mutations in the PINK1 and PARK2 genes. Loss-of-function mutations, especially prevalent in early-onset PD, lead to the accumulation of dysfunctional mitochondria, resulting in a bioenergetic collapse and excessive reactive oxygen species production [44–46]. In turn, this fuels a vicious cycle of oxidative stress and chronic neuroinflammation, which is marked by the activation of microglial NLRP3 inflammasome and increased pro-inflammatory cytokine levels, which further accelerate dopaminergic neuron loss [47, 48]. Interestingly, recent innovations in CRISPR technology enable multiple precision intervention approaches for PD. Base editing allows precise correction of single-nucleotide mutations without double-strand breaks [49, 50], whereas dCas9-VPR systems accelerate endogenous gene expression in cases of haploinsufficiency [51]. Additionally, combinatorial sgRNA approaches enhance allelic coverage and improve specificity by minimizing off-target activity [52]. In patient-derived midbrain organoids, targeted editing of PINK1 and PARK2 not only reinstates mitophagy activity but also restores mitochondrial membrane potential and reduces pathological α-synuclein accumulation [53]. These effects are especially compelling when directed at vulnerable dopaminergic neurons. In vivo, AAV-mediated delivery of CRISPR components to the substantia nigra can lead to both anatomical preservation and improved motor function, highlighting the therapeutic potential of gene correction in mitigating neurodegeneration at its source [54].

Although these findings demonstrate CRISPR's potential in disease-modifying, several critical challenges also remain for clinical translation. For example, the BBB poses a major delivery hurdle, though advances in engineered AAV serotypes and lipid nanoparticle carriers can offer promising solutions [55]. Hence, ensuring long-term genomic safety of permanent edits in non-dividing neurons is essential, with emerging epigenetic modulation techniques presenting opportunities for transient yet lasting therapeutic effects. Future development may also focus on combining gene editing with complementary treatments such as NAD+ boosters to enhance mitophagy or LRRK2 inhibitors to target complementary pathogenic mechanisms and pathways [8, 56–58].

Furthermore, the hexanucleotide repeat expansion in C9ORF72 represents the most common genetic cause of ALS and FTD, driving neurodegeneration through two primary mechanisms: (a) loss of protein function and (b) toxic gain-of-function from repeat-derived RNA and dipeptide repeat proteins [17, 55, 59–61]. Recent advances in gene editing using CRISPR-Cas9 and antisense oligonucleotides have potential in selectively targeting and disrupting repeat expansions and preserving wild-type C9ORF72 function. Correcting this mutation not only restores physiological autophagy processes but also reduces the accumulation of neurotoxic dipeptide repeat proteins such as poly-GA and poly-GR. Mechanistically, loss of C9ORF72 function impairs autophagy–lysosomal pathways, resulting in defective clearance of protein aggregates [17]. Concurrently, dipeptide repeat proteins directly inhibit the proteasome and sequester which are essential autophagy adaptors such as p62/SQSTM1, perpetuating a cycle of proteotoxic stress [62]. CRISPR-mediated excision of pathogenic repeats or oligonucleotide-mediated suppression of repeat-containing transcripts restores autophagic flux in patient-derived neurons, correlating with reduced dipeptide repeat protein accumulation and enhanced neuronal survival [63]. Additionally, emerging techniques such as targeted RNA base editing (e.g., REPAIRx) or small-molecule repeat-stabilizing compounds that stabilize repeats present promising alternatives to reduce repeat-associated toxicity without permanent changes to the genome [64].

However, significant challenges remain in affected CNS regions and in minimizing off-target effects of repeat-targeting therapies. Moreover, since autophagy hyperactivation in some C9ORF72 models may paradoxically promote dipeptide repeat protein aggregation, therapeutic approaches should carefully balance restoring cellular homeostasis without triggering excessive autophagic stress [65]. Future directions could involve combinatorial treatments, such as coupling repeat-targeting CRISPR with TFEB overexpression to synergistically enhance lysosomal clearance, or nanoparticle-based delivery of systems to co-administer autophagy modulators alongside gene-editing tools. By addressing both root cause and downstream proteostatic dysfunction, these integrated approaches offer promising potential to halt disease progression [3, 21].

Delivering CRISPR-Cas9 therapeutics to neurons remains one of the challenges in neurogenetic medicine due to the requirement for a balance between efficiency, specificity, and safety. Although viral vectors, especially engineered AAVs, led the field due to their natural neurotropism and long-lasting expression, their limitations are evident in clinical translation. Table 15.1 highlights key concerns such as immunogenicity ceiling, stringent cargo size limits, and enduring concerns about off-target effects from persistent nuclease activity [66]. These constraints underscore the growing need for alternative delivery platforms that can meet the unique

Table 15.1 Comparison of gold nanoparticle-Cas9 delivery versus viral vectors

Feature	Gold nanoparticle-Cas9 delivery	Viral vectors (AAV/LV)
Immunogenicity	Low	Moderate-high
Cargo capacity	High (~20–100 nm range)	Limited (~4.7 kb for AAV)
Targeting precision	Tunable surface chemistry	Dependent on capsid tropism
Integration risk	None (non-viral)	Low (AAV)/High (LV)

AAV adeno-associated virus vector, *LV* lentiviral vector

demands of CNS gene editing [66]. Interestingly, gold nanoparticle conjugates offer a highly promising non-viral alternative, leveraging their unique physicochemical properties to overcome many of the limitations associated with viral vectors. Their key advantage offers the potential for dynamic surface engineering as gold nanoparticles can be precisely functionalized with neuron-targeting peptides (e.g., rabies virus glycoprotein derivatives), BBB penetrating motifs like Angiopep-2, and nuclear localization signals in a modular fashion. Additionally, gold nanoparticles exhibit a transient activity window, typically under 72 h, providing a significant safety benefit by limiting off-target effects while still enabling permanent genomic edits [66]. The materials science revolution is significantly expanding the therapeutic potential of gold nanoparticles, enabling sophisticated design features such as:

- Charge-tunable surfaces that optimize efficient neuronal uptake with rapid systemic clearance
- Photothermal-responsive release systems that allow precise spatiotemporal control of cargo delivery, and
- Hybrid lipid-gold nanostructures that integrate the membrane fusion capabilities of lipid carriers with the structural stability of inorganic nanoparticles.

Dual-cut sgRNA strategies enhance CRISPR-Cas9 precision by requiring two adjacent DNA cuts for effective gene editing, minimizing off-target effects critical in polygenic disorders where unintended edits could disrupt multiple disease-associated loci. Approaches such as paired nickases (Cas9-D10A), double-nicking with offset sgRNAs, and dCas9-FokI fusions improve specificity by ensuring dual-guided breaks occur only when both guides bind correctly, while single off-target nicks are efficiently repaired. These approaches are particularly valuable for correcting risk alleles in complex diseases like diabetes or schizophrenia, where genetic networks demand high fidelity. However, challenges remain in optimizing sgRNA design, delivery, and spacing for therapeutic applications. Future advancements in Cas9 engineering and base or prime editing could further increase the safety and precision of dual-cut platforms for therapeutic use for polygenic disorders.

15.5 Case Study: Base Editing to Restore Lysosome Function and Lower CX3CL1 Signaling

Base editing of gene regulator network mutations in FTD patients microglia is a significant advancement in precision neurotherapeutics. Adenine base editors restored progranulin to 60–70% of normal levels, mitigating the overexpression hazards associated with viral treatments. This adjustment enhanced lysosomal function and decreased CX3CL1–CX3CR1 signaling [67, 68]. In gene regulator network mutant animals, CD11b-targeted lipid nanoparticle administration can result in cerebral microglia, leading to a reduction in lipofuscin, improved localization of TDP-43, and enhanced cognitive function. Therapeutic effects depend on microglia-specific editing and the secretion of progranulin. Compared to AAV, base editing may maintain endogenous regulation and allow allele-specific correction. Challenges remain in optimizing delivery and managing microglial turnover. Still, this strategy offers a promising, targeted treatment for gene regulator network—FTD and a platform for broader interventions in microglia-driven neurodegeneration [69].

15.6 Optogenetic Control of Microglia: Spatiotemporal Modulation of Neuroimmunity

Optogenetics provides precise spatiotemporal control of microglial activity, enabling localized modulation of neuroinflammation without systemic immunosuppression. Light-induced Ca^{2+} influx suppresses nuclear factor kappa B (NF-κB)–mediated pro-inflammatory signaling using channelrhodopsin-2 driven by the microglia-specific TMEM1 promoter. This targeted activation reduces pyroptosis through caspase-1 inhibition and enhances Aβ clearance via improved phagocytosis in AD models. Unlike broad pharmacological inhibitors, this strategy preserves essential microglial roles in synaptic pruning and injury response [70, 71]. Interestingly, in traumatic brain injury models, optogenetic modulation of microglia enables reversible suppression of neuroinflammation, protecting neuronal viability without the risks of chronic immunosuppression associated with systemic anti-inflammatory drugs [72]. Combining this approach with autophagy enhancers like trehalose further boosts therapeutic effects by promoting protein aggregate clearance, simultaneously reducing inflammation. Such dual strategies hold particular promise for diseases like PD, where α-synuclein aggregates and microglial activation perpetuate a damaging feedback loop. However, clinical translation will require overcoming current hardware constraints to enable effective and practical application [72]. However, the main challenge is with biocompatibility of fiber-optic implants, as chronic intracranial hardware can cause glial scarring and increase infection risk. To avoid permanent implants, less invasive alternatives such as near–infrared–activated upconversion nanoparticles or sonogenetics are being explored. Additionally, long-term safety needs to be evaluated as prolonged optogenetic stimulation could affect microglial metabolism or inadvertently activate neighboring neurons. Overcoming these obstacles is crucial to translating optogenetics into a viable clinical approach for precision neuroimmunomodulation [73].

15.7 Immune and Delivery Challenges in Next-Gen Neurotherapeutics

The transition of CRISPR-based neurotherapeutics from preclinical success to clinical application faces significant immune and biological hurdles. A major challenge is the immunity against delivery vectors, and studies show humans harbor neutralizing antibodies against AAV9, as primate models reveal anti-Cas9 T-cell responses may compromise efficacy or trigger adverse effects [73]. The BBB adds further complexity, as its integrity and transporter expression, such as LRP1, play important roles. For example, AD patients exhibit reduced LRP1 levels, hindering vector delivery to brain. These biological variables highlight the need for patient stratification and adaptable delivery approaches tailored to individual immune status and BBB characteristics [74]. Nonetheless, significant regulatory hurdles persist, particularly due to limitations in preclinical modeling. Existing animal systems do not fully capture the complex neuroinflammatory–autophagy interactions observed in humans, highlighting the need for advanced models such as humanized microglia in cerebral organoids or CRISPR-engineered primates. Bridging the translational gaps requires an integrated approach combining immunology, nanotechnology, and computational biology. For example, machine learning algorithms that predict AAV serotype compatibility based on a patient's HLA profile could personalize vector selection, and nanoparticle coatings engineered with BBB-shuttling peptides like Angiopep-2 may bypass LRP1-mediated transport. Additionally, transient CRISPR systems can further reduce prolonged immune exposure. Additionally, regulatory agencies must evolve by establishing guidelines for neuro-specific off-target assessment and immune monitoring during clinical trials. Collaborative efforts between academia and industry, exemplified by initiatives like the NIH Somatic Cell Genome Editing program, are essential to systematically overcome these barriers and translate advances in autophagy modulation and gene editing into safe, effective treatments for neurodegenerative diseases [75].

Emerging technologies are expanding the multimodal neurotherapeutics landscape. Sonogenetics leverages non-invasive stimulation of autophagy in deep brain regions, offering a way to bypass the BBB and achieve targeted modulation of neurodegeneration-associated pathways. Next, CRISPR activation or interference allows transcriptional fine-tuning, either upregulating autophagy genes or suppressing neuroinflammatory mediators to rebalance cellular homeostasis. These approaches offer spatiotemporal precision, allowing dynamic, localized intervention while minimizing off-target effects and enhancing therapeutic specificity [73]. Importantly, artificial intelligence-driven approaches will play a pivotal role in optimizing both delivery and personalized neurotherapies. Machine learning models predict BBB-penetrant peptide sequences for nanoparticle efficacy to enhance delivery efficiency. It may further uncover patient-specific therapeutic combinations by integrating genetic, proteomic, and clinical data. As these technologies converge, the integration of gene editing, nanotechnology, neuromodulation, and computational biology will accompany a multimodal neurotherapeutics transforming how we treat neurodegenerative diseases from monotherapy to systems-level intervention [76].

15.8 Integrating Autophagy, Gene Editing, and Neuroinflammation

Nanoparticles enable dual drug delivery across the BBB via receptor-mediated transcytosis, though limited endosomal escape restricts cytosolic access. Interestingly, AAV-mediated TFEB overexpression enhances autophagy and reduces neuroinflammation by modulating lysosomal–mTOR signaling. Furthermore, CRISPR correction of the gene prevents inflammasome activation, shifting focus to disease-modifying therapies. Additionally, optogenetic autophagy control offers precise regulation but faces translational barriers due to invasive light delivery, highlighting the need for less invasive stimulation methods [73]. Despite advances, key uncertainties persist. Although autophagy is generally neuroprotective, in contexts like C9ORF72-linked ALS, excessive flux may worsen neuroinflammation or aggregation. Age-related lysosomal decline may also limit nanoparticle or AAV efficacy in older patients [77]. These challenges highlight the need for tailored, context-specific strategies, as universal autophagy modulation is unlikely to suit all neurodegenerative diseases. The interplay between autophagy and neuroinflammation is important to link TFEB lysosomal biogenesis with NLRP3 suppression, which suggests autophagy inducers act partly through immunomodulation [78]. CRISPR correction of genes restores mitophagy and lowers mitochondrial reactive oxygen species-driven inflammasome activation to connect proteostasis, metabolism, and inflammation. However, the long-term efficacy of such interventions, chiefly amid aging-related decline, remains uncertain and demands further validation. Future strategies may benefit from multimodal approaches, such as combining AAV-TFEB delivery with CRISPR correction or nanoparticle-based autophagy inducers with optogenetic activation, to synergistically target multiple disease pathways. Approaches should overcome barriers like BBB penetration and cell-type specificity, and enhance delivery platforms such as sonogenetics. Next, artificial intelligence-driven modeling may further optimize therapies by integrating variables like aging and disease progression. This will shift toward systems-level interventions that address neurodegeneration as a dynamic, multifactorial network.

15.9 Conclusion

Neurodegenerative diseases demand integrated strategies targeting interconnected pathologies. This chapter underscores the necessity of combining nanoparticle delivery, viral gene therapy, CRISPR editing, and optogenetics to overcome biological barriers and achieve precise interventions. Nanoparticles enable dual cargo delivery across the BBB, whereas AAV-mediated TFEB amplification enhances lysosomal clearance and suppresses inflammation. CRISPR corrects causal mutations, and optogenetics offers spatiotemporal control of microglial responses. Critically, synergies emerge when these modalities converge, such as nanoparticle-facilitated CRISPR delivery with TFEB overexpression, yielding enhanced aggregate clearance and immunomodulation. Persistent challenges include delivery inefficiency, immunogenicity, and off-target effects. Future progress hinges on multimodal integration, patient stratification, and emerging innovations like sonogenetics, artificial intelligence-optimized carriers, and transient editing systems. By addressing both molecular etiology and dynamic disease networks, these approaches herald a shift from symptomatic management toward precision disease modification.

References

1. Barmaki H, Nourazarian A, Khaki-Khatibi F. Proteostasis and neurodegeneration: a closer look at autophagy in Alzheimer's disease. Front Aging Neurosci. 2023;15:1281338.
2. Nayak V, et al. Regulation of neuroinflammation in Alzheimer's disease via nanoparticle-loaded phytocompounds with anti-inflammatory and autophagy-inducing properties. Phytomedicine. 2024;122:155150.
3. Gomez Limia C, et al. Emerging perspectives on gene therapy delivery for neurodegenerative and neuromuscular disorders. J Pers Med. 2022;12(12):1979.
4. Voicu V, et al. Decoding neurodegeneration: a comprehensive review of molecular mechanisms, genetic influences, and therapeutic innovations. Int J Mol Sci. 2023;24(16):13006.
5. Paul A, Collins MG, Lee HY. Gene therapy: the next-generation therapeutics and their delivery approaches for neurological disorders. Front Genome Ed. 2022;4:899209.
6. Zha S, et al. Functionalized nanomaterials capable of crossing the blood-brain barrier. ACS Nano. 2024;18(3):1820–45.
7. Zhao N, et al. Microglia-targeting nanotherapeutics for neurodegenerative diseases. APL Bioeng. 2020;4(3):030902.
8. Clark EH, et al. Targeting mitophagy in Parkinson's disease. J Biol Chem. 2021;296:100209.
9. Masaldan S, Callegari S, Dewson G. Therapeutic targeting of mitophagy in Parkinson's disease. Biochem Soc Trans. 2022;50(2):783–97.
10. Zhao YJ, et al. Neurotropic virus infection and neurodegenerative diseases: potential roles of autophagy pathway. CNS Neurosci Ther. 2024;30(6):e14548.
11. Rahman MU, et al. CRISPR-Cas9-based technology and its relevance to gene editing in Parkinson's disease. Pharmaceutics. 2022;14(6):1252.
12. Cho WH, Barcelon E, Lee SJ. Optogenetic glia manipulation: possibilities and future prospects. Exp Neurobiol. 2016;25(5):197–204.
13. Liang Y, et al. Optogenetic neuromodulation in inflammatory pain. Neuroscience. 2024;536: 104–18.
14. Taha EA, Lee J, Hotta A. Delivery of CRISPR-Cas tools for in vivo genome editing therapy: trends and challenges. J Control Release. 2022;342:345–61.
15. Hakim CH, et al. Cas9-specific immune responses compromise local and systemic AAV CRISPR therapy in multiple dystrophic canine models. Nat Commun. 2021;12(1):6769.
16. Foley RA, et al. Delivering the CRISPR/Cas9 system for engineering gene therapies: recent cargo and delivery approaches for clinical translation. Front Bioeng Biotechnol. 2022;10: 973326.
17. Beckers J, Tharkeshwar AK, Van Damme P. C9orf72 ALS-FTD: recent evidence for dysregulation of the autophagy-lysosome pathway at multiple levels. Autophagy. 2021;17 (11):3306–22.
18. Cortes CJ, La Spada AR. TFEB dysregulation as a driver of autophagy dysfunction in neurodegenerative disease: molecular mechanisms, cellular processes, and emerging therapeutic opportunities. Neurobiol Dis. 2019;122:83–93.

19. Cunningham KM, et al. TFEB/Mitf links impaired nuclear import to autophagolysosomal dysfunction in C9-ALS. elife. 2020;9:e59419.
20. Decressac M, et al. TFEB-mediated autophagy rescues midbrain dopamine neurons from alpha-synuclein toxicity. Proc Natl Acad Sci USA. 2013;110(19):E1817–26.
21. Diab R, Pilotto F, Saxena S. Autophagy and neurodegeneration: unraveling the role of C9ORF72 in the regulation of autophagy and its relationship to ALS-FTD pathology. Front Cell Neurosci. 2023;17:1086895.
22. Tang X, et al. Divergence, convergence, and therapeutic implications: a cell biology perspective of C9ORF72-ALS/FTD. Mol Neurodegener. 2020;15(1):34.
23. Pouliopoulos AN, et al. Non-invasive optogenetics with ultrasound-mediated gene delivery and red-light excitation. Brain Stimul. 2022;15(4):927–41.
24. Liu R, Zuo R, Hudalla GA. Harnessing molecular recognition for localized drug delivery. Adv Drug Deliv Rev. 2021;170:238–60.
25. Jimenez A, et al. Nanotechnology to overcome blood-brain barrier permeability and damage in neurodegenerative diseases. Pharmaceutics. 2025;17(3):281.
26. Pupyshev AB, et al. Therapeutic activation of autophagy by combined treatment with rapamycin and trehalose in a mouse MPTP-induced model of Parkinson's disease. Pharmacol Biochem Behav. 2019;177:1–11.
27. Zhang D, et al. Drug-loaded PEG-PLGA nanoparticles for cancer treatment. Front Pharmacol. 2022;13:990505.
28. Huang N, et al. PLGA nanoparticles modified with a BBB-penetrating peptide co-delivering Abeta generation inhibitor and curcumin attenuate memory deficits and neuropathology in Alzheimer's disease mice. Oncotarget. 2017;8(46):81001–13.
29. Gu Z, et al. TFEB in Alzheimer's disease: from molecular mechanisms to therapeutic implications. Neurobiol Dis. 2022;173:105855.
30. Lin J, et al. TFEB agonist clomiphene citrate activates the autophagy-lysosomal pathway and ameliorates Alzheimer's disease symptoms in mice. J Biol Chem. 2024;300(12):107929.
31. Molle LM, et al. Nanoparticles for vaccine and gene therapy: overcoming the barriers to nucleic acid delivery. Wiley Interdiscip Rev Nanomed Nanobiotechnol. 2022;14(6):e1809.
32. Yang Z, et al. pH and redox dual-responsive multifunctional gene delivery with enhanced capability of transporting DNA into the nucleus. Colloids Surf B Biointerfaces. 2017;153:111–22
33. Wang S. pH-responsive amphiphilic carboxylate polymers: design and potential for endosomal escape. Front Chem. 2021;9:645297.
34. Arotcarena ML, et al. Transcription factor EB overexpression prevents neurodegeneration in experimental synucleinopathies. JCI Insight. 2019;4(16):e129719.
35. Kilpatrick K, et al. Genetic and chemical activation of TFEB mediates clearance of aggregated alpha-synuclein. PLoS One. 2015;10(3):e0120819.
36. Xia Q, et al. TDP-43 loss of function increases TFEB activity and blocks autophagosome-lysosome fusion. EMBO J. 2016;35(2):121–42.
37. Ivankovic D, et al. Mitochondrial and lysosomal biogenesis are activated following PINK1/Parkin-mediated mitophagy. J Neurochem. 2016;136(2):388–402.
38. Nezich CL, et al. MiT/TFE transcription factors are activated during mitophagy downstream of Parkin and Atg5. J Cell Biol. 2015;210(3):435–50.
39. Chan KY, et al. Engineered AAVs for efficient noninvasive gene delivery to the central and peripheral nervous systems. Nat Neurosci. 2017;20(8):1172–9.
40. Deverman BE, et al. Cre-dependent selection yields AAV variants for widespread gene transfer to the adult brain. Nat Biotechnol. 2016;34(2):204–9.
41. Galvan A, et al. Intracerebroventricular administration of AAV9-PHP.B SYN1-EmGFP induces widespread transgene expression in the mouse and monkey central nervous system. Hum Gene Ther. 2021;32(11–12):599–615.
42. Jiang S, Bhaskar K. Degradation and transmission of tau by autophagic-endolysosomal networks and potential therapeutic targets for tauopathy. Front Mol Neurosci. 2020;13:586731.

43. Wang H, et al. TFEB overexpression in the P301S model of tauopathy mitigates increased PHF1 levels and lipofuscin puncta and rescues memory deficits. eNeuro. 2016;3(2): ENEURO.0042-16.2016.
44. Charmpilas N, Fang EF, Palikaras K. Mitophagy and neuroinflammation: a compelling interplay. Curr Neuropharmacol. 2023;21(7):1477–81.
45. Sliter DA, et al. Parkin and PINK1 mitigate STING-induced inflammation. Nature. 2018;561 (7722):258–62.
46. Xiao B, et al. Reactive oxygen species trigger Parkin/PINK1 pathway-dependent mitophagy by inducing mitochondrial recruitment of Parkin. J Biol Chem. 2017;292(40):16697–708.
47. Mouton-Liger F, et al. Parkin deficiency modulates NLRP3 inflammasome activation by attenuating an A20-dependent negative feedback loop. Glia. 2018;66(8):1736–51.
48. Xu Y, et al. PINK1-mediated mitophagy protects against hepatic ischemia/reperfusion injury by restraining NLRP3 inflammasome activation. Free Radic Biol Med. 2020;160:871–86.
49. Ryu SM, et al. Adenine base editing in mouse embryos and an adult mouse model of Duchenne muscular dystrophy. Nat Biotechnol. 2018;36(6):536–9.
50. Zeng J, et al. Therapeutic base editing of human hematopoietic stem cells. Nat Med. 2020;26(4): 535–41.
51. Savell KE, et al. A neuron-optimized CRISPR/dCas9 activation system for robust and specific gene regulation. eNeuro. 2019;6(1):ENEURO.0495-18.2019.
52. Gisler S, et al. Multiplexed Cas9 targeting reveals genomic location effects and gRNA-based staggered breaks influencing mutation efficiency. Nat Commun. 2019;10(1):1598.
53. Chung SY, et al. Parkin and PINK1 patient iPSC-derived midbrain dopamine neurons exhibit mitochondrial dysfunction and alpha-synuclein accumulation. Stem Cell Reports. 2016;7(4): 664–77.
54. Li H, et al. Co-editing PINK1 and DJ-1 genes via adeno-associated virus-delivered CRISPR/ Cas9 system in adult monkey brain elicits classical Parkinsonian phenotype. Neurosci Bull. 2021;37(9):1271–88.
55. Salomonsson SE, Clelland CD. Building CRISPR gene therapies for the central nervous system: a review. JAMA Neurol. 2024;81(3):283–90.
56. Lautrup S, et al. NAD(+) in brain aging and neurodegenerative disorders. Cell Metab. 2019;30 (4):630–55.
57. Aman Y, et al. The NAD(+)-mitophagy axis in healthy longevity and in artificial intelligence-based clinical applications. Mech Ageing Dev. 2020;185:111194.
58. Ilieva NM, et al. LRRK2 kinase inhibition protects against Parkinson's disease-associated environmental toxicants. Neurobiol Dis. 2024;196:106522.
59. Kempthorne L, et al. Dual-targeting CRISPR-CasRx reduces C9orf72 ALS/FTD sense and antisense repeat RNAs in vitro and in vivo. Nat Commun. 2025;16(1):459.
60. McCallister TX, et al. A high-fidelity CRISPR-Cas13 system improves abnormalities associated with C9ORF72-linked ALS/FTD. Nat Commun. 2025;16(1):460.
61. Braems E, Swinnen B, Van Den Bosch L. C9orf72 loss-of-function: a trivial, stand-alone or additive mechanism in C9 ALS/FTD? Acta Neuropathol. 2020;140(5):625–43.
62. Lim J, et al. Proteotoxic stress induces phosphorylation of p62/SQSTM1 by ULK1 to regulate selective autophagic clearance of protein aggregates. PLoS Genet. 2015;11(2):e1004987.
63. Gittings LM, Sattler R. Recent advances in understanding amyotrophic lateral sclerosis and emerging therapies. Fac Rev. 2020;9:12.
64. Liu Y, et al. REPAIRx, a specific yet highly efficient programmable A > I RNA base editor. EMBO J. 2020;39(22):e104748.
65. Boivin M, et al. Reduced autophagy upon C9ORF72 loss synergizes with dipeptide repeat protein toxicity in G4C2 repeat expansion disorders. EMBO J. 2020;39(4):e100574.
66. Cavazza A, et al. Advanced delivery systems for gene editing: a comprehensive review from the GenE-HumDi COST Action Working Group. Mol Ther Nucleic Acids. 2025;36(1):102457.
67. Arrant AE, et al. Restoring neuronal progranulin reverses deficits in a mouse model of frontotemporal dementia. Brain. 2017;140(5):1447–65.

68. Krabbe G, et al. Microglial NFkappaB-TNFalpha hyperactivation induces obsessive-compulsive behavior in mouse models of progranulin-deficient frontotemporal dementia. Proc Natl Acad Sci USA. 2017;114(19):5029–34.
69. Sevigny J, et al. Progranulin AAV gene therapy for frontotemporal dementia: translational studies and phase 1/2 trial interim results. Nat Med. 2024;30(5):1406–15.
70. Liu J, et al. Microglial TMEM119 binds to amyloid-beta to promote its clearance in an Abeta-depositing mouse model of Alzheimer's disease. Immunity. 2025;58(7):1830–1846.e7.
71. Hyung S, Park JH, Jung K. Application of optogenetic glial cells to neuron-glial communication. Front Cell Neurosci. 2023;17:1249043.
72. Delaney SL, et al. Optogenetic modulation for the treatment of traumatic brain injury. Stem Cells Dev. 2020;29(4):187–97.
73. Gropman AL, Uittenbogaard MN, Chiaramello AE. Challenges and opportunities to bridge translational to clinical research for personalized mitochondrial medicine. Neurotherapeutics. 2024;21(1):e00311.
74. Thapar N, et al. Application of CRISPR/Cas9 in the management of Alzheimer's disease and Parkinson's disease: a review. Ann Med Surg (Lond). 2024;86(1):329–35.
75. Lau CH, Suh Y. In vivo genome editing in animals using AAV-CRISPR system: applications to translational research of human disease. F1000Res. 2017;6:2153.
76. Noury H, et al. AI-driven innovations in smart multifunctional nanocarriers for drug and gene delivery: a mini-review. Crit Rev Oncol Hematol. 2025;210:104701.
77. Nassif M, Woehlbier U, Manque PA. The enigmatic role of C9ORF72 in autophagy. Front Neurosci. 2017;11:442.
78. Chen J, et al. p38-TFEB pathways promote microglia activation through inhibiting CMA-mediated NLRP3 degradation in Parkinson's disease. J Neuroinflammation. 2021;18(1):295.

Lifestyle, Diet, and the Gut–Brain Axis in Neuroprotection

16

Preeti Sharma, Neha Verma, Anushka Verma, Ankita Wal, Sudhir Kumar, and Prabhat Upadhyay

Abstract

Neurodegenerative disorders, including Alzheimer's, Parkinson's, Huntington's disease, and amyotrophic lateral sclerosis, are marked by the gradual degeneration of neurons, ultimately resulting in significant impairments in cognition and motor function. Despite growing knowledge of their molecular mechanisms, including oxidative stress, mitochondrial dysfunction, and protein aggregation, available treatments are largely symptomatic and fail to halt disease progression. Emerging evidence highlights the critical role of lifestyle factors in modulating neurodegeneration through the gut–brain axis linking gut microbiota, immune system, and central nervous system. Diet strongly influences gut microbial composition, producing metabolites like short-chain fatty acids and neurotransmitter precursors that affect brain function. Disruption of this axis due to poor nutrition, inactivity, or stress contributes to neuroinflammation and cognitive decline. This chapter explores how dietary patterns such as the Mediterranean diet support neuroprotection by enhancing GABAergic signaling, reducing

P. Sharma · N. Verma · A. Wal
Pranveer Singh Institute of Technology (Pharmacy), Bhauti, Kanpur, Uttar Pradesh, India
e-mail: preeti.sharma@psit.ac.in; neha.verma@psit.ac.in; ankita.wal@psit.ac.in

A. Verma
Faculty of Pharmaceutical Sciences, Rama University, Madhana, Kanpur, Uttar Pradesh, India

S. Kumar
Department of Microbiology and Immunology, School of Medicine, Emory University, Atlanta, GA, USA
e-mail: sudhir.kumar@emory.edu

P. Upadhyay (✉)
Vincent Center for Reproductive Biology, Department of Obstetrics and Gynaecology, Massachusetts General Hospital, Harvard Medical School, Boston, MA, USA

P. Khanal et al. (eds.), *Neuroinflammation and Autophagy in Neurodegeneration*,
https://doi.org/10.1007/978-981-95-3945-1_16

oxidative stress, and promoting neurotrophic factors. It also discusses the promise of bioactive phytochemicals, functional foods, and microbiota-targeted therapies. Integrating diet, exercise, and microbial modulation offers a holistic, non-pharmacological strategy to delay onset, alleviate symptoms, and improve quality of life in aging populations at risk for neurodegenerative disorders.

Keywords

Dietary intervention · Gut · brain axis · Microbiota modulation · Neuroprotection

16.1 Introduction

Neurodegenerative diseases are marked by the gradual loss of neurons and encompass a range of progressive neuropsychiatric conditions such as Alzheimer's disease (AD), multiple sclerosis (MS), Parkinson's disease (PD), amyotrophic lateral sclerosis (ALS), Huntington's disease (HD), and other related disorders [1]. Neurological diseases typically occur in later life due to progressive neuron loss and synaptic network dysfunction. Recent research identified autophagy as a crucial intracellular neuroprotective process that maintains neuronal integrity. Autophagy is a conserved lysosomal pathway that degrades and removes damaged organelles and misfolded proteins. The accumulation of β-amyloid, α-synuclein, and huntingtin aggregates is a characteristic feature of diseases arising from defective or reduced autophagic flux. The enhancement of autophagy has been shown to reduce neuroinflammation. Moreover, it may mitigate oxidative damage and restore neuronal plasticity [2–4], and is the primary cause of physical and cognitive impairment globally. It has been indicated that neurodegenerative diseases have a restricted survival rate. For example, AD is recognized as the sixth largest cause of mortality globally. As of 2021, approximately 11.77 million people worldwide were living with Parkinson's disease (PD), marking a dramatic rise from just 3.15 million cases in 1990, a nearly 275% increase in prevalence. Disability-adjusted life years (DALYs) attributable to PD climbed even more sharply, reaching about 7.47 million in 2021, a 162% increase since 1990. Moreover, the total number of patients has significantly increased during the previous 30 years, and the burden of chronic neurodegenerative disorders is anticipated to at least double in the next two decades. The transformation is primarily driven by an increasing aging population, which presents an important barrier to ensuring accessible neurological care for all individuals. This neuronal degeneration results in several neuropsychiatric disorders and enduring impairment in a person. The molecular pathways responsible for neuronal loss and neurodegeneration have garnered significant research during the last decade. For example, neurodegenerative diseases possess distinct pathologies, but they involve common mechanisms such as protein aggregation, oxidative damage, neuroinflammation, apoptosis, and mitochondrial dysfunction, leading to neuronal loss [5–7]. Recent investigations have concentrated on formulating neuroprotective interventions to address each of these processes. It comprises neuroprotective compounds to safeguard against neuronal damage resulting from acute conditions or associated with chronic neurodegenerative diseases [8, 9]. Thus, it is important to understand that the early mitigation of neurodegenerative hallmarks (e.g., inflammation, oxidative stress, immune dysfunction, protein aggregation, and apoptosis) could delay neuronal dysfunction. The underlying problem is that commercially accessible medications may mitigate symptoms of several neurodegenerative disorders, although their prolonged usage often correlates with severe adverse effects. Furthermore, none of these therapeutic agents halt the advancement of the degenerative process [10–12].

Recent epidemiological and clinical research indicates that signaling pathways are involved in autophagy control, with dietary elements like resveratrol, curcumin, and intermittent fasting influencing these pathways. Lifestyle modifications, such as dietary adjustments and increased physical activity, provide a viable approach to mitigate and avoid cognitive decline and dementia. Next, autophagy activation may serve as a promising therapeutic strategy in the early stages of neurodegeneration, while also working in concert with gut–brain axis mechanisms to maintain central nervous system (CNS) homeostasis [13, 14]. These lifestyle variables interact with biological systems associated with dementia, including oxidative damage, inflammation of the nervous system, mitochondrial impairment, and neurotrophic signaling networks [15, 16]. Research highlights the critical role of the gut–brain axis in enabling these effects. The gut–brain axis is a two-way communication system that includes the central and enteric nervous systems, the immune system, endocrine signals, and the gut flora [17]. Dietary components and lifestyle habits strongly influence the structure and function of gut microbiota. This produces metabolites that affect brain homeostasis, and disruption of this axis resulting from insufficient diet, inactivity, or chronic stress increases neuroinflammation, impaired neurogenesis, and accelerated cognitive decline [18–20]. Alternative dietary strategies and eating patterns are increasingly emerging as advantageous for cognitive function, realizing lifestyle changes, including nutrition, as a feasible adjunctive treatment to mitigate cognitive loss. The Mediterranean and MIND diets have biological effects that protect brain function, and similar consistent physical exercise has been shown to promote regeneration and elevate brain-derived neurotrophic factor (BDNF) levels, which is important in sustaining neuronal plasticity and viability [21–23]. In this chapter, we explore the role of nutrition and lifestyle in supporting brain health through neuroprotection through gut–brain axis modulation.

16.2 The Gut–Brain Axis: A Neuroprotective Pathway

The gut–brain axis represents a complex, bidirectional communication system linking the gastrointestinal tract and the central nervous system through neural, immune, and endocrine pathways. Disruptions in this axis have been observed in several neurodegenerative disorders, such as AD, PD, and ALS, suggesting its possible role in disease pathogenesis. The gut microbiota is a crucial element of the gut–brain axis, and gut dysbiosis has been linked to dementia. Moreover, the gut microbiota may affect CNS homeostasis by modifying the immune system and directly controlling the metabolites that impact the neurological and endocrine systems [24]. For example, gut microbiota influences the development, function, and diseases of the CNS and enteric nervous systems by interacting and activating pattern recognition receptors [24–29]. Clinical, epidemiological, and immunological studies highlight that enteric microbiota profoundly affect gut–brain assembly, including psychological behavior, emotional regulation, neuromuscular function, and hypothalamic–pituitary–adrenal (HPA) axis regulation. Hence, it is important to clarify the mechanisms behind these effects on both emotional and cognitive brain centers, and it has been found that these variations in microbiota are associated with changes in communication networks. A wide range of mood disorders, such as anxiety, depression, and autism spectrum disorders, are closely linked to functional disturbances in the gastrointestinal system. Conversely, gastrointestinal conditions like irritable bowel syndrome (IBS) and inflammatory bowel disease (IBD) are often accompanied by psychological comorbidities, which have been associated with changes in the gut microbiota [30–35].

The gut microbiota plays a vital role within the gut–brain axis, and its imbalance, known as dysbiosis, has been associated with neurodegenerative processes. It can influence CNS homeostasis by modulating immune responses and regulating the production of metabolites that impact both neurological and endocrine functions [36, 37]. Structural components of gut bacteria, such as lipopolysaccharides, typically provide low-level stimulation to the innate immune system. However, disruptions like dysbiosis, small intestinal bacterial overgrowth, or increased intestinal permeability can lead to excessive immune activation, triggering systemic and CNS inflammation. Bacterial proteins may mimic host antigens, provoking abnormal adaptive immune responses. In addition, bacterial enzymes produce both harmful metabolites, such as D-lactic acid and ammonia, and beneficial ones, like short-chain fatty acids. Gut microbes also possess receptors for host hormones, influencing their own growth and pathogenic potential. Notably, certain gut bacteria can stimulate afferent neurons of the enteric nervous system, transmitting signals to the brain via the vagus nerve. They also impact sleep patterns and modulate the hypothalamic–pituitary–adrenal (HPA) axis, thereby affecting cognition, memory, and emotional regulation [38–41].

16.3 Dietary Intervention in Modulating the Inhibitory Neurotransmitter

The main inhibitory neurotransmitter is gamma-aminobutyric acid (GABA). It is a key player in mediating and regulating a wide range of CNS functions and is widely distributed and abundant in the brain and spinal cord. GABA and glutamate have inhibitory and excitatory effects, respectively, and are balanced in the mature brain; this homeostasis is essential for homeostatic brain function. GABAergic transmission dysregulation is linked to multiple neuropsychiatric and neurodegenerative conditions, such as AD, PD, anxiety, depression, and epilepsy [42–46]. Although traditional pharmacological treatments like benzodiazepines, barbiturates, and

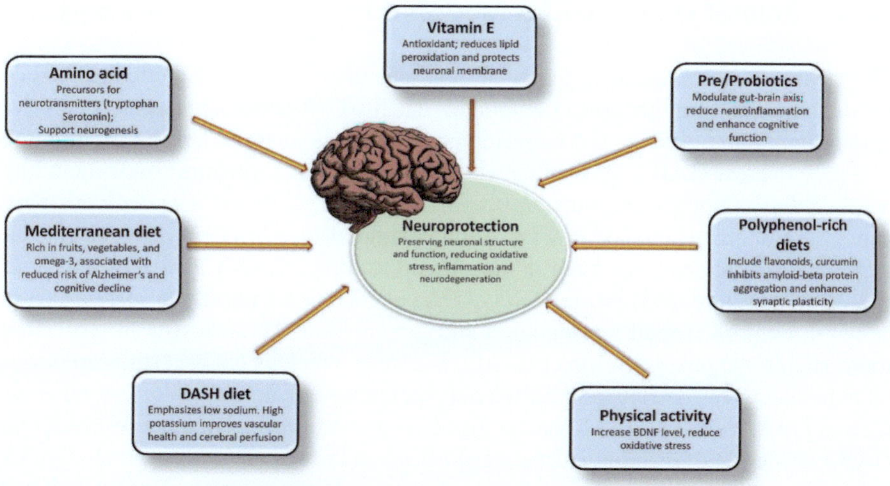

Fig. 16.1 Dietary and lifestyle interventions contribute to neuroprotection by preserving neuronal structure and function. Amino acids act as precursors for neurotransmitters and play a crucial role in supporting neurogenesis. Vitamin E functions as a potent antioxidant. Pre- and probiotics modulate gut–brain axis, enhancing cognitive function. Polyphenol-rich diets, including compounds such as flavonoids and curcumin, inhibit Aβ aggregation and promote synaptic plasticity. Physical activity enhances BDNF levels and lowers oxidative stress. The DASH diet increases vascular health and cerebral perfusion. Meanwhile, the Mediterranean diet has been associated with a reduced risk of AD and cognitive decline

GABA analogs are pharmacologically effective, they exhibit side effects like tolerance, dependency, sedation, and withdrawal symptoms [47]. As a result, non-pharmacological methods such as dietary changes have drawn attention as sustainable and safer ways to increase GABAergic activity. Numerous dietary sources and bioactive compounds have potential in influencing GABAergic function by either enhancing GABA synthesis, inhibiting its degradation, or modulating GABA-A receptor activity. For example, GABA-enriched foods such as germinated brown rice, fermented dairy products, and GABA tea have been shown to improve mood, sleep, and blood pressure control [48]. Furthermore, as positive allosteric modulators of GABA-A receptors, natural phytochemicals demonstrate sedative and anxiolytic effects [49–51]. Interestingly, endogenous GABA synthesis and receptor function are supported by dietary precursors such as glutamate and vitamins B6 and D. Certain probiotics and prebiotics have been shown to influence gut microbiota production of GABA (Fig. 16.1), highlighting the gut–brain axis as a pathway through which dietary components exert neuromodulatory effects [24, 52–54].

16.4 Role of Diet in GABA Modulation

The synthesis of neurotransmitters and brain chemistry is fundamentally influenced by diet. Dietary polyphenolic compounds such as vitamins and amino acids act either as modulators of the GABA receptor or as precursors for GABA biosynthesis. The precursor of GABA, glutamate, is found in large quantities in foods including whole grains, legumes, and nuts. Furthermore, the enzyme glutamate decarboxylase converts glutamate into GABA by using vitamin B6 as a coenzyme. Moreover, vitamin B6 deficiency has been demonstrated to affect GABA synthesis and raise anxiety and seizures [55]. Moreover, GABA can be found naturally in some foods or added through microbial fermentation and germination. This consists of fermented foods such as yogurt, kefir, kimchi, natto, and miso, whereas grain-germinated foods include buckwheat, barley, and brown rice, including *Lactobacillus*-fermented black tea and GABA oolong tea. However, processing conditions have a major impact on GABA content. For example, lactic acid bacteria in fermented foods promote GABA synthesis, whereas germination activates glutamate decarboxylase, which transforms glutamate to GABA. Thus, adequate protein intake (glutamate precursors) and micronutrients like vitamin B6, magnesium, and zinc are crucial for optimal GABA synthesis. Certain dietary polyphenols and terpenoids modulate GABA receptors allosterically, mimicking the effects of pharmaceutical GABA agonists with fewer side effects. Flavonoids such as apigenin (found in parsley and chamomile), kaempferol (in broccoli and spinach), and baicalin (in *Scutellaria baicalensis*) have been shown to bind the benzodiazepine site of GABA receptors [56–58]. These findings suggest the therapeutic potential of diet-derived compounds in sleep disorders and anxiety.

The gut–brain axis has emerged as a key pathway through which diet influences GABA metabolism. Certain gut bacteria, predominantly *Lactobacillus* and *Bifidobacterium* species, produce GABA through glutamate decarboxylation. Diets rich in prebiotics and fermented foods can enhance the abundance and activity of these GABA-producing microbes [18, 59, 60]. Furthermore, microbial-derived GABA stimulates gut vagal afferents, influencing mood and anxiety via signaling to the central nervous system, even without crossing the blood-brain barrier [61].

Beyond specific nutrients or foods, overall dietary patterns also influence GABAergic function. For example, diets rich in whole grains, fibers, fruits, and the Mediterranean diet are associated with enhanced mental well-being and balanced neurotransmission. These diets support GABA biosynthesis, promote gut microbial diversity, and supply essential cofactors. Conversely, high-fat, high-sugar western diets have been linked to reduced GABA receptor expression and increased anxiety-like behaviors [62, 63]. Moreover, GABA can be added to a variety of foods through fermentation and processing, or it can be found naturally. Seaweeds, fermented milk, soy products, GABA-enriched tea, and germinated brown rice are a few examples. Consumption of these foods can raise GABA levels and reduce anxiety, hypertension, and insomnia symptoms [64–66]. Emerging research highlights the effect of the gut–brain axis in modulating GABA activity. Certain gut-resident bacteria, including *Lactobacillus* and *Bifidobacterium* species, produce GABA through glutamate decarboxylation, and dietary components such as prebiotics and fermented foods have been shown to support these GABA-producing microbes [60, 64, 67–69]. Various phytochemicals act as natural modulators of GABA receptors or enzymes involved in GABA metabolism. Flavonoids, terpenoids, and lignans enhance GABAergic transmission by modulating GABA-A receptor subtypes or inhibiting GABA transaminase. A compound derived from hops, Humulone, has been shown to reduce the time it takes to fall asleep by enhancing the activity of GABA-A receptors, thereby promoting calming and sleep-inducing effects through increased inhibitory signaling in the brain [70].

Although GABA does not efficiently cross the blood-brain barrier, oral GABA and dietary GABAergic compounds may exert central effects through vagal afferent signaling, gut hormones, and immune modulation. It has been shown that GABA supplementation modulates stress hormones, improves mood, and suppresses food intake through peripheral mechanisms. In rodent models, co-administration of GABA with GABA transaminase inhibitors produced weight-loss effects similar to GLP-1 analogs like semaglutide [71–73].

16.5 Lifestyle Modification Engaging Neuroprotection

A growing public health concern in aging populations is the prevalence of neurodegenerative diseases, and current pharmaceutical treatments are effective in managing symptoms, but they are not very effective at intervening in disease progression. Because of this constraint, there is now more interest in lifestyle changes that affect brain aging and provide neuroprotective advantages, such as diet, exercise, cognitive training, sleep control, and stress management. Through multiple mechanisms that affect inflammation, oxidative stress, synaptic plasticity, and mitochondrial function, integrative lifestyle changes may lower neurodegenerative diseases or slow their progression [74]. A move toward preventive measures that go beyond symptomatic treatment is required due to the growing worldwide burden of neurodegenerative diseases, including AD, MS, PD, and other types of dementia. The World Health Organization estimates that if effective interventions are not found, the number of people with dementia, which currently stands at over 55 million, will triple by 2050. There is currently no cure or way to considerably slow the progression of neurodegenerative diseases, even though pharmacological developments have provided some relief in symptom control. Due to this, it is clear that modifiable lifestyle factors are effective as non-invasive neuroprotective measures. By lowering risk factors such as oxidative stress, chronic inflammation, mitochondrial dysfunction, vascular impairment, and protein misfolding, which are key contributors to neurodegenerative diseases, lifestyle changes can enhance neuroprotection. Unlike targeted drug therapies, lifestyle interventions exert broad, systemic effects on brain function through mechanisms like gut–brain axis regulation, synaptic plasticity, and epigenetic modulation.

Among various lifestyle factors, physical activity is the most consistently associated with improved cognitive outcomes across the lifespan. Regular aerobic exercise promotes BDNF release, which is a key molecule involved in synaptic plasticity, neurogenesis, and neuronal survival. BDNF levels decline with age and in neurodegenerative diseases, but can be upregulated following consistent physical activity [75, 76]. In addition, exercise enhances cerebral blood flow, reduces oxidative stress, and promotes neurovascular coupling, which contribute to improved cognitive performance. A meta-analysis of randomized controlled studies showed both aerobic and resistance exercises significantly enhanced memory, executive function, and processing speed in older adults [77–79]. Furthermore, neuroimaging studies revealed that physical activity leads to increases in hippocampal volume. These structural changes were positively correlated with functional outcomes, including better memory consolidation and slower rates of cognitive decline.

16.6 Nutritional Patterns and Cognitive Resilience

Dietary patterns significantly influence brain function through multiple biological pathways, including oxidative stress regulation, inflammation modulation, lipid metabolism, and mitochondrial efficiency. The Mediterranean diet is rich in omega-3 fatty acids, monounsaturated fats, polyphenols, and plant-based foods, and has consistently been associated with lower rates of cognitive decline and reduced incidence of dementia. The MIND diet is a hybrid of Mediterranean diet and DASH diet that was designed for neuroprotection. Longitudinal studies showed that adherence to MIND diet is associated with a 53% reduction in AD risk, even when adherence is moderate [80, 81]. Moreover, activities such as cognitive training, social interaction, and lifelong learning help strengthen cognitive reserve, improving the brain's capacity to adapt and preserve cognitive function in the face of aging and neurodegenerative changes [82].

16.7 Conclusion

The complex and diverse nature of neurodegenerative diseases needs comprehensive and preventative strategies that go beyond typical pharmaceutical therapies. This chapter highlights the critical role of lifestyle modifications, including diet, exercise, and gut microbiota control, in modulating neuroprotective pathways. The emerging link between gut–brain axis and CNS function emphasizes the need to maintain microbial balance through prebiotics, probiotics, and GABA-rich diets. Dietary patterns demonstrated considerable benefits in enhancing GABAergic signaling, lowering oxidative stress, and reducing neuroinflammation, which are major contributors to dementia. Furthermore, bioactive phytochemicals and functional foods showed promising results in modulating GABAergic activity and improving psychological well-being, suggesting a possible alternative to prescription GABA agonists. The joint effect of gut microbiota on neurotransmitter homeostasis and cognitive performance suggests new therapeutic options in the development of personalized dietary regimens. Further study should focus on longitudinal clinical investigations to identify causal relationships between eating habits, microbiota composition, and neurodegenerative disorders.

References

1. Wilson DM, Cookson MR, Van Den Bosch L, Zetterberg H, Holtzman DM, Dewachter I. Hallmarks of neurodegenerative diseases. Cell. 2023;186(4):693–714.
2. Li Y-Y, Qin Z-H, Sheng R. The multiple roles of autophagy in neural function and diseases. Neurosci Bull. 2024;40(3):363–82.
3. Nah J, Yuan J, Jung Y-K. Autophagy in neurodegenerative diseases: from mechanism to therapeutic approach. Mol Cells. 2015;38(5):381–9.
4. Talebi M, Mohammadi Vadoud SA, Haratian A, Talebi M, Farkhondeh T, Pourbagher-Shahri AM, Samarghandian S. The interplay between oxidative stress and autophagy: focus on the development of neurological diseases. Behav Brain Funct. 2022;18(1):3.
5. Jellinger KA. Basic mechanisms of neurodegeneration: a critical update. J Cell Mol Med. 2010;14(3):457–87.
6. Dash UC, Bhol NK, Swain SK, Samal RR, Nayak PK, Raina V, et al. Oxidative stress and inflammation in the pathogenesis of neurological disorders: mechanisms and implications. Acta Pharm Sin B. 2025;15(1):15–34.
7. Lin M-M, Liu N, Qin Z-H, Wang Y. Mitochondrial-derived damage-associated molecular patterns amplify neuroinflammation in neurodegenerative diseases. Acta Pharmacol Sin. 2022;43(10):2439–47.
8. Zhou L, Tan F, Zhang X, Li Y, Yin W. Neuroprotection and mechanisms of ginsenosides in nervous system diseases: progress and perspectives. IUBMB Life. 2024;76(11):862–82.
9. Karvandi MS, Sheikhzadeh Hesari F, Aref AR, Mahdavi M. The neuroprotective effects of targeting key factors of neuronal cell death in neurodegenerative diseases: the role of ER stress, oxidative stress, and neuroinflammation. Front Cell Neurosci. 2023;17:1105247.
10. Lamptey RN, Chaulagain B, Trivedi R, Gothwal A, Layek B, Singh J. A review of the common neurodegenerative disorders: current therapeutic approaches and the potential role of nanotherapeutics. Int J Mol Sci. 2022;23(3):1851.
11. Jiang Q, Liu J, Huang S, Wang X-Y, Chen X, Liu G-H, et al. Antiageing strategy for neurodegenerative diseases: from mechanisms to clinical advances. Signal Transduct Target Ther. 2025;10(1):76.
12. Durães F, Pinto M, Sousa E. Old drugs as new treatments for neurodegenerative diseases. Pharmaceuticals. 2018;11(2):44.
13. Mitra S, Munni YA, Dash R, Sadhu T, Barua L, Islam MA, et al. Gut microbiota in autophagy regulation: new therapeutic perspective in neurodegeneration. Life. 2023;13(4):957.
14. Yao P, Han H. Advances in autophagy–lysosomal pathway and neurodegeneration via brain–gut axis. Biomedicine. 2025;13(6):1390.
15. Dominguez LJ, Veronese N, Vernuccio L, Catanese G, Inzerillo F, Salemi G, Barbagallo M. Nutrition, physical activity, and other lifestyle factors in the prevention of cognitive decline and dementia. Nutrients. 2021;13(11):4080.
16. Xue B, Waseem SMA, Zhu Z, Alshahrani MA, Nazam N, Anjum F, et al. Brain-derived neurotrophic factor: a connecting link between nutrition, lifestyle, and Alzheimer's disease. Front Neurosci. 2022;16:925991.
17. Suganya K, Koo B-S. Gut–brain axis: role of gut microbiota on neurological disorders and how probiotics/prebiotics beneficially modulate microbial and immune pathways to improve brain functions. Int J Mol Sci. 2020;21(20):7551.
18. Berding K, Vlckova K, Marx W, Schellekens H, Stanton C, Clarke G, et al. Diet and the microbiota–gut–brain axis: sowing the seeds of good mental health. Adv Nutr. 2021;12(4): 1239–85.

19. Hasan A, Scuderi SA, Capra AP, Giosa D, Bonomo A, Ardizzone A, Esposito E. An updated and comprehensive review exploring the gut–brain axis in neurodegenerative disorders and neurotraumas: implications for therapeutic strategies. Brain Sci. 2025;15(6):654.
20. Abildinova GZ, Benberin VV, Vochshenkova TA, Afshar A, Mussin NM, Kaliyev AA, et al. The gut-brain-metabolic axis: exploring the role of microbiota in insulin resistance and cognitive function. Front Microbiol. 2024;15:1463958.
21. Timlin D, McCormack JM, Kerr M, Keaver L, Simpson EE. The MIND diet, cognitive function, and well-being among healthy adults at midlife: a randomised feasibility trial. BMC Nutr. 2025;11(1):59.
22. Andreu-Caravaca L, Ramos-Campo DJ, Moncada-Jiménez J, Abellán-Aynés O, Rubio-Arias JÁ. Immediate and short-term effect of physical exercise on BDNF in multiple sclerosis patients: a systematic review and meta-analysis. Sci Rep. 2025;15(1):19696.
23. Petersson SD, Philippou E. Mediterranean diet, cognitive function, and dementia: a systematic review of the evidence. Adv Nutr. 2016;7(5):889–904.
24. Loh JS, Mak WQ, Tan LKS, Ng CX, Chan HH, Yeow SH, et al. Microbiota–gut–brain axis and its therapeutic applications in neurodegenerative diseases. Signal Transduct Target Ther. 2024;9(1):37.
25. Kabouridis PS, Pachnis V. Emerging roles of gut microbiota and the immune system in the development of the enteric nervous system. J Clin Invest. 2015;125(3):956–64.
26. Yuan C, He Y, Xie K, Feng L, Gao S, Cai L. Review of microbiota gut brain axis and innate immunity in inflammatory and infective diseases. Front Cell Infect Microbiol. 2023;13:1282431.
27. Keogh CE, Rude KM, Gareau MG. Role of pattern recognition receptors and the microbiota in neurological disorders. J Physiol. 2021;599(5):1379–89.
28. Ma Q, Xing C, Long W, Wang HY, Liu Q, Wang R-F. Impact of microbiota on central nervous system and neurological diseases: the gut-brain axis. J Neuroinflammation. 2019;16(1):53.
29. Kasarello K, Cudnoch-Jedrzejewska A, Czarzasta K. Communication of gut microbiota and brain via immune and neuroendocrine signaling. Front Microbiol. 2023;14:1118529.
30. Grau-Del Valle C, Fernández J, Solá E, Montoya-Castilla I, Morillas C, Bañuls C. Association between gut microbiota and psychiatric disorders: a systematic review. Front Psychol. 2023;14:1215674.
31. Ahmed GK, Ramadan HK-A, Elbeh K, Haridy NA. Bridging the gap: associations between gut microbiota and psychiatric disorders. Middle East Curr Psychiatry. 2024;31(1):2.
32. Młynarska E, Barszcz E, Budny E, Gajewska A, Kopeć K, Wasiak J, et al. The gut–brain–microbiota connection and its role in autism spectrum disorders. Nutrients. 2025;17(7):1135.
33. Simpson CA, Mu A, Haslam N, Schwartz OS, Simmons JG. Feeling down? A systematic review of the gut microbiota in anxiety/depression and irritable bowel syndrome. J Affect Disord. 2020;266:429–46.
34. Marano G, Rossi S, Sfratta G, Traversi G, Lisci FM, Anesini MB, et al. Gut microbiota: a new challenge in mood disorder research. Life. 2025;15(4):593.
35. Mayer EA, Ryu HJ, Bhatt RR. The neurobiology of irritable bowel syndrome. Mol Psychiatry. 2023;28(4):1451–65.
36. Park KJ, Gao Y. Gut-brain axis and neurodegeneration: mechanisms and therapeutic potentials. Front Neurosci. 2024;18:1481390.
37. Schaible P, Henschel J, Erny D. How the gut microbiota impacts neurodegenerative diseases by modulating CNS immune cells. J Neuroinflammation. 2025;22(1):60.
38. Galland L. The gut microbiome and the brain. J Med Food. 2014;17(12):1261–72.
39. Salami M, Soheili M. The microbiota-gut-hippocampus axis. Front Neurosci. 2022;16:1065995.
40. Mahbub NU, Islam MM, Hong S-T, Chung H-J. Dysbiosis of the gut microbiota and its effect on α-synuclein and prion protein misfolding: consequences for neurodegeneration. Front Cell Infect Microbiol. 2024;14:1348279.

41. Bertollo AG, Santos CF, Bagatini MD, Ignácio ZM. Hypothalamus-pituitary-adrenal and gut-brain axes in biological interaction pathway of the depression. Front Neurosci. 2025;19: 1541075.
42. Teleanu RI, Niculescu A-G, Roza E, Vladâcenco O, Grumezescu AM, Teleanu DM. Neurotransmitters—key factors in neurological and neurodegenerative disorders of the central nervous system. Int J Mol Sci. 2022;23(11):5954.
43. Li Y, Sun H, Chen Z, Xu H, Bu G, Zheng H. Implications of GABAergic neurotransmission in Alzheimer's disease. Front Aging Neurosci. 2016;8:31.
44. Alharbi B, Al-Kuraishy HM, Al-Gareeb AI, Elekhnawy E, Alharbi H, Alexiou A, et al. Role of GABA pathway in motor and non-motor symptoms in Parkinson's disease: a bidirectional circuit. Eur J Med Res. 2024;29(1):205.
45. Bi D, Wen L, Wu Z, Shen Y. GABAergic dysfunction in excitatory and inhibitory (E/I) imbalance drives the pathogenesis of Alzheimer's disease. Alzheimers Dement. 2020;16(9): 1312–29.
46. Govindpani K, Calvo-Flores Guzmán B, Vinnakota C, Waldvogel HJ, Faull RL, Kwakowsky A. Towards a better understanding of GABAergic remodeling in Alzheimer's disease. Int J Mol Sci. 2017;18(8):1813.
47. Vinkers CH, Olivier B. Mechanisms underlying tolerance after long-term benzodiazepine use: a future for subtype-selective GABAA receptor modulators? Adv Pharmacol Pharm Sci. 2012;2012(1):416864.
48. Almutairi S, Sivadas A, Kwakowsky A. The effect of oral GABA on the nervous system: potential for therapeutic intervention. Nutraceuticals. 2024;4(2):241–59.
49. Alexeev M, Grosenbaugh DK, Mott DD, Fisher JL. The natural products magnolol and honokiol are positive allosteric modulators of both synaptic and extra-synaptic GABAA receptors. Neuropharmacology. 2012;62(8):2507–14.
50. Ríos J-L, Schinella GR, Moragrega I. Phenolics as GABAA receptor ligands: an updated review. Molecules. 2022;27(6):1770.
51. Manayi A, Nabavi SM, Daglia M, Jafari S. Natural terpenoids as a promising source for modulation of GABAergic system and treatment of neurological diseases. Pharmacol Rep. 2016;68(4):671–9.
52. Ross K. Psychobiotics: are they the future intervention for managing depression and anxiety? A literature review. Explore. 2023;19(5):669–80.
53. Liwinski T, Lang UE, Brühl AB, Schneider E. Exploring the therapeutic potential of gamma-aminobutyric acid in stress and depressive disorders through the gut–brain axis. Biomedicine. 2023;11(12):3128.
54. Ķimse L, Reinis A, Miķelsone-Jansone L, Gintere S, Krūmiņa A. A narrative review of psychobiotics: probiotics that influence the gut–brain axis. Medicina. 2024;60(4):601.
55. Sousou JM, Griffith EM, Marsalisi C, Reddy P. Pyridoxine deficiency and neurologic dysfunction: an unlikely association. Cureus. 2023;15(10):e47647.
56. Wasowski C, Marder M. Flavonoids as GABAA receptor ligands: the whole story? J Exp Pharmacol. 2012;4:9–24.
57. Gazola AC, Costa GM, Castellanos L, Ramos FA, Reginatto FH, Lima TC, Schenkel EP. Involvement of GABAergic pathway in the sedative activity of apigenin, the main flavonoid from Passiflora quadrangularis pericarp. Rev Bras. 2015;25(2):158–63.
58. Çiçek SS. Structure-dependent activity of natural GABA (A) receptor modulators. Molecules. 2018;23(7):1512.
59. Kaur S, Sharma P, Mayer MJ, Neuert S, Narbad A, Kaur S. Beneficial effects of GABA-producing potential probiotic Limosilactobacillus fermentum L18 of human origin on intestinal permeability and human gut microbiota. Microb Cell Factories. 2023;22(1):256.
60. Braga JD, Yang Y, Nagao T, Kato N, Yanaka N, Nishio K, et al. Fructooligosaccharides and Aspergillus enzymes increase brain GABA and homocarnosine by modulating microbiota in adolescent mice. npj Sci Food. 2025;9(1):48.

61. Ramadan YN, Alqifari SF, Alshehri K, Alhowiti A, Mirghani H, Alrasheed T, et al. Microbiome gut-brain-axis: impact on brain development and mental health. Mol Neurobiol. 2025;62:1–21.
62. Sandoval-Salazar C, Ramírez-Emiliano J, Trejo-Bahena A, Oviedo-Solís CI, Solís-Ortiz MS. A high-fat diet decreases GABA concentration in the frontal cortex and hippocampus of rats. Biol Res. 2016;49(1):15.
63. Xia G, Han Y, Meng F, He Y, Srisai D, Farias M, et al. Reciprocal control of obesity and anxiety–depressive disorder via a GABA and serotonin neural circuit. Mol Psychiatry. 2021;26(7):2837–53.
64. Icer MA, Sarikaya B, Kocyigit E, Atabilen B, Çelik MN, Capasso R, et al. Contributions of gamma-aminobutyric acid (GABA) produced by lactic acid bacteria on food quality and human health: current applications and future prospects. Foods. 2024;13(15):2437.
65. Patil SB, Khan MK. Germinated brown rice as a value added rice product: a review. J Food Sci Technol. 2011;48(6):661–7.
66. Tsou S-F, Hsu H-Y, Chen S-D. Effects of different pretreatments on the GABA content of germinated brown rice. Appl Sci. 2024;14(13):5771.
67. Langa S, Santos S, Flores JA, Peirotén Á, Rodríguez S, Curiel JA, Landete JM. Selection of GABA-producing lactic acid bacteria strains by polymerase chain reaction using novel gadB and gadC multispecies primers for the development of new functional foods. Int J Mol Sci. 2024;25(24):13696.
68. Park S-Y, Lee J-W, Lim S-D. The probiotic characteristics and GABA production of Lactobacillus plantarum K154 isolated from kimchi. Food Sci Biotechnol. 2014;23(6):1951–7.
69. Braga JD, Thongngam M, Kumrungsee T. Gamma-aminobutyric acid as a potential postbiotic mediator in the gut–brain axis. npj Sci Food. 2024;8(1):16.
70. Benkherouf AY, Eerola K, Soini SL, Uusi-Oukari M. Humulone modulation of GABAA receptors and its role in hops sleep-promoting activity. Front Neurosci. 2020;14:594708.
71. Geisler CE, Ghimire S, Bruggink SM, Miller KE, Weninger SN, Kronenfeld JM, et al. A critical role of hepatic GABA in the metabolic dysfunction and hyperphagia of obesity. Cell Rep. 2021;35(13):109301.
72. Nagao T, Braga JD, Chen S, Thongngam M, Chartkul M, Yanaka N, Kumrungsee T. Synergistic effects of peripheral GABA and GABA-transaminase inhibitory drugs on food intake control and weight loss in high-fat diet-induced obese mice. Front Pharmacol. 2024;15:1487585.
73. Chuong V, Farokhnia M, Khom S, Pince CL, Elvig SK, Vlkolinsky R, et al. The glucagon-like peptide-1 (GLP-1) analogue semaglutide reduces alcohol drinking and modulates central GABA neurotransmission. JCI Insight. 2023;8(12):e170671.
74. Raza ML, Bhojani A, Batool SK, Zehra D. Non pharmacoligical approaches for neurodegenerative diseases: a narrative review. Exp Gerontol. 2024;198:112620.
75. Zuccato C, Cattaneo E. Brain-derived neurotrophic factor in neurodegenerative diseases. Nat Rev Neurol. 2009;5(6):311–22.
76. Walsh EI, Smith L, Northey J, Rattray B, Cherbuin N. Towards an understanding of the physical activity-BDNF-cognition triumvirate: a review of associations and dosage. Ageing Res Rev. 2020;60:101044.
77. Smith PJ, Blumenthal JA, Hoffman BM, Cooper H, Strauman TA, Welsh-Bohmer K, et al. Aerobic exercise and neurocognitive performance: a meta-analytic review of randomized controlled trials. Psychosom Med. 2010;72(3):239–52.
78. Dhahbi W, Briki W, Heissel A, Schega L, Dergaa I, Guelmami N, et al. Physical activity to counter age-related cognitive decline: benefits of aerobic, resistance, and combined training—a narrative review. Sports Med-Open. 2025;11(1):56.
79. Sun G, Ding X, Zheng Z, Ma H. Effects of exercise interventions on cognitive function in patients with cognitive dysfunction: an umbrella review of meta-analyses. Front Aging Neurosci. 2025;17:1553868.

80. Liu X, Morris MC, Dhana K, Ventrelle J, Johnson K, Bishop L, et al. Mediterranean-DASH Intervention for Neurodegenerative Delay (MIND) study: rationale, design and baseline characteristics of a randomized control trial of the MIND diet on cognitive decline. Contemp Clin Trials. 2021;102:106270.
81. Morris MC, Tangney CC, Wang Y, Sacks FM, Bennett DA, Aggarwal NT. MIND diet associated with reduced incidence of Alzheimer's disease. Alzheimers Dement. 2015;11(9): 1007–14.
82. Wang H-X, MacDonald SW, Dekhtyar S, Fratiglioni L. Association of lifelong exposure to cognitive reserve-enhancing factors with dementia risk: a community-based cohort study. PLoS Med. 2017;14(3):e1002251.

Biomarkers to Personalized Therapies in Neurodegenerative Medicine

17

Asif Shaik, Krishna Shevate, Hannah Lalengzuali Fanai, and Jagdish Chand

Abstract

In this chapter, we discuss how the concept of biomarkers will change the way neurodegenerative diseases are diagnosed, stratified, and treated, and how neuroinflammatory disorders and integrated autophagy insights can be considered in personalizing treatment. The symptom-based methods can be ineffective in treating diseases such as Alzheimer's disease and Parkinson's disease since they do not consider the biological variability of diseases. Fluid-based biomarkers and neuroimaging can detect diseases earlier and track changes over time. The molecular subtyping and personalized intervention are achievable through a multi-omics strategy. The increased efficacy to track disease in real-time and precision of therapies are possible with the advent of new technologies like artificial intelligence-based analytics, CRISPR-based diagnostics, and wearable biosensors. To further the goals of equitable solutions and being potentially scalable, some of the tools of the next generation, such as liquid biopsies and nanotech sensors, are promising, despite barriers to validation, control, and access globally. Filling the gaps between biological knowledge and clinical use, biomarkers are at the center of a precision medicine age, which allows the practice of pre-symptomatic treatment and disease-modifying treatments. This chapter highlights how such collaboration needs to offset biomarker innovations into

A. Shaik · K. Shevate · H. L. Fanai
Department of Pharmaceutical Chemistry, JSS College of Pharmacy, JSS Academy of Higher Education & Research, Ooty, The Nilgiris, Tamil Nadu, India

J. Chand (✉)
Department of Pharmaceutical Chemistry, JSS College of Pharmacy, JSS Academy of Higher Education & Research, Ooty, The Nilgiris, Tamil Nadu, India

Silicon Script Sciences Private Limited, Bharatpur, Ghorahi, Dang, Nepal

P. Khanal et al. (eds.), *Neuroinflammation and Autophagy in Neurodegeneration*,
https://doi.org/10.1007/978-981-95-3945-1_17

patient-specific care pathways accessible to patients to make a difference in the outcomes of the neurodegenerative diseases.

Keywords

Biomarkers · Multi-omics integration · Neurodegenerative diseases · Neuroinflammation · Precision medicine

17.1 Introduction to Biomarker-Driven Precision Medicine

Since the early days of defining the course of neurodegenerative disease treatment, the paradigm has changed dramatically, as historical frameworks have shifted toward a more refined, precision medicine armamentarium based on biomarkers [1]. Due to the biological diversity of such diseases as Parkinson's disease (PD) and Alzheimer's disease (AD), traditional approaches that focus on clinical signs, like tremor in PD or memory loss in AD, over and over again fail to reflect the biological diversity of these diseases [2]. This simple categorization has resulted in over-generalized treatment, which in many cases does not work because of the variability in disease mechanisms among individuals [3]. This intervention most often occurs late, whereas the symptoms can be delayed until the neuronal degeneration is considerable [4]. As a result, it has been observed that there is a rush to redefine diagnosis and treatment in molecular biological terms. The definitive moment in this transformation is the introduction of the biomarkers to diagnostic algorithms [5]. Biomarkers offer real-time information on pathophysiological processes like neuroinflammation and dysfunctional autophagy, which are now known to be core elements of the neurodegenerative disease inception and development [6]. On the basis of this mechanistic understanding, the patient can be stratified according to the molecular drivers of pathology and receive early intervention and rational choice of therapy. As an example, levels of phosphorylated tau in cerebrospinal fluid (CSF) and of neurofilament light chain in serum are no longer only useful in order to make a diagnosis but also in terms of monitoring the dynamics of disease progression [7]. Neuroimaging positron emission tomography (PET) scans, like the ones with a tracer specific to inflammation or autophagy, provide a spatial resolution to their pathology much earlier than it can be detected [8]. However, biomedical translation comes with difficulties in translating biomarkers into regular clinical applications. The standardization of the assay protocols, the inter-cohort validation, and regulatory recognition are still major bottlenecks. In addition, the technological cost of some biomarker systems makes them unaffordable, especially to low-resource countries, and hence, this enhances global health inequities [9]. Moreover, with the multi-omics strategies, these obstacles are to be managed by combining genomics, proteomics, and metabolomics to obtain more specific as well as scalable composite biomarker signatures [10]. This suggests that biomarker identification and pattern recognition can now be achieved at a faster rate. Additionally, a higher degree of precision in medical diagnostics becomes feasible in practical environments with the help of artificial intelligence. Biomarker-guided care has the potential to save the economic costs of ineffective treatments and a long period of disease progression [11]. Advances in precision medicine allow healthcare systems to maximize effective resource use by eliminating trial-and-error that exists to a greater or lesser extent in the current iteration of therapeutic regimens. In such a way, evolution toward the concept of precision medicine in neurodegenerative disease is not just a scientific breakthrough but a prerequisite for clinical practices reorientation, intended to challenge the complexity of devastating diseases and treat them in a proactive, patient-centered approach.

17.2 Biomarker Discovery and Validation

17.2.1 Fluid-Based Biomarkers

Discovery and confirmation of fluid-based biomarkers can be described as a breakthrough in the development of precision medicine for neurodegenerative disease [12]. Biomarkers identified from CSF and blood can provide novel insights into the biochemical environment in the central nervous system [13]. The CSF biomarkers landscape currently encompasses phosphorylated tau isoforms and the expected amyloid-β $(A\beta)_{42/40}$ ratio to represent the AD pathology, and cytokine levels can represent underlying neuroinflammatory states implicated in neurodegenerative disease [14]. Nonetheless, the utilization of CSF biomarkers is limited by several as yet unresolved challenges in translation [15]. For example, ethical questions are prompted because lumbar puncture is invasive, especially when used in asymptomatic individuals or those who need to be sampled longitudinally. Moreover, pre-analytical variability in pre-analytical processing, storage, and assay systems is a factor in the disparity between research centers [16], diminishing the reproducibility of research and clinical trustworthiness of results. The barriers can be overcome by standardization, such as coordinated procedures and the production of standard materials. Such frameworks are already being adopted by international consortia and regulatory bodies [17]. To overcome some of these constraints, efforts have been directed at blood-based biomarkers, which have a much less invasive and much more scalable approach [18]. Serum neurofilament light chain has shown a significant potential as a dynamic measure of axon degeneration in multiple neurodegenerative diseases based on the severity and progression [19, 20]. Similarly, there is an increased use of glial fibrillary acidic protein (GFAP) as a measure that indirectly determines astroglia activation [21] when induced by neuroinflammation and reactive gliosis [22]. Although such blood-based biomarkers are easier to collect and have the potential to broaden the screening to a large population [23], peripheral confounders and low analyte concentration may affect their sensitivity and specificity. Longitudinal studies that contain sufficiently large and heterogeneous cohorts are required to consider variability in demographics and comorbidity to validate these biomarkers [24]. However, the problem of low abundance of biomarkers in blood can be solved by ultra-sensitive detection platforms, including single-molecule array (Simoa) technology [25], leading to a higher fidelity of diagnostic procedures. Moreover, ethical frameworks need to be reconsidered in using biomarker data, especially with the increasing use of predictive testing [24]. Therefore, fluid-based biomarker discovery and validation, which is not easy to accomplish, has the promise of bringing a paradigm shift to how neurodegenerative disease are diagnosed and monitored, and eventually opening the precise opportunity of earlier intervention and improved outcomes in patients.

17.2.2 Neuroimaging Biomarkers

Use of neuroimaging biomarkers has transformed the diagnostic and mechanistic concepts of neurodegenerative disorders since they enable non-invasive, spatially and temporally resolved visualization of pathological mechanisms [26]. It involves the application of PET tracers, more specifically Aβ and tau protein tracers, which emerged as the hallmarks of AD [27]. The presence of amyloid can be detected before cognitive manifestation due to new technologies like imaging with [^{11}C] Pib and [^{18}F] florbetapir. It is seen that the tau-specific PET ligands, such as [^{18}F] flortaucipir, indicate the stage and disease progression, with correlation to clinical severity [28]. These conventional targets, however, only form a part of the neuropathological topography, and new PET ligands are already being developed to provide images of neuropathological components of autophagy, including LC3B and p62 [29]. These novel tracers further broaden the application of imaging biomarkers to include maladaptive clearance of proteins, not only in proteinopathies. Although PET imaging holds promise, there still exist limitations. For example, PET imaging is expensive and not always accessible with limited resources. Certain measures that strive to democratize access are the development of long-lived radioisotopes, easier synthesis procedures, and the inter-institutional tracer-sharing programs [30]. In addition to PET, magnetic resonance imaging (MRI) is also a widely used method. However, advanced imaging methods, such as diffusion tensor imaging (DTI) and functional MRI, provide structural and functional data on the integrity of white matter and functional dynamics of brain networks. DTI can trace axonal damage and abnormalities of anatomical shrinkage. The functional MRI can detect aberrant patterns of connectivity in resting state or task-based paradigm, and this provides insight into dysfunction in the cognitive network that is usually evident at disease onset. These techniques are non-invasive, reversible, and do not involve radiation, which is good for longitudinal studies [31, 32]. Combining PET and MRI multimodal imaging is one of the prevailing approaches to address the correlation of molecular pathology and structural and functional alterations in the brain. Such a synergistic approach permits obtaining a full image of disease processes to relate metabolic abnormalities observed by PET to microstructural and functional changes. Nonetheless, multimodal platforms demand multiple instruments and require knowledge of image co-registration and merging data. In order to manage such limitations, there is a rising use of machine learning algorithms to automate the analysis process to determine a disease-specific imaging signature. Neuroimaging biomarkers streamline diagnostics, guide the choice of treatment, and evaluate its effectiveness by linking the molecular understanding with anatomical and functional output, which needs to be developed further, integrated into clinical practice, and better neurological advancements have to be made to deal with neurodegenerative diseases.

17.3 Stratifying Patients and Monitoring Progression

Patient stratification has become one of the critical solutions in the implementation of effective and individualized care in neurodegenerative diseases. As opposed to considering these disorders as unitary, homogenous ones, precision medicine has realized that diseases such as AD and PD are made of multiple biological subtypes with different and unique pathological courses. In AD, imaging and fluid biomarkers showed that in some people, there is an apparent tau pathology preceding amyloid accumulation, termed the tau-first subtype, whereas in others, it is another amyloid-first pathway. Such variation is important, as the therapeutic response to tau or anti-amyloid-based treatment is likely to depend on both the time and space predominance of these proteins. Similarly, PD can no longer be classified as a single pathway disorder. It is possible to split these into inflammatory fall or metabolic fall, depending on disorder-specific neurodegeneration, time of symptoms appearance, and responsiveness to treatments [33, 34]. The issue is with classical categorization of all patients into macro-diagnostic categories, making treatment responses inconsistent. The current solution to this would be to employ biomarker-driven stratification of clinical trials rather than the combined approach, which may be randomized to test the potential effectiveness of a treatment or intervention by biasing it to participants who are more likely to respond to it. As an illustration, in any trials of tau treatment, the individuals will be enrolled with proven tau accumulation opposed to general AD cohorts, which increases the odds of showing efficacy. This focused strategy increases the probability of success in trials, as well as minimizing the cost and shortening the regulatory approval schedules of drug. Nevertheless, the extensive application of biomarker-directed trials is under threat due to inaccessibility of advanced diagnostics and the necessity to develop uniform enrolment conditions across study sites [35, 36]. The second new area of interest is the tracking and early diagnosis of those subjects who are at risk of developing neurodegenerative disease, even before the symptom onset. An understanding of fluid and imaging biomarkers of neuroinflammation or protein aggregation in genetically predisposed individuals, for example, APOEε4 or LRRK2, can indicate the presence of the disease decades before it is shown in symptoms [37, 38]. Identification of such changes presents a precondition for preventive interventions, though the moral aspect of pre-symptomatic diagnosis has not been closed yet. A careful and informed consent is required to balance the benefits and drawbacks, with improved monitoring on the one hand and psychological and social repercussions on the other hand. Last but not least, precise patient classification and early disease monitoring are revolutionizing how we see and treat neurodegeneration and assisting the entire profession in moving toward treatments that are as unique as the patient.

17.4 Multi-omics Integration for Subtype Identification

The merging of multi-omics technologies has inserted a paradigm-shift perspective into our approach toward neurodegenerative diseases, having the potential to lead to the identification of unique biological groupings that are not necessarily identified through single modality analysis. Genomics has been critical in doing so with the utilization of polygenic risk scores, combining the consequences of many genetic variants to predict the likelihood of developing a disorder. Availability of variants like APOEε4, which has been closely linked with AD, as well as TREM2 that regulates microglial activation, has highlighted the neuroinflammation risk [39, 40]. Moreover, genes that contribute to autophagy process, like ATG5 and ATG7, are gaining significance in neuronal proteostasis. Once combined, these generic elements assist in the design of patient-specific pathways of fragility that can be further classified. In addition to DNA, proteomics introduces a completely new layer of information due to dynamic nature of proteomics, representing the ongoing biological activity. CSF proteome profiling has revealed protein fingerprints of different levels and types of neurodegeneration. Biomarkers that can distinguish rapidly progressive disease variants from slower, indolent forms include elevated inflammatory proteins and reduced levels of enzymes required for autophagy [41]. These proteomic profiles not only confirm evidence of a genomic discovery but also offer potential therapeutic targeting biomarkers. Nevertheless, true potential of multi-omics can be identified as data integration, and here the main challenges arise. Integrating large, complex genomics, proteomics, and imaging datasets requires advanced computational tools to harmonize highly diverse data formats and scales, addressing sparsity, batch effects, and sampling disparities. Machine learning algorithms, dimensionality reduction strategies, and data harmonization pipelines that reveal latent patterns across modalities are some of the novel approaches [42]. This framework is further complemented by adopting a systems medicine strategy as it circumvents the problem of tackling each condition individually by leveraging network analysis to derive hub proteins and central molecular pathways that can serve between neuroinflammation and autophagy (Fig. 17.1).

17.5 Emerging Technologies for Biomarker Innovation

17.5.1 Artificial Intelligence and Machine Learning

Artificial intelligence and machine learning have served as accelerators to identify and use biomarkers in neurodegenerative conditions. The traditional statistics tools are generally poor at handling the high-dimensional, multifaceted, and large amounts of information that are produced by the modern multi-omics experiments. But

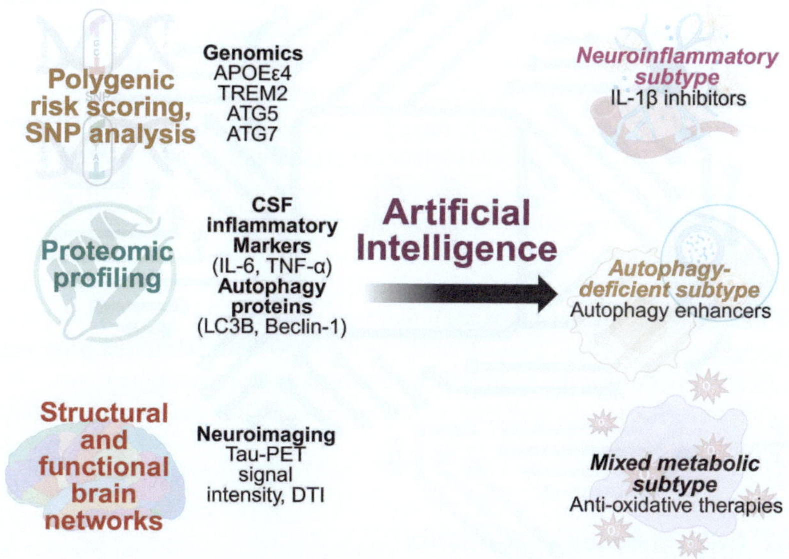

Fig. 17.1 A multi-omics approach to the stratification of patients with neurodegenerative diseases into biologically relevant subtypes. Clustering algorithms based on artificial intelligence are used to harmonize genomic, proteomic, and imaging data, identifying molecular signatures that can be used to guide precision therapies. (Created using BioRender. https://www.biorender.com/)

artificial intelligence has brought forth the possibility of uncovering the hidden patterns and relationships. Deep learning models can group patients with molecular similarity, and this approach can reveal novel subtypes of disease that would normally be masked. As an example, combining gene expression, proteomic alterations, and imaging biomarkers, these models can identify patient groups with similar disease trajectories, even though the clinical manifestations of patients may not be similar. Such reclassification is essential to develop more effective and individualized therapeutic approaches, which can substitute an archaic one-size-fits-all approach [43–45]. These benefits notwithstanding, the use of artificial intelligence in biomarker science is not a smooth journey. However, one of the problems is data heterogeneity. The data generated by various platforms are in different formats, and the quality also varies, which causes inconsistent model training. In addition to that, plenty of omics datasets are characterized by limited sample size [46], particularly rare forms of neurodegeneration, potentially restricting the applicability of artificial intelligence models. In addressing these challenges, it is necessary to come up with effective data pre-processing techniques and also to consider data federated learning techniques, which enable multi-institutional collaboration without causing any violation of patient privacy [47]. The other strong example of machine learning implementation is in predictive analytics, which relies on longitudinal biomarker data to predict the disease path [48]. These models have the potential to examine decades of clinical and biomarker data to understand when a person at risk of developing AD may transition to symptomatic or how fast an individual with very early PD may progress. Predictive analytics has incalculable clinical utility, as it is possible to perform proactive interventions and recruit as informed participants in trials. Nevertheless, there is an issue regarding model interpretability. The clinicians might hesitate to execute predictions when the explanation thereof is unclear. To this end, new artificial intelligence systems are being developed with explainable frameworks to enable better insights into how they predict and make their predictions more clinically actionable. Artificial intelligence and machine learning contribute to observing the complexity of neurodegenerative diseases, and are in the bigger picture of biomarker innovation that they enable and surpass what used to be feasible in early detection, patient stratification, and therapeutic decision-making. There is great hope that the field will be converted to precautionary care, with the help of these technologies, and the overall outcome will be better management and prevention of diseases.

17.5.2 CRISPR-Based Diagnostics

CRISPR diagnostics are rapidly becoming a disruptive technology in the biomarker discovery of neurodegenerative disease, but these diagnostics provide unprecedented sensitivity, specificity, and flexibility. Two gene-editing systems, CRISPR-Cas12 and Cas13, have been hijacked to achieve ultrasensitive nucleic-acid-based detection in blood and other biofluids. These systems take advantage of collateral cleavage activity, whereby Cas12 seeks out DNA and Cas13 targets RNA, and is able to produce very specific signals even when the disease-related genetic material is present in very low quantities. Such ability renders them worthwhile in neurodegeneration, where early biomarkers could be present in low concentrations many years before the onset of clinical symptoms [49]. Moreover, it is now possible to detect fragments of mutant transcripts or microRNAs that cause problems with autophagy or neuroinflammation much earlier than can be done with the current diagnostic methods. Even so, the application potential in diagnostics is evident, but in applying the CRISPR platforms clinically, hurdles exist. The major obstacles are stability and reproducibility of system in sample, which has background noise variability. Additionally, advances in gene editing are being used in preclinical models to alter pathways involved in neurodegeneration involving autophagy dysfunction. Precision editing of autophagy-related genes like ATG5 or BECN1 has the potential to correct cell clearance mechanisms and can be used to slow or even reverse disease progression. In this regard, the benefit of CRISPR is to achieve long-term, targeted, heritable changes in genome. However, targeted delivery into brain regions and cell types is a challenging step in the delivery of CRISPR components to brain [50]. The existing options include engineered viral vectors and non-viral delivery [51], off-target effects, and immune responses. As technology gets more refined, there is a high probability that the future of personalized, precision neurology will have CRISPR-based tools as its focus.

17.5.3 Wearable Real-Time Monitoring

Wearable real-time monitoring technologies can continuously detect and track changes in neurodegenerative diseases as they occur in real-life situations, and they are revolutionizing how we observe and track these conditions. Digital phenotyping using consumer and medical-grade smartwatches is undeniably one of the most promising applications. These devices can passively evaluate non-motor indicators of PD, like heart rate variability and poor sleep, as well as motor symptoms, like bradykinesia, tremor, and walk abnormalities. Wearable devices record the dynamics of symptoms over time, unlike episodic clinical care, which presents a more comprehensive and dynamic picture of symptoms as they shift throughout the day. This enables not only better diagnosis, but also adjustment of treatment according to real-world data on a patient-by-patient basis. Nevertheless, the quality and the interpretation of data are inconsistent when different devices and platforms are used. Incomplete connections of data due to imperfections in accuracy of photographic recorders, and noncompliance can compromise the data. Developers are working on the optimization of signal processing algorithms and incorporation of machine learning to minimize noise and identify patterns that are clinically relevant [52, 53]. In addition to surface wearables, even biosensor arrays are being worked upon so that they can monitor the molecular-level biochemical changes. Biosensors that can monitor cytokine concentration continuously in the CSF are the state of the art in this technology. Hopefully, such devices will provide real-time data on processes of neuroinflammation, which are represented earlier before structural alterations take place. Such a type of biosensor can be considered as an early-warning signal [54]. However, key problems are associated with biocompatibility and long-lasting stability of materials that are intended to be used as implants for the safe transfer of sensitive data. Such constraints are to be gradually addressed through bioengineered wireless transmissions to inch ahead toward real neurochemical recording. This live data can be more useful when it is correlated with cloud-based platforms, which can help in aggregating and analyzing on the same scale. On the one hand, these platforms are assuming the role of helping to track individual patient trajectories [55]; on the other hand, this can lead to population-level discovery of biomarkers by the finding of common patterns in thousands of people. The problem is that the data systems should be secure and interoperable enough to ensure patient privacy [56], but allow massive analysis. A cross-section of healthcare providers, technology companies, and rulemaking organizations is now working together to dictate ethical standards in the management of data and cloud inhabitants. Finally, wearable gadgets and on-demand observations are transitioning precision neurology out of clinics and into everyday lives of patients.

17.6 Translating Biomarkers to Bespoke Therapies

Biomarker research in neurodegenerative diseases can enable personalized treatments tailored to a patient's molecular profile. For example, rising GFAP levels indicate astrocyte-driven neuroinflammation and help identify patients who may benefit from anti-inflammatory biologics [57]. Recent studies supported this method, demonstrating that the enrichment of trial inclusion was favorable to patients with high GFAP readings and positive Aβ status [58, 59]. The resulting IL-1β-specific biologics resulted in massive reductions in inflammatory rates, slowed cognitive deterioration, and reduced trial costs and cohort summaries by more than 50%. This example is extended by using the distinctive application of IL-2 modulation in mild to severe AD in regulating T-cells, CSF-$A\beta_{42}$, and GFAP [60]. Such therapeutic benefits were associated with less cognitive impairment and substantial patient stratification, which makes the application of individual immunomodulation intervention feasible in clinical practice. Meantime, pharmacogenomics has been used in progressively crucial ways to streamline the safety and dosage of medications. Polymorphisms of cytochrome P450 enzymes, for example, CYP2D6 [61], have a significant effect on autophagy-targeting drugs. The other frontier of customized therapeutics is the adaptive clinical trial model involving real-time biomarker feedback to perform dynamic alterations in treatment regimen [62]. This strategy can be used to optimize an autophagy-enhancing drug, achieving more effective results and decisions in a shorter time. However, this model is still developing, but more effective studies can present a more responsive design to patients. Overall, progressing-GFAP-driven biologic targeting, CYP2D6-based dosing, and biomarker-informed adaptive trials are creating a new chapter to address neurodegeneration that is choreographed at the molecular level, and not just by general diagnosis.

17.7 Challenges and Future Directions

In the ever-changing biomarker landscape in neurodegenerative diseases, there are various issues to be addressed so that these advancements can be scientifically reliable. The problem of biomarker validation can be listed as one of the most common obstacles [63]. Although some promising candidates have been generated by multi-omics and neuroimaging studies, achieving consistency in different patient populations, different platforms, and different stages of the disease is a complex process. Biomarker expression can also be greatly influenced by biological variants across ethnic groups, environmental exposures, and even comorbid conditions [64]. In addition, sample processing discrepancies, assay calibration errors, and analysis methods also lead to discrepancies in results. In order to get out of discovery and into clinical adoption, more harmonized protocols and multicentric validation studies are needed, with standardization and reproducibility as priorities. To worsen the situation, there are no clear regulatory frameworks for approving dynamic biomarkers [65]. The regulatory frameworks that were traditionally made are based on the concept of static diagnostics and are frequently not sufficiently suited to test machine learning models, which can learn over time. Moreover, regulatory lag establishes a barrier to the translation of artificial intelligence-enabled biomarkers in clinical workflow, though it shows good technical performance [66]. Regulatory bodies are just starting to research adaptive models capable of measuring algorithmic learning and real-time evolution of a biomarker, but implementation is in progress. In long term, these tools would require collaborative work among data scientists, clinicians, and policymakers so that they can be safe and effective at the same time. However, limited access to advanced biomarker technologies in low- and middle-income countries hinders diagnosis and skews research toward populations with stronger healthcare infrastructure. These differences can be mitigated through inexpensive methods of diagnostics, education programs of local scientists, and sharing infrastructure solutions to the precision medicine. In future, biomarker tools of next generation are serving as new hand-held pathways to do away with these barriers. The less invasive way to monitor neurodegenerative activity can be the use of liquid biopsies to measure autophagy flux in blood or CSF. Likewise, real-time nanotechnology-based sensors can be developed to allow continuous detection of disease-related proteins and metabolites. These technologies are under development, but they indicate a better move toward dynamic biomarker systems. To be truly transformative, validation, regulation, and access issues need to be addressed in collaboration. It is only after that time that biomarker science will have the optimum efficacy to redefine how neurodegenerative diseases are diagnosed and treated globally.

17.8 Conclusion

To sum up, biomarkers are transforming the management of neurodegenerative diseases into a patient-focused, biology-driven model. They aid early, precise diagnosis by detecting molecular signatures of neuroinflammation, autophagy dysfunction, and protein aggregation. Combining fluid markers like phosphorylated tau and neurofilament light chain with PET and advanced MRI can track disease progression even before symptoms appear. Multi-omics platforms and artificial intelligence-driven analytics can classify diseases into molecular subtypes, predict aggressive courses, and guide personalized treatments. Emerging tools such as CRISPR-based tests, nanotechnology sensors, and wearable biosensors enable minimally invasive and real-time monitoring. However, challenges remain, including biomarker validation across populations, regulatory adaptation for artificial intelligence diagnostics, and limited access to high-end technologies in low-resource settings. Broader collaboration is essential to make biomarker-driven precision neurology accessible, enabling earlier interventions, safer therapies, and meaningful disease modification.

References

1. Menendez-Gonzalez M. Implementing a tridimensional diagnostic framework for personalized medicine in neurodegenerative diseases. Alzheimers Dement. 2025;21:e14591.
2. Chopade P, Chopade N, Zhao Z, Mitragotri S, Liao R, Chandran Suja V. Alzheimer's and Parkinson's disease therapies in the clinic. Bioeng Transl Med. 2023;8:e10367.
3. Kearney BE, Lanius RA. The brain-body disconnect: a somatic sensory basis for trauma-related disorders. Front Neurosci. 2022;16:1015749.
4. Martin J, Clarke J, Howard R, Waltho M. Palliative care in neurology. In: Howard R, Kullmann D, Werring D, Zandi M, editors. Neurology [Internet]. 1st ed. Wiley; 2024 [cited 2025 July 30]. p. 1203–18. Available from: https://onlinelibrary.wiley.com/doi/10.1002/9781119715672.ch35.
5. Chudzik A, Śledzianowski A, Przybyszewski AW. Machine learning and digital biomarkers can detect early stages of neurodegenerative diseases. Sensors. 2024;24:1572.
6. Giorgi C, Bouhamida E, Danese A, Previati M, Pinton P, Patergnani S. Relevance of autophagy and mitophagy dynamics and markers in neurodegenerative diseases. Biomedicine. 2021;9:149.
7. Park Y, Kc N, Paneque A, Cole PD. Tau, glial fibrillary acidic protein, and neurofilament light chain as brain protein biomarkers in cerebrospinal fluid and blood for diagnosis of neurobiological diseases. Int J Mol Sci. 2024;25:6295.
8. Bao W, Xie F, Zuo C, Guan Y, Huang YH. PET neuroimaging of Alzheimer's disease: radiotracers and their utility in clinical research. Front Aging Neurosci. 2021;13:624330.
9. Burkhart MC, Lee LY, Vaghari D, Toh AQ, Chong E, Chen C, et al. Unsupervised multimodal modeling of cognitive and brain health trajectories for early dementia prediction. Sci Rep. 2024;14:10755.
10. La Cognata V, Morello G, Cavallaro S. Omics data and their integrative analysis to support stratified medicine in neurodegenerative diseases. Int J Mol Sci. 2021;22:4820.
11. Elhage A, Cohen S, Cummings J, Van Der Flier WM, Aisen P, Cho M, et al. Defining benefit: clinically and biologically meaningful outcomes in the next-generation Alzheimer's disease clinical care pathway. Alzheimers Dement. 2025;21:e14425.
12. El Abiad E, Al-Kuwari A, Al-Aani U, Al Jaidah Y, Chaari A. Navigating the Alzheimer's biomarker landscape: a comprehensive analysis of fluid-based diagnostics. Cells. 2024;13:1901.
13. Alcolea D, Beeri MS, Rojas JC, Gardner RC, Lleó A. Blood biomarkers in neurodegenerative diseases: implications for the clinical neurologist. Neurology. 2023;101:172–80.
14. Sun Q, Ni J, Wei M, Long S, Li T, Fan D, et al. Plasma β-amyloid, tau, neurodegeneration biomarkers and inflammatory factors of probable Alzheimer's disease dementia in Chinese individuals. Front Aging Neurosci. 2022;14:963845.
15. Agnello L, Gambino CM, Ciaccio AM, Masucci A, Vassallo R, Tamburello M, et al. Molecular biomarkers of neurodegenerative disorders: a practical guide to their appropriate use and interpretation in clinical practice. Int J Mol Sci. 2024;25:4323.
16. Hansson O, Batrla R, Brix B, Carrillo MC, Corradini V, Edelmayer RM, et al. The Alzheimer's Association international guidelines for handling of cerebrospinal fluid for routine clinical measurements of amyloid β and tau. Alzheimers Dement. 2021;17:1575–82.
17. Giangrande C, Delatour V, Andreasson U, Blennow K, Gobom J, Zetterberg H. Harmonization and standardization of biofluid-based biomarker measurements for AT(N) classification in Alzheimer's disease. Alzheimer's Dement Diagn Assess Dis Monit. 2023;15:e12465.
18. Schindler SE, Galasko D, Pereira AC, Rabinovici GD, Salloway S, Suárez-Calvet M, et al. Acceptable performance of blood biomarker tests of amyloid pathology — recommendations from the Global CEO Initiative on Alzheimer's Disease. Nat Rev Neurol. 2024;20:426–39.
19. Bar-Or A, Nicholas J, Feng J, Sorrell F, Cascione M. Exploring the clinical utility of neurofilament light chain assays in multiple sclerosis management. Neurol Neuroimmunol Neuroinflamm. 2025;12:e200427.

20. Beydoun MA, Noren Hooten N, Georgescu MF, Beydoun HA, Eid SM, Fanelli-Kuczmarski MT, et al. Serum neurofilament light chain as a prognostic marker of all-cause mortality in a national sample of US adults. Eur J Epidemiol. 2024;39:795–809.
21. Agnello L, Gambino CM, Ciaccio AM, Salemi G, Brighina F, Ragonese P, et al. The value of serum glial fibrillary acidic protein as a biomarker of astrogliosis in different neurological diseases. Clin Chim Acta. 2025;572:120248.
22. Marksteiner J, Humpel C. Glial fibrillary acidic protein as a biomarker for diagnosis of Alzheimer's disease in cerebrospinal fluid, plasma and saliva measured with Lumipulse technology: a narrative review. NeuroMarkers. 2025;2:100038.
23. Leipp F, Vialaret J, Mohaupt P, Coppens S, Jaffuel A, Niehoff A-C, et al. Glial fibrillary acidic protein in Alzheimer's disease: a narrative review. Brain Commun. 2024;6:fcae396.
24. Lanz TA, Ruprecht K, Somps CJ, Göpfert JC, Kam-Thong T, Tekle F, et al. Longitudinal evaluation of serum neurofilament light levels in normal healthy volunteers: defining a threshold of concern. J Neurol. 2025;272:512.
25. Böken D, Xia Z, Lam JYL, Fertan E, Wu Y, English EA, et al. Ultrasensitive protein aggregate quantification assays for neurodegenerative diseases on the simoa platform. Anal Chem. 2025;97:290–9.
26. Bronte A, Prieto E, Quincoces G, Erro E, Arbizu J. The basics of PET molecular imaging in neurodegenerative disorders with dementia and/or parkinsonism. Eur Radiol. 2025;35:4621–34.
27. Rabinovici GD, Knopman DS, Arbizu J, Benzinger TLS, Donohoe KJ, Hansson O, et al. Updated appropriate use criteria for amyloid and tau PET: a report from the Alzheimer's Association and Society for Nuclear Medicine and Molecular Imaging Workgroup. Alzheimers Dement. 2025;21:e14338.
28. Boccalini C, Peretti DE, Mathoux G, Iaccarino L, Ribaldi F, Scheffler M, et al. Early-phase 18F-Flortaucipir tau-PET as a proxy of brain metabolism in Alzheimer's disease: a comparison with 18F-FDG-PET and early-phase amyloid-PET. Eur J Nucl Med Mol Imaging. 2025;52:1958–69.
29. Tutas J, Tolve M, Özer-Yildiz E, Ickert L, Klein I, Silverman Q, et al. Autophagy regulator ATG5 preserves cerebellar function by safeguarding its glycolytic activity. Nat Metab. 2025;7:297–320.
30. Coupé P, Mansencal B, Morandat F, Morell-Ortega S, Villain N, Manjón JV, et al. petBrain: a new pipeline for amyloid, tau tangles and neurodegeneration quantification using PET and MRI [Internet]. arXiv. 2025 [cited 2025 July 30]. Available from: https://arxiv.org/abs/2506.03217.
31. Bergamino M, Zhu S, Shill HA, Stokes AM. Investigating white matter alterations in Parkinson's disease using multi-shell free-water DTI and NODDI: insights into neurodegeneration and levodopa effects. Front Neurol. 2025;16:1605753.
32. Zhang R, Aksman L, Wijesinghe D, Ringman JM, Wang DJJ, Jann K, et al. A longitudinal study of functional brain complexity in progressive Alzheimer's disease. Alzheimer's Dement Diagn Assess Dis Monit. 2025;17:e70059.
33. Aksman LM, Oxtoby NP, Scelsi MA, Wijeratne PA, Young AL, Alves IL, et al. A data-driven study of Alzheimer's disease related amyloid and tau pathology progression. Brain. 2023;146:4935–48.
34. Gao S, Wang Z, Huang Y, Yang G, Wang Y, Yi Y, et al. Early detection of Parkinson's disease through multiplex blood and urine biomarkers prior to clinical diagnosis. NPJ Parkinsons Dis. 2025;11:35.
35. Penny LK, Lofthouse R, Arastoo M, Porter A, Palliyil S, Harrington CR, et al. Considerations for biomarker strategies in clinical trials investigating tau-targeting therapeutics for Alzheimer's disease. Transl Neurodegener. 2024;13:25.
36. Duan X-P, Qin B-D, Jiao X-D, Liu K, Wang Z, Zang Y-S. New clinical trial design in precision medicine: discovery, development and direction. Signal Transduct Target Ther. 2024;9:57.

37. Mishra S, Blazey TM, Holtzman DM, Cruchaga C, Su Y, Morris JC, et al. Longitudinal brain imaging in preclinical Alzheimer disease: impact of APOE ε4 genotype. Brain. 2018;141:1828–39.
38. Zarkali A, Thomas GEC, Zetterberg H, Weil RS. Neuroimaging and fluid biomarkers in Parkinson's disease in an era of targeted interventions. Nat Commun. 2024;15:5661.
39. Venkatesh R, Cardone KM, Bradford Y, Moore AK, Kumar R, Moore JH, et al. Integrative multi-omics approaches identify molecular pathways and improve Alzheimer's Disease risk prediction [Internet]. Neurology. 2025 [cited 2025 July 30]. Available from: http://medrxiv.org/lookup/doi/10.1101/2025.05.31.25328688.
40. D'Aoust T, Clocchiatti-Tuozzo S, Rivier CA, Mishra A, Hachiya T, Grenier-Boley B, et al. Polygenic score integrating neurodegenerative and vascular risk informs dementia risk stratification. Alzheimers Dement. 2025;21:e70014.
41. Fleming A, Bourdenx M, Fujimaki M, Karabiyik C, Krause GJ, Lopez A, et al. The different autophagy degradation pathways and neurodegeneration. Neuron. 2022;110:935–66.
42. Ren F, Wei J, Chen Q, Hu M, Yu L, Mi J, et al. Artificial intelligence-driven multi-omics approaches in Alzheimer's disease: progress, challenges, and future directions. Acta Pharm Sin B. 2025;15(9):4327–85.
43. Ozdemir C, Olaimat MA, Vashishath Y, Bozdag S, Initiative ADN. IGCN: Integrative Graph Convolution Networks for patient level insights and biomarker discovery in multi-omics integration [Internet]. arXiv. 2024 [cited 2025 July 31]. Available from: https://arxiv.org/abs/2401.17612.
44. Zhang Y, Yu L, Lv Y, Yang T, Guo Q. Artificial intelligence in neurodegenerative diseases research: a bibliometric analysis since 2000. Front Neurol. 2025;16:1607924.
45. Aborode AT, Emmanuel OA, Onifade IA, Olotu E, Otorkpa OJ, Mehmood Q, et al. The role of machine learning in discovering biomarkers and predicting treatment strategies for neurodegenerative diseases: a narrative review. NeuroMarkers. 2025;2:100034.
46. Koksalmis GH, Soykan B, Brattain LJ, Huang H-H. Artificial intelligence for personalized prediction of Alzheimer's disease progression: a survey of methods, data challenges, and future directions [Internet]. arXiv. 2025 [cited 2025 July 31]. Available from: https://arxiv.org/abs/2504.21189.
47. Ouyang X, Shuai X, Li Y, Pan L, Zhang X, Fu H, et al. ADMarker: a multi-modal federated learning system for monitoring digital biomarkers of Alzheimer's disease [Internet]. arXiv. 2023 [cited 2025 July 31]. Available from: https://arxiv.org/abs/2310.15301.
48. Ghosh D, Pal S, Lutz M, Luo S. Ensemble survival analysis for preclinical cognitive decline prediction in Alzheimer's disease using longitudinal biomarkers [Internet]. arXiv. 2025 [cited 2025 July 31]. Available from: https://arxiv.org/abs/2503.16645.
49. Bagi M, Jamalzadegan S, Steksova A, Wei Q. CRISPR–Cas based platforms for RNA detection: fundamentals and applications. Chem Commun. 2025. https://doi.org/10.1039/D5CC03257A.
50. Barmaki H, Nourazarian A, Khaki-Khatibi F. Proteostasis and neurodegeneration: a closer look at autophagy in Alzheimer's disease. Front Aging Neurosci. 2023;15:1281338.
51. Klinkovskij A, Shepelev M, Isaakyan Y, Aniskin D, Ulasov I. Advances of genome editing with CRISPR/Cas9 in neurodegeneration: the right path towards therapy. Biomedicine. 2023;11:3333.
52. Adams JL, Kangarloo T, Gong Y, Khachadourian V, Tracey B, Volfson D, et al. Using a smartwatch and smartphone to assess early Parkinson's disease in the WATCH-PD study over 12 months. NPJ Parkinsons Dis. 2024;10:112.
53. Bougea A. Application of wearable sensors in Parkinson's disease: state of the art. J Sens Actuator Netw. 2025;14:23.
54. Campuzano S, Yáñez-Sedeño P, Pingarrón JM. Revisiting electrochemical biosensing in the 21st century society for inflammatory cytokines involved in autoimmune, neurodegenerative, cardiac, viral and cancer diseases. Sensors. 2020;21:189.

55. Ianculescu M, Băjenaru L, Vasilevschi A-M, Gheorghe-Moisii M, Gheorghe C-G. Advancing neurodegenerative disease management: technical, ethical, and regulatory insights from the NeuroPredict platform. Future Internet. 2025;17:320.
56. Bradshaw A, Hughes N, Vallez-Garcia D, Chokoshvili D, Owens A, Hansen C, et al. Data sharing in neurodegenerative disease research: challenges and learnings from the innovative medicines initiative public-private partnership model. Front Neurol. 2023;14:1187095.
57. Kim KY, Shin KY, Chang K-A. GFAP as a potential biomarker for Alzheimer's disease: a systematic review and meta-analysis. Cells. 2023;12:1309.
58. Lista S, Imbimbo BP, Grasso M, Fidilio A, Emanuele E, Minoretti P, et al. Tracking neuroinflammatory biomarkers in Alzheimer's disease: a strategy for individualized therapeutic approaches? J Neuroinflammation. 2024;21:187.
59. Bellaver B, Povala G, Ferreira PCL, Bauer-Negrini G, Lussier FZ, Leffa DT, et al. Plasma GFAP for populational enrichment of clinical trials in preclinical Alzheimer's disease. Alzheimers Dement. 2025;21:e70209.
60. Duggan MR, Morgan DG, Price BR, Rajbanshi B, Martin-Peña A, Tansey MG, et al. Immune modulation to treat Alzheimer's disease. Mol Neurodegener. 2025;20:39.
61. Ahmad SR, Zeyaullah M, Khan MS, AlShahrani AM, Altijani AAG, Ali H, et al. Pharmacogenomics for neurodegenerative disorders – a focused review. Front Pharmacol. 2024;15:1478964.
62. Noh M-Y, Kwon HS, Kwon M-S, Nahm M, Jin HK, Bae J, et al. Biomarkers and therapeutic strategies targeting microglia in neurodegenerative diseases: current status and future directions. Mol Neurodegener. 2025;20:82.
63. Mattsson-Carlgren N, Palmqvist S, Blennow K, Hansson O. Increasing the reproducibility of fluid biomarker studies in neurodegenerative studies. Nat Commun. 2020;11:6252.
64. Myrou A, Barmpagiannos K, Ioakimidou A, Savopoulos C. Molecular biomarkers in neurological diseases: advances in diagnosis and prognosis. Int J Mol Sci. 2025;26:2231.
65. U.S. Food and Drug Administration. Considerations for the use of artificial intelligence to support regulatory decision-making for drug and biological products [Internet]. U.S. Department of Health and Human Services; 2025. Report No.: FDA-2024-D-4689. Available from: https://www.fda.gov/regulatory-information/search-fda-guidance-documents/considerations-use-artificial-intelligence-support-regulatory-decision-making-drug-and-biological.
66. Wu K, Wu E, Rodolfa K, Ho DE, Zou J. Regulating AI adaptation: an analysis of AI medical device updates. 2024 [cited 2025 July 31]. Available from: https://arxiv.org/abs/2407.16900.

MIX
Papier aus verantwortungsvollen Quellen
Paper from responsible sources
FSC® C105338

If you have any concerns about our products,
you can contact us on
ProductSafety@springernature.com

In case Publisher is established outside the EU,
the EU authorized representative is:
Springer Nature Customer Service Center GmbH
Europaplatz 3, 69115 Heidelberg, Germany

Printed by Libri Plureos GmbH
in Hamburg, Germany